Lecture Notes in Physics

The Lecture Notes in Physics

The series Lecture Notes in Physics (LNP), founded in 1969, reports new developments in physics research and teaching – quickly and informally, but with a high quality and the explicit aim to summarize and communicate current knowledge in an accessible way. Books published in this series are conceived as bridging material between advanced graduate textbooks and the forefront of research and to serve three purposes:

- to be a compact and modern up-to-date source of reference on a well-defined topic

- to serve as an accessible introduction to the field to postgraduate students and nonspecialist researchers from related areas

- to be a source of advanced teaching material for specialized seminars, courses and schools

Both monographs and multi-author volumes will be considered for publication. Edited volumes should, however, consist of a very limited number of contributions only. Proceedings will not be considered for LNP.

Volumes published in LNP are disseminated both in print and in electronic formats, the electronic archive being available at springerlink.com. The series content is indexed, abstracted and referenced by many abstracting and information services, bibliographic networks, subscription agencies, library networks, and consortia.

Proposals should be sent to a member of the Editorial Board, or directly to the managing editor at Springer:

Christian Caron
Springer Heidelberg
Physics Editorial Department I
Tiergartenstrasse 17
69121 Heidelberg / Germany
christian.caron@springer.com

A. Jüngel

Transport Equations
for Semiconductors

 Springer

Ansgar Jüngel
TU Wien
Inst. Analysis und Scientific Computing
Wiedner Hauptstr. 8-10
1040 Wien
Austria
juengel@anum.tuwien.ac.at

Jüngel, A., *Transport Equations for Semiconductors*, Lect. Notes Phys. 773 (Springer, Berlin Heidelberg 2009), DOI 10.1007/978-3-540-89526-8

ISBN 978-3-540-89525-1   e-ISBN 978-3-540-89526-8

DOI 10.1007/978-3-540-89526-8

Lecture Notes in Physics ISSN 0075-8450   e-ISSN 1616-6361

Library of Congress Control Number: 2008940023

Cover design: Integra Software Services Pvt Ltd., Pondicherry

Printed on acid-free paper

9 8 7 6 5 4 3 2 1

springer.com

Preface

The modern computer and telecommunication industry relies heavily on the use of semiconductor devices. The first semiconductor device (a germanium transistor) was built in 1947 by Bardeen, Brattain, and Shockley, who were awarded the Nobel prize in 1956. In the following decades, a lot of different devices for special applications have been invented, for instance, light-emitting diodes, metal-oxide semiconductor transistors, semiconductor lasers, solar cells, and single-electron transistors.

A fundamental fact of the success of the semiconductor technology is that the device length is much smaller than that of previous electronic devices (like tube transistors). The first transistor of Bardeen, Brattain, and Shockley had a characteristic length (the emitter–collector length) of 20 μm, compared to the size of a few centimeter of a tube transistor. The first Intel processor 4004, built in 1971, consisted of 2250 transistors, each of them with a characteristic length of 10 μm. This length could be reduced to 45 nm for transistors in actual processors. Modern quantum-based devices (like tunneling diodes) have structures of only a few nanometer length.

The main objective of this book is the derivation of transport equations describing the electron flow through a semiconductor device due to the application of a voltage. Depending on the device structure, the main transport phenomena may be very different, caused by diffusion, drift, scattering, or quantum-mechanical effects. The choice of the model equations depends on certain key parameters, such as the number of free electrons in the device, the mean free path of the charge carriers (i.e., the average distance between two consecutive collisions for a particle), the device dimension, and the ambient temperature.

Usually, a large number of electrons is flowing through a device such that a particle-like description using kinetic or fluid-type equations seems to be appropriate. On the other hand, electrons in a semiconductor crystal are quantum mechanical objects such that a wave-like description using the Schrödinger equation or the density-matrix formalism is necessary. For this reason, we have to devise different models which are able to describe the important physical phenomena for a particular situation or for a particular device. Moreover, since in some cases we are not interested in all the available physical information, we need simpler models which help to reduce the computation costs in the numerical simulations.

This leads to a *hierarchy* of semiconductor models. Roughly speaking, we distinguish four classes of semiconductor models: (i) microscopic semi-classical, (ii) macroscopic semi-classical, (iii) microscopic quantum, and (iv) macroscopic quantum model equations.

The first model class involves kinetic equations in which the quantum mechanical description is incorporated only in a semi-classical way. The electrons are specified by a distribution function, which has a probabilistic interpretation, depending on the phase space variables and the time. The evolution equations are of kinetic type, such as the Liouville, Vlasov, and Boltzmann equation.

When collisions become dominant in the semiconductor domain, i.e., when the mean free path is much smaller than the characteristic device size, a fluid dynamical description is appropriate. This description takes into account quantities which are averages of the distribution function of the Boltzmann equation over the momentum or energy space, like the particle, current, and energy densities. This leads to the second model class, which consists of semi-classical diffusive or hyperbolic moment equations, such as the drift-diffusion, energy-transport, hydrodynamic, and spherical harmonics expansion (SHE) models.

The semi-classical description is reasonable if the carriers can be treated as particles. An important parameter, which measures the validity of this description, is the de Broglie wavelength corresponding to a thermal average carrier. In physical situations, in which the electric potential varies rapidly on the scale of the de Broglie length or in which the mean free path is much larger than the de Broglie length, quantum mechanical models are more appropriate. In the third model class, three formulations of quantum microscopic models are presented: the Schrödinger, the density-matrix, and the kinetic Wigner formulation.

When the modeling of both quantum effects and collisions is important, but a computationally less expensive macroscopic description of the transport phenomena is needed, averaged quantum models can be formulated. This leads to the model class of macroscopic quantum models. These models are, in some sense, quantum analogues of the semi-classical equations, such as the quantum drift-diffusion, quantum energy-transport, and quantum hydrodynamic equations. Similar to their semi-classical analogues, they are derived from Wigner–Boltzmann equations.

The model hierarchies determine the structure of this book, which consists of five parts. The first part provides a short introduction to some basic notions of semiconductor physics and explains the strategy of the derivation of macroscopic models, starting from kinetic equations. The four model classes are then presented in the remaining four parts. Figure 1 gives an overview of the model hierarchy.

This book is not intended to be comprehensive, and there are a lot of important models which are not discussed, like hybrid models, lattice heat flow equations, transport in magnetic fields, subband models, quantum SHE equations, models for spintronics, and many others. The emphasis is placed on transport fundamentals and concepts of macroscopic modeling. For details on the crystal structure of semiconductors and their properties and on specific semiconductor devices, we refer to the

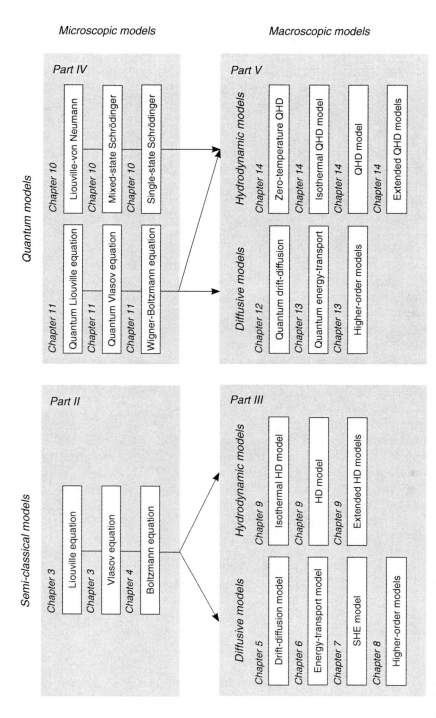

Fig. 1 Overview of the presented model classes. SHE, spherical harmonics expansion; HD, hydrodynamic; QHD, quantum hydrodynamic

physical and engineering literature. Furthermore, in order to make this book accessible to a wide range of readers, such as researchers and graduate students from physics, engineering, and applied mathematics, not the most general situations are considered and mathematical details are usually replaced by references to the corresponding literature.

In this spirit, the derivations of the model equations are purely formal, although in several instances some mathematical properties are mentioned. It turned out to be convenient to summarize the results in the form of lemmas, propositions, and theorems as it is common in mathematics. However, a "proof" of a lemma, proposition, or theorem is not a proof in the strict mathematical sense, since the underlying function spaces and regularity assumptions are generally not specified.

Some existing material on semiconductor modeling was employed in this book. For instance, the book of Brennan [1] was used for the chapter on semiconductor physics. The classical book of Markowich, Ringhofer, and Schmeiser [2] was an important source for the presentation of semi-classical and quantum kinetic theory. The works of Poupaud [3] were additionally employed for the description of kinetic equations, the papers of Arnold [4] were very useful for microscopic quantum modeling, and finally, Ben Abdallah [5] contributed to the derivation of boundary conditions for the Schrödinger equation. The chapters on semi-classical macroscopic models are based on several sources. For instance, the derivation of the low- and high-field drift-diffusion models from the Boltzmann equation was done by Poupaud in [6, 7]. The derivation of energy-transport and SHE equations from the Boltzmann equation is due to Degond and coworkers [8, 9]. An extension of the energy-transport model was derived by Grasser and coworkers [10]. Hydrodynamic models and their extensions were studied intensively by Anile, Romano, and coworkers [11, 12]. Furthermore, the derivation of quantum fluid models from a Wigner–Boltzmann equation was initiated by Degond and Ringhofer [13], and quantum drift-diffusion and quantum energy-transport models have been derived [14]. I'm very grateful to the above mentioned authors for many fruitful discussions and various suggestions on the material. Moreover, I want to express my gratitude to (in alphabetic order) Kazuo Aoki (Kyoto), José Antonio Carrillo (Barcelona), Irene Gamba (Austin), Thierry Goudon (Lille), Hans-Christoph Kaiser (Berlin), Florian Méhats (Rennes), Paola Pietra (Pavia), and Riccardo Sacco (Milan) for valuable discussions.

I would like to express my thanks to Bertram Düring, Jan Haškovec, Stefan Krause, Peter Kristöfel, and Daniel Matthes from Vienna for many corrections and suggestions. Finally, I acknowledge partial support of the German Science Foundation (DFG), the Austrian Science Fund (FWF), the German Academic Exchange Service (DAAD), the European Science Foundation (ESF), and the Austrian Exchange Service (ÖAD).

Vienna, Austria   *Ansgar Jüngel*

References

1. K. Brennan. *The Physics of Semiconductors.* Cambridge University Press, Cambridge, 1999.
2. P. Markowich, C. Ringhofer, and C. Schmeiser. *Semiconductor Equations.* Springer, Vienna, 1990.
3. F. Poupaud. On a system of nonlinear Boltzmann equations of semiconductors physics. *SIAM J. Appl. Math.* 50 (1990), 1593–1606.
4. A. Arnold. Mathematical properties of quantum evolution equations. In: G. Allaire, A. Arnold, P. Degond, and T. Hou (eds.), *Quantum Transport – Modelling, Analysis and Asymptotics*, Lecture Notes Math. 1946, 45–110. Springer, Berlin, 2008.
5. N. Ben Abdallah. On a multidimensional Schrödinger-Poisson scattering model for semiconductors. *J. Math. Phys.* 41 (2000), 4241–4261.
6. F. Poupaud. Diffusion approximation of the linear semiconductor Boltzmann equation: analysis of boundary layers. *Asympt. Anal.* 4 (1991), 293–317.
7. F. Poupaud. Runaway phenomena and fluid approximation under high fields in semiconductor kinetic theory. *Z. Angew. Math. Mech.* 72 (1992), 359–372.
8. N. Ben Abdallah and P. Degond. On a hierarchy of macroscopic models for semiconductors. *J. Math. Phys.* 37 (1996), 3308–3333.
9. P. Degond, C. Levermore, and C. Schmeiser. A note on the energy-transport limit of the semiconductor Boltzmann equation. In: N. Ben Abdallah et al. (eds.), *Proceedings of Transport in Transition Regimes* (Minneapolis, 2000), IMA Math. Appl. 135, 137–153. Springer, New York, 2004.
10. T. Grasser. Non-parabolic macroscopic transport models for semiconductor device simulation. *Physica A* 349 (2005), 221–258.
11. A. Anile and V. Romano. Non parabolic transport in semiconductors: closure of the moment equations. *Continuum Mech. Thermodyn.* 11 (1999), 307–325.
12. A. Anile, V. Romano, and G. Russo. Extended hydrodynamical model of carrier transport in semiconductors. *SIAM J. Appl. Math.* 61 (2000), 74–101.
13. P. Degond and C. Ringhofer. Quantum moment hydrodynamics and the entropy principle. *J. Stat. Phys.* 112 (2003), 587–628.
14. P. Degond, F. Méhats, and C.Ringhofer. Quantum energy-transport and drift-diffusion models. *J. Stat. Phys.* 118 (2005), 625–665.

Contents

Part III Macroscopic Semi-Classical Models

Part IV Microscopic Quantum Models

List of Symbols

$A^\top$	transpose of the matrix A
B	Brillouin zone
$C(x)$	doping profile
D	Wigner–Seitz cell
$\mathscr{D}$	diffusion matrix, $\mathscr{D} = (D_{ij})$
D_{ij}	diffusion coefficients
e	macroscopic energy
E	electric field, $E = -\nabla_x V$
$E(t)$	energy at time t
E_c	conduction band energy
E_g	energy gap
E_v	valence band energy
E_F	Fermi energy
$E(k)$	energy band
$f(x,k,t)$	distribution function
$F_{1/2}$	Fermi integral of index $1/2$
$\mathscr{F}(f)$	Fourier transform of f
$\hbar$	reduced Planck constant, $\hbar = h/2\pi$
H	Hamilton operator
Id	identity matrix
$\mathrm{Im}(z)$	imaginary part of the complex number z
J, J_n	electron current density
J_e	energy current density
k	pseudo-wave vector
k_B	Boltzmann constant
$L^2(\mathbb{R}^3)$	set of all real functions f with $\|f\|_{L^2}^2 = \int_{\mathbb{R}^3} f^2 \, \mathrm{d}x < \infty$
$L^2(\mathbb{R}^3;\mathbb{C})$	set of all complex functions f with $\|f\|_{L^2}^2 = \int_{\mathbb{R}^3} |f|^2 \, \mathrm{d}x < \infty$
m	electron rest mass
m^*, m_e^*	effective electron mass
m_h^*	effective hole mass

m_i	ith moment
M	Maxwellian distribution
n	electron density
n_i	intrinsic density
N_c	effective density of states of the conduction band
N_v	effective density of states of the valence band
$N(E)$	density of states of energy E
$N(Q)$	kernel (null space) of the operator Q
$\mathcal{O}(x)$	term of order x
p	hole density, crystal momentum $p = \hbar k$
P	quantum momentum operator, stress tensor
q	elementary charge, heat flux
$Q(f), Q(w)$	collision operator
R	vorticity matrix, $R_{j\ell} = \partial u_j / \partial x_\ell - \partial u_\ell / \partial x_j$
$R(n,p)$	recombination-generation rate
$R(Q)$	range of the operator Q
$s(x,k,k')$	scattering rate
S	entropy, phase function
t	time
T	temperature
T_L	lattice temperature
Tr	trace of a matrix
u	macroscopic velocity
U_T	thermal voltage, $U_T = k_B T_L / q$
$v(k)$	mean velocity, $v(k) = \nabla_k E(k) / \hbar$
V	electric potential
V_{bi}	built-in potential
V_{ext}	external potential
V_L	lattice potential
$\text{vol}(\Omega)$	volume or measure of the set Ω
$w(x,p,t)$	Wigner function
W	energy relaxation term
$W(\widehat{\rho})$	Wigner–Weyl transform of $\widehat{\rho}$
x	spatial variable
α	nonparabolicity parameter, scaled mean free path
Γ	Gamma function
δ_{ij}	Kronecker delta
Δ	Laplace operator
ε	energy, scaled Planck constant
ε_0	permittivity of vacuum
ε_s	semiconductor permittivity
$\eta(x)$	exterior unit normal vector at the boundary point x
θ	temperature tensor
$\theta[V]$	potential operator

κ	weight function in moment method
λ_i	Lagrange multiplier, occupation probability
λ_D	scaled Debye length
μ	chemical potential
μ_0	(low-field) mobility
$\mu(E)$	high-field mobility at electric field E
$\rho(x,y,t)$	density matrix function
$\widehat{\rho}$	density matrix operator
$\sigma(x,k,k')$	scattering cross-section
τ	relaxation time, typical time
ψ	wave function
ω	collision frequency

Part I
Introduction

The following two chapters are of introductory nature. First, we present a summary of basic notions and definitions from semiconductor physics. Only those subjects relevant to the subsequent chapters are included here. Second, we explain the strategy of deriving macroscopic model equations from the microscopic Boltzmann equation by assuming dominant scattering. Here, we distinguish the diffusion scaling, leading to diffusive models which are mathematically of parabolic type, and the hydrodynamic scaling, leading to hydrodynamic models which are mathematically of hyperbolic type.

Chapter 1
Basic Semiconductor Physics

In this chapter we present a short summary of the physics and main properties of semiconductors. We refer to [1–6] for introductory textbooks of solid-state and semiconductor physics and to [7–12] for more advanced expositions.

1.1 Semiconductor Crystals

What is a semiconductor? Historically, the term "semiconductor" has been used to denote solid materials whose conductivity is much larger than that of insulators but much smaller than that of metals, measured at room temperature. A modern and more precise definition is that a semiconductor is a solid with an *energy gap* larger than zero and smaller than a few electron volt (up to about 3 or 4 eV; see [13]). Metals do not have an energy gap, whereas it is usually larger than a few electron volt in insulators. In order to give a meaning to the notion "energy gap", we shall review some facts about the crystal structure of solids.

An ideal solid is made of an infinite three-dimensional array of atoms arranged in a lattice

$$L = \{n_1 a_1 + n_2 a_2 + n_3 a_3 : n_1, n_2, n_3 \in \mathbb{Z}\} \subset \mathbb{R}^3,$$

where $a_1, a_2, a_3 \in \mathbb{R}^3$ are the basis vectors of L, called *primitive vectors* of the lattice (see Fig. 1.1). The set L is called the *Bravais lattice*. The periodic structure of the lattice is specified in the following definitions [3, 14]:

1. The *reciprocal lattice* (or dual lattice) L^* of L is defined by

$$L^* = \{n_1 a_1^* + n_2 a_2^* + n_3 a_3^* : n_1, n_2, n_3 \in \mathbb{Z}\} \subset \mathbb{R}^3,$$

   where the *primitive vectors* $a_1^*, a_2^*, a_3^* \in \mathbb{R}^3$ are the dual basis, satisfying

$$a_m \cdot a_n^* = 2\pi \delta_{mn} \quad \text{for all } m, n = 1, 2, 3. \tag{1.1}$$

Jüngel, A.: *Basic Semiconductor Physics*. Lect. Notes Phys. **773**, 3–44 (2009)
DOI 10.1007/978-3-540-89526-8_1   © Springer-Verlag Berlin Heidelberg 2009

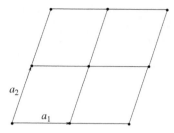

Fig. 1.1 Illustration of a two-dimensional lattice L

2. A connected set $D \subset \mathbb{R}^3$ is called a *primitive cell* of L (or L^*) if the volume of D equals the volume of the parallelepiped spanned by the basis vectors of L (or L^*),

$$\mathrm{vol}(D) = a_1 \cdot (a_2 \times a_3) \quad (\text{or} \quad \mathrm{vol}(D) = a_1^* \cdot (a_2^* \times a_3^*)),$$

and if the whole space $\mathbb{R}^3$ is covered by the union of translates of D by the primitive vectors. Here, the symbol "$\times$" denotes the vector product in $\mathbb{R}^3$.
3. The special primitive cell

$$D = \left\{ x \in \mathbb{R}^3 : x = \sum_{n=1}^{3} \alpha_n a_n, \; \alpha_n \in \left[-\frac{1}{2}, \frac{1}{2} \right] \right\},$$

which consists of all points being closer to the origin than to any other point of the lattice, is called the *Wigner–Seitz cell*.
4. The Wigner–Seitz cell of the reciprocal lattice is called the (first) *Brillouin zone* (see Fig. 1.2):

$$B = \left\{ k \in \mathbb{R}^3 : k = \sum_{n=1}^{3} \beta_n a_n^*, \; \beta_n \in \left[-\frac{1}{2}, \frac{1}{2} \right] \right\}.$$

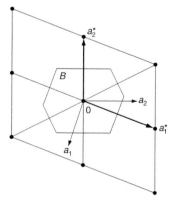

Fig. 1.2 The primitive vectors of a two-dimensional lattice L and its reciprocal lattice L^* and the Brillouin zone B

We give some explanations of the above definitions. What is the meaning of the reciprocal lattice? The reciprocal lattice vectors and the direct lattice vectors can be seen as conjugate variables, like time and frequency are conjugate variables in signal analysis. In fact, let $x \in L$ and $k \in L^*$ be given by

$$x = \sum_{m=1}^{3} \alpha_m a_m \quad \text{and} \quad k = \sum_{n=1}^{3} \beta_n a_n^*,$$

where $\alpha_m, \beta_n \in \mathbb{Z}$. Then, by (1.1),

$$e^{ik \cdot x} = \exp\left(i \sum_{m,n=1}^{3} 2\pi \delta_{mn} \alpha_m \beta_n\right) = \exp\left(2\pi i \sum_{m=1}^{3} \alpha_m \beta_m\right) = 1. \qquad (1.2)$$

As the position vector x has the dimension of length, k has the dimension of inverse length and therefore, k is called a wave vector. (More precisely, k is called a *pseudo-wave vector*; see below.)

Physically, the reciprocal lattice appears in X-ray diffraction experiments with crystals. It can be shown that the intensity peaks of the reflected X-rays are obtained when the change in the wave vector $\triangle k$ of the X-ray wave is an element of the reciprocal lattice [3, p. 404]. This allows one to determine the structure of the crystal lattice.

The primitive vectors a_ℓ^* of the Brillouin zone can be computed from the vectors a_m by

$$a_\ell^* = 2\pi \frac{a_m \times a_n}{a_1 \cdot (a_2 \times a_3)},$$

where (ℓ, m, n) is $(1,2,3)$, $(2,3,1)$, or $(3,1,2)$. If A, A^* denote the 3×3 matrices whose columns are the vectors $a_n = (a_{1n}, a_{2n}, a_{3n})^\top$, $a_n^* = (a_{1n}^*, a_{2n}^*, a_{3n}^*)^\top$, respectively, the relation (1.1) implies that

$$(A^\top A^*)_{mn} = \sum_{j=1}^{3} a_{jm} a_{jn}^* = a_m \cdot a_n^* = 2\pi \delta_{mn}$$

and thus $A^\top A^* = 2\pi \text{Id}$, where Id is the identity matrix of $\mathbb{R}^{3 \times 3}$. Hence,

$$A^* = 2\pi (A^\top)^{-1} = 2\pi (A^{-1})^\top. \qquad (1.3)$$

Graphically, the Brillouin zone can be constructed as follows. Draw arrows from a lattice point of L^* to its nearest neighbors and determine the midpoints of the arrows. Then the planes through these points perpendicular to the arrows form the surface of the (bounded) Brillouin zone. In two space dimensions, the Brillouin zone is a hexagon or a square (see Fig. 1.2). In three space dimensions, the zone is a polyhedron (e.g., a "capped" octahedron; see Fig. 1.3).

Lemma 1.1. *The volumes of a primitive cell D and its Brillouin zone B are related by the equation*

$$\text{vol } B = \frac{(2\pi)^3}{\text{vol } D}.$$

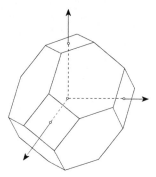

Fig. 1.3 Brillouin zone of semiconductors like silicon, germanium, gallium arsenide, etc.

Proof. With the above notations, we have

$$\text{vol}\, D = a_1 \cdot (a_2 \times a_3) = \det(a_1, a_2, a_3) = \det A,$$

and hence, by (1.3),

$$\text{vol}\, B = \det(a_1^*, a_2^*, a_3^*) = \det A^* = \det(2\pi(A^{-1})^\top) = (2\pi)^3 \det(A^{-1})$$
$$= \frac{(2\pi)^3}{\det A} = \frac{(2\pi)^3}{\text{vol}\, D},$$

finishing the proof.   □

1.2 The Schrödinger Equation

In the previous section we have introduced the semiconductor solid by its crystalline structure, i.e., by the nuclei lying at lattice points. In fact, the crystal consists of the nuclei, the core electrons, and the valence electrons. Their state has to be described by quantum mechanics. More precisely, the state of a quantum particle is represented by a complex-valued wave function $\phi(x,t)$, where $x \in \mathbb{R}^3$ and $t \in \mathbb{R}$. The dynamics of the wave function is given by the *Schrödinger equation*

$$i\hbar \partial_t \phi = H\phi, \quad x \in \mathbb{R}^3, \, t > 0, \quad \phi(\cdot, 0) = \phi_I, \tag{1.4}$$

where $\partial_t = \partial/\partial t$ and H is the so-called *Hamilton operator*. For instance, the Hamilton operator of a single electron with mass m moving in an electric potential $V(x)$ reads as

$$H = -\frac{\hbar^2}{2m}\Delta - qV(x), \quad x \in \mathbb{R}^3, \tag{1.5}$$

where $\Delta = \sum_{j=1}^3 \partial^2/\partial x_j^2$ is the Laplace operator in $\mathbb{R}^3$.

Stationary states can be obtained from the ansatz $\phi(x,t) = e^{-iEt/\hbar}\psi(x)$, where E is a real number. Inserting this ansatz into (1.4) and dividing by $e^{-iEt/\hbar}$ gives the *stationary* Schrödinger equation

$$H\psi = E\psi \qquad (1.6)$$

or, in the case of a single electron,

$$-\frac{\hbar^2}{2m}\Delta\psi - qV(x)\psi = E\psi, \quad x \in \mathbb{R}^3.$$

Thus, the quantum state is stationary if ψ is an eigenfunction and E is an eigenvalue of H. Physically, E describes the energy of the system if it is in the eigenstate ψ. The set of all possible energy values is represented by the spectrum of the Hamiltonian H.

The solution ϕ of (1.4) with the Hamiltonian (1.5) can be interpreted as follows. We take the derivative

$$\partial_t|\phi|^2 = (\partial_t\overline{\phi})\phi + \overline{\phi}(\partial_t\phi) = -\frac{i\hbar}{2m}\Delta\overline{\phi}\phi + \frac{i\hbar}{2m}\overline{\phi}\Delta\phi$$

$$= -\frac{i\hbar}{2m}\mathrm{div}\,(\nabla\overline{\phi}\phi - \overline{\phi}\nabla\phi) = -\frac{\hbar}{m}\mathrm{div}\,\mathrm{Im}(\overline{\phi}\nabla\phi),$$

where $\overline{z}$ denotes the conjugate of the complex number $z \in \mathbb{C}$, $\mathrm{Im}(z)$ is its imaginary part, and $\mathrm{div}\,u = \sum_{j=1}^3 \partial u_j/\partial x_j$ is the divergence of a vector field $u = (u_1, u_2, u_3)$. Introducing the variables

$$n = |\phi|^2, \quad J = -\frac{q\hbar}{m}\mathrm{Im}(\overline{\phi}\nabla\phi),$$

we arrive at the conservation law

$$\partial_t n - \frac{1}{q}\mathrm{div}\,J = 0,$$

expressing the conservation of the integral $\int_{\mathbb{R}^3} n\,dx$. According to the pioneering works of Einstein, Planck, etc., we may interpret n as the *electron density* and J as the *electron current density* J. The integral $\int_\Omega |\phi(x,t)|^2\,dx$ is the probability to find the electron at time t in the domain Ω.

We illustrate the stationary Schrödinger equation and its solutions by two simple examples.

Example 1.2 (State of a free-electron). Consider a free-electron in a one-dimensional vacuum, i.e., $V(x) = 0$ for all $x \in \mathbb{R}$. We need to solve the Schrödinger equation

$$-\frac{\hbar^2}{2m}\psi'' = E\psi \quad \text{in } \mathbb{R}. \qquad (1.7)$$

A computation shows that eigenfunctions are given by

$$\psi_k(x) = Ae^{ikx} + Be^{-ikx}, \quad x \in \mathbb{R},$$

where $k^2 = 2mE/\hbar^2$, with eigenvalues

$$E = E(k) = \frac{\hbar^2 k^2}{2m}, \quad k \in \mathbb{R}.$$

Thus, the eigenvalue problem (1.7) has infinitely many bounded solutions parametrized by $k \in \mathbb{R}$ and corresponding to different real-valued energies $E(k)$. The functions $e^{\pm ikx}$ are called *plane waves*. Thus, the eigenstates of a free particle are plane waves. □

Example 1.3 (Infinite square-well potential). We consider an electron in an infinite square-well potential. This is a one-dimensional structure of length L with a vanishing potential inside the well and an infinite potential outside. As the potential is confining the electron to the inner region, we have to solve the Schrödinger equation (1.6) in the interval $(0,L)$ with boundary conditions

$$\psi(0) = \psi(L) = 0$$

and potential $V(x) = 0$ for $x \in (0,L)$. The general solution of (1.6) is

$$\psi(x) = Ae^{(a+ik)x} + Be^{-(a+ik)x},$$

where $A, B \in \mathbb{C}$ and a and k are real numbers such that $-(a+ik)^2 = 2mE/\hbar^2$. Using the boundary conditions, it is not difficult to see that they can only be satisfied if $a = 0$ and $\sin(kL) = 0$. Hence, the eigenfunctions are given by

$$\psi_k(x) = A\left(e^{ikx} - e^{-ikx}\right) = C\sin(kx), \quad \text{where } k = \frac{n\pi}{L}, \, n \in \mathbb{Z},$$

and $C = 2iA$, and the eigenvalues are

$$E(k) = \frac{\hbar^2 k^2}{2m}.$$

The integration constant C can be determined by assuming that

$$\int_0^L |\psi_k(x)|^2 \, dx = 1$$

holds, stating that the probability of finding the electron in the square well is equal to one. A simple computation shows that $C = \sqrt{2L}$. The system only allows *discrete* energy states. In particular, the parameter k can only take discrete values. □

1.3 Electrons in a Periodic Potential

The semiconductor solid can be described by ions (nuclei and core electrons) and valence electrons. These electrons are responsible for the electronic properties of the solid. The evolution of their state is quantum mechanically given by the Hamiltonian which takes into account the relevant physical phenomena, like ion vibrations, electron–ion interactions, and electron–electron scattering. We assume that the ions are fixed and in equilibrium such that we can neglect lattice vibrations and their interaction with the electrons (see [2, 11] for lattice dynamics and electron–phonon interactions).

Let the state of the ion–electron system be described by the wave function $\psi(x)$, where $x = (x_1, \ldots, x_M)^\top \in \mathbb{R}^{3M}$ is the vector of all possible positions $x_j \in \mathbb{R}^3$ of the M electrons. Then, the Hamiltonian of the quantum system (see Sect. 1.2) consists of the kinetic energy part, the electron–ion interactions, and the electron–electron interactions,

$$H = -\frac{\hbar^2}{2m} \sum_{j=1}^{M} \Delta_j + H_{\text{ei}} + H_{\text{ee}},$$

where Δ_j is the Laplace operator acting on the x_j variable only. In the following, we will derive explicit expressions for H_{ei} and H_{ee}.

The lattice ions generate a periodic electrostatic potential V_{ei},

$$V_{\text{ei}}(x+y) = V_{\text{ei}}(x) \quad \text{for } x \in \mathbb{R}^3, \ y \in L$$

(recall that L is the Bravais lattice; see Sect. 1.1), which is the superposition of the Coulomb potentials

$$V_j(x) = \frac{Q}{4\pi\varepsilon_0 |x - R_j|}$$

of the crystal ions located at R_j, i.e.,

$$V_{\text{ei}}(x) = \sum_{j=1}^{M_i} \frac{Q}{4\pi\varepsilon_0 |x - R_j|}, \quad x \in \mathbb{R}^3$$

(see Fig. 1.4). Here, Q is the ion charge, ε_0 the permittivity, and M_i the number of ions. The lattice potential describes the interaction of a single electron with the ions. It is periodic with respect to the lattice. Hence, the electron–ion Hamiltonian is given by

$$H_{\text{ei}} = -q \sum_{\ell=1}^{M} V_{\text{ei}}(x_\ell) = -\sum_{\ell=1}^{M} \sum_{j=1}^{M_i} \frac{qQ}{4\pi\varepsilon_0 |x_\ell - R_j|}.$$

The electron–electron interactions are modeled by

$$V_{\text{ee}}(x) = -\frac{1}{2} \sum_{j,\ell=1, j\neq\ell}^{M} \frac{q}{4\pi\varepsilon_0 |x_j - x_\ell|}, \quad x \in \mathbb{R}^{3M},$$

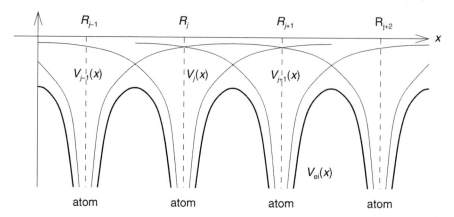

Fig. 1.4 Potentials $V_j(x)$ of a single ion at $x = R_j$ and net potential $V_{ei}(x)$ of a one-dimensional crystal lattice

and the Hamiltonian is given by $H_{ee} = -qV_{ee}(x)$. The factor $\frac{1}{2}$ takes into account that the sum counts each interaction twice.

Thus, the Hamiltonian of the system reads as

$$H = \sum_{j=1}^{M} \left(-\frac{\hbar^2}{2m}\Delta_j - qV_{ei}(x_j) \right) - qV_{ee}(x).$$

The solution of the eigenvalue problem $H\psi = E\psi$ is computationally very expensive, due to the presence of the potentials and the large number of electrons. In the following, we simplify the problem by making two approximations. First, we replace the electron–electron interactions by an effective single-particle potential. This reduces the $3M$-dimensional problem to a three-dimensional one (Hartree–Fock approximation). Second, the solution of the Schrödinger equation in the whole space $\mathbb{R}^3$ is reduced to the solution in a primitive cell of the lattice (Bloch decomposition).

Hartree–Fock approximation. The reduction to a single-particle potential is based on the following idea. If the electron–electron interactions can be neglected, the Schrödinger equation is the sum of single-particle Schrödinger equations. Consequently, the wave function ψ can be written as the product of the single-particle wave functions. Even in the presence of electron–electron interactions, one may try the product ansatz

$$\psi(x) = \prod_{j=1}^{M} \psi_j(x_j). \tag{1.8}$$

This approximation of the wave function is called the *Hartree approximation*. The single-particle wave functions ψ_j are determined by assuming that they minimize the energy $(\psi, H\psi)_{L^2} = \int_{\mathbb{R}^{3N}} \overline{\psi}H\psi\,dx$ under the constraint of normalized wave functions,

$$\min_{\psi}(\psi, H\psi)_{L^2} \quad \text{subject to } \|\psi_j\|_{L^2}^2 = 1 \text{ for all } j, \tag{1.9}$$

where $\|\psi_j\|_{L^2}^2 = \int_{\mathbb{R}^3} |\psi_j|^2 \, dx$ and $\overline{\psi}$ denotes the complex conjugate of ψ. The minimum is taken over all wave functions satisfying (1.8).

Proposition 1.4 (Hartree equation). *A necessary condition for the solution of the constrained minimization problem* (1.9) *is given by the solution of the Schrödinger eigenvalue problem*

$$-\frac{\hbar^2}{2m}\Delta_j\psi_j - qV_{ei}(x_j)\psi_j - qV_{H,j}(x)\psi_j = E_j\psi_j \quad in \ \mathbb{R}^3, \tag{1.10}$$

the so-called Hartree equation, *where the* Hartree potential $V_{H,j}$ *is defined by*

$$V_{H,j}(x) = -q\sum_{\ell\neq j}\int_{\mathbb{R}^3} \frac{|\psi_\ell|^2 \, dx_\ell}{4\pi\varepsilon_0|x - x_\ell|}, \quad j = 1,\ldots,M.$$

Proof. The constrained minimization problem (1.9) can be solved by the method of Lagrange multipliers. We define

$$F(\psi_1,\ldots,\psi_M,E_1,\ldots,E_M) = (\psi,H\psi)_{L^2} - \sum_{j=1}^{M} E_j\left(\|\psi_j\|_{L^2}^2 - 1\right),$$

where E_j are the Lagrange multipliers. A computation, using (1.8) and the normalization $\|\psi_j\|_{L^2}^2 = 1$, shows that the expectation value $(\psi,H\psi)_{L^2}$ can be written as

$$(\psi,H\psi)_{L^2} = \sum_{j=1}^{M}\int_{\mathbb{R}^3} \overline{\psi_j}\left(-\frac{\hbar^2}{2m}\Delta_j - qV_{ei}(x_j)\right)\psi_j \, dx_j$$

$$+ \frac{q^2}{2}\sum_{j,\ell=1,\, j\neq\ell}^{M}\int_{\mathbb{R}^6} \frac{|\psi_j(x_j)|^2|\psi_\ell(x_\ell)|^2}{4\pi\varepsilon_0|x_j - x_\ell|} \, dx_j \, dx_\ell.$$

Then, a necessary condition for the solution of (1.9) reads as

$$0 = \frac{\partial F}{\partial \psi_j}(\phi) = \left(\left(-\frac{\hbar^2}{2m}\Delta_j - qV_{ei} + q^2\sum_{\ell\neq j}\int_{\mathbb{R}^3}\frac{|\psi_\ell(x_\ell)|^2 \, dx_\ell}{4\pi\varepsilon_0|x_j - x_\ell|} - E_j\right)\psi_j,\phi\right)_{L^2}$$

for all functions ϕ. This gives (1.10). □

The above approach has a drawback. By the Pauli principle, the total wave function of an electron ensemble has to be anti-symmetric (with respect to the spatial and spin variables). This is not necessarily the case if the above product ansatz is employed. To overcome this limitation, we construct a properly symmetrized wave function by a linear combination of products of the type $\psi_1(x_{j_1})\cdots\psi_N(x_{j_N})$. More precisely, ψ is given by the *Slater determinant*, and x_{j_k} includes spatial and spin variables. By computing the necessary condition for the corresponding constrained minimization problem, we obtain a Schrödinger equation similar to (1.10) augmented by the additional term $qV_{ex,j}\psi_j$. Instead of going into the details, we only present the result and refer to [3, Sect. 7.2] for a definition of the Slater determinant and the computations:

$$V_{ex,j}\psi_j(x) = -q \sum_{\ell \neq j,\|} \int_{\mathbb{R}^3} \frac{\overline{\psi_j(x')}\psi_\ell(x')}{4\pi\varepsilon_0 |x-x'|} \, dx' \psi_\ell(x).$$

The summation is over all states ℓ with parallel spin. This exchange term comes from the fact that the Slater determinant is the sum of weighted and signed products of single-particle functions with interchanged coordinates. Notice that the expression $V_{ex,j}\psi_j$ is not a multiplication of two functions but $V_{ex,j}$ is a nonlocal nonlinear integral operator since ψ_j appears under the integral.

The Schrödinger equation with the above effective potential can be reformulated by introducing the electron density and exchange particle density, respectively,

$$n(x) = \sum_{j=1}^{M} |\psi_j(x)|^2, \quad n_{ex,j}(x,x') = \sum_{\ell,\|} \frac{\overline{\psi_j(x')}\psi_\ell(x')\overline{\psi_j(x)}\psi_\ell(x)}{\overline{\psi_j(x)}\psi_j(x)}.$$

Then, the so-called *Hartree–Fock equation* reads as

$$E_j\psi_j = -\frac{\hbar^2}{2m}\Delta\psi_j - q\left(V_{ei}(x) + V_{H,j}(x) + V_{ex,j}(x)\right)\psi_j$$

$$= -\frac{\hbar^2}{2m}\Delta\psi_j - qV_{ei}(x)\psi_j + q^2 \int_{\mathbb{R}^3} \frac{n(x') - n_{ex,j}(x,x')}{4\pi\varepsilon_0 |x-x'|} \, dx' \psi_j.$$

The exchange potential depends on the state number j. A function independent of the state number is obtained by replacing the exchange density $n_{ex,j}$ by the average

$$\bar{n}_{ex}(x,x') = \frac{1}{M}\sum_{j=1}^{M} n_{ex,j}(x,x').$$

Then, introducing the effective single-particle potential

$$V_{eff}(x) = -q \int_{\mathbb{R}^3} \frac{n(x') - \bar{n}_{ex}(x,x')}{4\pi\varepsilon_0 |x-x'|} \, dx'$$

.

and the total effective potential $V_L = V_{ei} + V_{eff}$, we obtain the modified Hartree–Fock equation

$$-\frac{\hbar^2}{2m}\Delta\psi_j - qV_L(x)\psi_j = E_j\psi_j, \quad x \in \mathbb{R}^3, \ j = 1, \dots, M. \tag{1.11}$$

This is a single-particle equation in $\mathbb{R}^3$ incorporating the many-body aspect in terms of the total effective potential V_L. For a discussion of the validity of the Hartree–Fock approximation, we refer to [8, Sect. 1.4].

Bloch decomposition. In a perfect periodic crystal, we expect that the single-electron effective potential V_L is periodic, too [2, p. 132]. Thus, one might hope that the whole-space Schrödinger problem (1.11) can be reduced to an eigenvalue problem on a cell of the lattice. The following result, due to Bloch [15], states that this is indeed possible.

Theorem 1.5 (Bloch). *Let V_L be a periodic potential, i.e., $V_L(x + y) = V_L(x)$ for all $x \in \mathbb{R}^3$ and $y \in L$ (the Bravais lattice). Then the eigenvalue problem for the Schrödinger operator*

$$H = -\frac{\hbar^2}{2m}\Delta - qV_L(x), \quad x \in \mathbb{R}^3,$$

can be reduced to an eigenvalue problem of the Schrödinger equation on the primitive cell D of the lattice, indexed by $k \in B$ (the Brillouin zone),

$$H\psi = E\psi \quad in\ D, \quad \psi(x + y) = e^{ik \cdot y}\psi(x), \quad x \in D, \ y \in L. \tag{1.12}$$

For each $k \in B$, there exists a sequence $E_n(k)$, $n \geq 1$, of eigenvalues with associated eigenfunctions $\psi_{n,k}$. The eigenvalues $E_n(k)$ are real functions of k and periodic and symmetric on B. The spectrum of H is given by the union of the closed intervals $\{E_n(k) : k \in \overline{B}\}$ for $n \geq 1$ (with $\overline{B}$ being the closure of B).

For a proof of the Bloch theorem, we refer to [16, 17], where also more properties on the energies $E_n(k)$ are stated. A simple proof for the one-particle Schrödinger equation can be found in [18, Sect. 7.1]. In the following, we give a (mathematically not rigorous) motivation of the above statement, which helps to understand the role of the vector k.

We consider the translation operator T_a, defined by $(T_a\psi)(x) = \psi(x + a)$ for $a \in L$, $x \in \mathbb{R}^3$, and functions $\psi \in L^2(\mathbb{R}^3)$. First, we claim that the eigenvalues of T_a are given by $e^{i\theta}$ for $\theta \in \mathbb{R}$. To see this, let ψ be an eigenfunction to the eigenvalue λ, i.e., $T_a\psi = \lambda\psi$. Then

$$|\lambda|^2 \|\psi\|_{L^2}^2 = \|\lambda\psi\|_{L^2}^2 = \|T_a\psi\|_{L^2}^2 = \int_{\mathbb{R}^3} |\psi(x + a)|^2 \mathrm{d}x = \|\psi\|_{L^2}^2,$$

and thus, $|\lambda| = 1$ or $\lambda = e^{i\theta}$ for some $\theta \in \mathbb{R}$.

The Hamiltonian H commutes with all the translation operators T_a since V_L is periodic:

$$
\begin{aligned}
(T_a H\psi)(x) &= -\frac{\hbar^2}{2m}\Delta\psi(x + a) - qV_L(x + a)\psi(x + a) \\
&= -\frac{\hbar^2}{2m}\Delta\psi(x + a) - qV_L(x)\psi(x + a) = (HT_a\psi)(x).
\end{aligned}
$$

Therefore, if ψ is an eigenfunction of H, it is also an eigenfunction of T_a for any $a \in L$ and vice versa. (For this statement some mathematical properties are needed, like the self-adjointness of H and T_a; see, e.g., [19].) Let ψ be such a simultaneous eigenvector of H and T_a for any $a \in L$. Hence, for all $j = 1, 2, 3$, there exists $\theta_j \in \mathbb{R}$ such that

$$T_{-a_j} \psi = e^{i\theta_j} \psi, \tag{1.13}$$

where a_1, a_2, and a_3 are the primitive vectors of the Bravais lattice L. We set

$$k_0 = -\frac{1}{2\pi} \sum_{\ell=1}^{3} \theta_\ell a_\ell^*, \tag{1.14}$$

where a_1^*, a_2^*, and a_3^* are the primitive vectors of L^*. Then (1.2) implies that

$$k_0 \cdot a_j = -\frac{1}{2\pi} \sum_{\ell=1}^{3} \theta_\ell a_\ell^* \cdot a_j = -\theta_j. \tag{1.15}$$

We define $\phi(x) = e^{-ik_0 \cdot x} \psi(x)$ for $x \in \mathbb{R}^3$. We claim that $\phi(x+y) = \phi(x)$ for all $x \in \mathbb{R}^3$ and $y \in L$. Since every $y \in L$ is a linear combination of the vectors a_j, it is sufficient to prove the periodicity for $y = a_j$. We obtain, using (1.13) and (1.15),

$$\phi(x) = e^{-ik_0 \cdot x} \psi(x) = e^{-ik_0 \cdot x} (T_{-a_j} \psi)(x + a_j) = e^{-ik_0 \cdot x} e^{i\theta_j} \psi(x + a_j)$$
$$= e^{-ik_0 \cdot x} e^{i\theta_j} e^{ik_0 \cdot (x+a_j)} \phi(x + a_j) = e^{i(\theta_j + k_0 \cdot a_j)} \phi(x + a_j) = \phi(x + a_j).$$

It remains to show that k_0 can be restricted to the Brillouin zone. We decompose $k_0 = k + \ell$, where $k \in B$ and $\ell \in L^*$ is a point in the reciprocal lattice closest to k (see Fig. 1.5). Then

$$\psi(x) = e^{ik_0 \cdot x} \phi(x) = e^{ik \cdot x} u(x), \quad x \in \mathbb{R}^3, \tag{1.16}$$

where $u(x) = e^{i\ell \cdot x} \phi(x)$ satisfies, in view of (1.2),

$$u(x+y) = e^{i\ell \cdot x} e^{i\ell \cdot y} \phi(x+y) = e^{i\ell \cdot x} \phi(x) = u(x)$$

for all $x \in \mathbb{R}^3$ and $y \in L$. Now, the representation (1.16) implies, for $x \in D$ and $y \in L$, that $\psi(x+y) = e^{ik \cdot (x+y)} u(x) = e^{ik \cdot y} \psi(x)$, which proves (1.12).

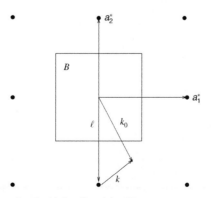

Fig. 1.5 Illustration of $k_0 = k + \ell$ with $k \in B$ and $\ell \in L^*$

Remark 1.6. The Brillouin zone in Theorem 1.5 is a nondiscrete set. In a semiconductor of finite size, however, only a finite number of values $k \in B$ is allowed. Suppose that ψ is a wave function satisfying the representation (1.16) and the boundary conditions $\psi(x + N_j a_j) = \psi(x)$ for all $x \in \mathbb{R}^3$, where a_j is a primitive vector and N_j is the number of primitive cells in the jth direction. Then (1.16) shows that

$$e^{ik \cdot x} u(x) = \psi(x) = \psi(x + N_j a_j) = e^{ik \cdot (x + N_j a_j)} u(x),$$

since $N_j a_j \in L$ and u is periodic on L. Thus, $N_j k \cdot a_j$ is a multiple of 2π and $k \cdot a_j = 2\pi n_j / N_j$ for some $n_j \in \mathbb{Z}$. Above we have decomposed k_0, defined in (1.14), as the sum of k and some vector $\ell \in L^*$. Thus, $k = k_0 - \ell$ and there exist some coefficients c_p such that $k = \sum_p c_p a_p^*$. The property $a_j \cdot a_p^* = 2\pi \delta_{jp}$ then yields

$$2\pi \frac{n_j}{N_j} = k \cdot a_j = \sum_{p=1}^{3} c_p a_p^* \cdot a_j = 2\pi c_j$$

and hence,

$$k = \sum_{p=1}^{3} \frac{n_j}{N_j} a_p^*, \quad n_j \in \mathbb{Z}, \quad -\frac{N_j}{2} \leq n_j \leq \frac{N_j}{2}. \tag{1.17}$$

Thus, k can take only a finite number of values. Typically, there are about 10^4 atoms in a semiconductor crystal of length 10^{-6} m in each direction. Thus, practically, k can take many numbers, and usually, it is assumed that k is not discrete but continuous as in a semiconductor occupying the whole space. This can be also justified by a scaling argument; see Sect. 1.6. □

Remark 1.7. The solutions ψ of the eigenvalue problem $H\psi = E_n \psi$ in $\mathbb{R}^3$ and $\psi_{n,k}$ of (1.12) are formally related by the expressions

$$\psi(x) = \frac{1}{\text{vol}(B)} \int_B \psi_{n,k}(x) \, dk \quad \text{and} \quad \psi_{n,k}(x) = \sum_{\ell \in L} e^{-ik \cdot \ell} \psi(x + \ell)$$

(see [20]). □

The above motivation shows that the eigenfunction $\psi_{n,k}$ of (1.12), parametrized by $n \in \mathbb{N}$ and $k \in B$, can be written as

$$\psi_{n,k}(x) = e^{ik \cdot x} u_{n,k}(x), \quad x \in D, \, k \in B,$$

and $u_{n,k}$ is periodic with respect to L (see (1.16)). In some sense, the so-called *Bloch functions* $\psi_{n,k}$ are plane waves which are modulated by a periodic function $u_{n,k}$ taking into account the influence of the crystal lattice. This explains why k is termed the *pseudo-wave vector*. It appears in modulated plane waves and is therefore not a real wave vector.

Which equation does $u_{n,k}$ solve? Inserting the above expression into the Schrödinger equation shows that $u_{n,k}$ is a solution of

$$-\frac{\hbar^2}{2m}(\Delta u_{n,k} + 2ik \cdot \nabla u_{n,k}) + \left(\frac{\hbar^2}{2m}|k|^2 - qV_L(x)\right)u_{n,k} = E_n(k)u_{n,k} \quad \text{in } D \quad (1.18)$$

with the periodic boundary conditions

$$u_{n,k}(x+y) = u_{n,k}(x), \quad x \in \mathbb{R}^3, \ y \in L. \quad (1.19)$$

The function $k \mapsto E_n(k)$ is called the *dispersion relation* and the set $\{E_n(k) : k \in B\}$ the *n*th *energy band*. It shows how the energy of the *n*th band depends on the (pseudo-) wave vector k. The union of ranges of E_n over $n \in \mathbb{N}$ is not necessarily the whole real line $\mathbb{R}$, i.e., there may exist energies E^* for which there is no number $n \in \mathbb{N}$ and no vector $k \in B$ such that $E_n(k) = E^*$. The connected components of the set of energies with this non-existence property are called *energy gaps*. We illustrate this property by the following example.

Example 1.8 (Kronig–Penney model). The Kronig–Penney model [21] is a simple model representing a one-dimensional single-crystal lattice (see [3, Sect. 8.2] or [10, Sect. 3.1.2]). The potential of the lattice atoms is modeled by the function

$$V_L(x) = \begin{cases} -V_0 & \text{if } -b < x \le 0, \\ 0 & \text{if } 0 < x \le a, \end{cases}$$

and V_L is extended to $\mathbb{R}$ with period $a+b$, $V_L(x) = V_L(x+a+b)$ for $x \in \mathbb{R}$, where a and $b > 0$ (see Fig. 1.6).

In order to solve the Schrödinger equation (1.12), we make the Bloch decomposition $\psi(x) = e^{ikx}u(x)$, where $u(x)$ is a $(a+b)$ periodic solution of (1.18),

$$-u'' - 2iku' + k^2u = \frac{2m}{\hbar^2}(E + qV_L)u \quad \text{in } \mathbb{R}. \quad (1.20)$$

We proceed as in [10, Sect. 3.1.2]. First we solve (1.20) in the interval $(0,a)$. Then $V_L(x) = 0$, $x \in (0,a)$, and the ansatz $u(x) = e^{i\gamma x}$ leads to

$$(\gamma^2 + 2k\gamma + k^2 - \alpha^2)e^{i\gamma x} = 0, \quad \text{where } \alpha = \frac{\sqrt{2mE}}{\hbar}.$$

The solutions of the quadratic equation in γ are given by $\gamma_{1/2} = -k \pm \alpha$ and therefore,

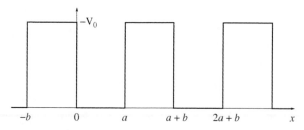

Fig. 1.6 The periodic square-well potential $V_L(x)$ of the Kronig–Penney model

$$u_1(x) = Ae^{i(\alpha-k)x} + Be^{-i(\alpha+k)x}, \quad x \in (0,a),$$

for some constants A and B.

In the interval $(-b,0)$ we make again the ansatz $u(x) = e^{i\gamma x}$ yielding

$$(\gamma^2 + 2k\gamma + k^2 - \beta^2)e^{i\gamma x} = 0, \quad \text{where } \beta = \frac{\sqrt{2m(E-qV_0)}}{\hbar}.$$

The solutions are $\gamma_{1/2} = -k \pm \beta$ and thus,

$$u_2(x) = Ce^{i(\beta-k)x} + De^{-i(\beta+k)x}, \quad x \in (-b,0),$$

where C and D are some constants. We notice that β is purely imaginary if $E < qV_0$, i.e., the electrons are bound within the crystal, and β is real if $E > qV_0$.

The constants A, B, C, and D are determined from the interface conditions. Assuming that u is continuously differentiable and periodic on $\mathbb{R}$,

$$u_1(0) = u_2(0), \qquad u_1'(0) = u_2'(0),$$
$$u_1(a) = u_2(-b), \qquad u_1'(a) = u_2'(-b),$$

we obtain the following four equations for the unknowns A, B, C, and D:

$$0 = A + B - C - D,$$
$$0 = (\alpha - k)A - (\alpha + k)B - (\beta - k)C + (\beta + k)D,$$
$$0 = Ae^{i(\alpha-k)a} + Be^{-i(\alpha+k)a} - Ce^{-i(\beta-k)b} - De^{i(\beta+k)b},$$
$$0 = (\alpha - k)Ae^{i(\alpha-k)a} - (\alpha + k)Be^{-i(\alpha+k)a} - (\beta - k)Ce^{-i(\beta-k)b}$$
$$+ (\beta + k)De^{i(\beta+k)b}.$$

This is a homogeneous linear system which has nontrivial solutions only if the determinant of the coefficient matrix vanishes. A lengthy calculation shows that this condition is equivalent to the equation

$$-\frac{\alpha^2 + \beta^2}{2\alpha\beta} \sin(\alpha a)\sin(\beta b) + \cos(\alpha a)\cos(\beta b) = \cos(k(a+b)), \qquad (1.21)$$

which relates the wave vector k to the energy E through the parameters α and β.

There are values of E for which there does not exist any k satisfying (1.21). In order to see this we assume that $E < qV_0$ such that β is purely imaginary and set $\beta = i\gamma$. Since $\sin(ix) = i\sinh(x)$ and $\cos(ix) = \cosh(x)$, (1.21) becomes

$$\frac{\gamma^2 - \alpha^2}{2\alpha\gamma} \sin(\alpha a)\sinh(\gamma b) + \cos(\alpha a)\cosh(\gamma b) = \cos(k(a+b)). \qquad (1.22)$$

Using $\lim_{\alpha \to 0} \sin(\alpha a)/\alpha = a$, we obtain for $E = 0$ (which implies that $\alpha = 0$):

$$\frac{\gamma a}{2} \sinh(\gamma b) + \cosh(\gamma b) = \cos(k(a+b)).$$

Since $\sinh(\gamma b) > 0$ and $\cosh(\gamma b) > 1$, the left-hand side is strictly larger than one and thus, this equation cannot have a solution. By continuity, there is no solution in a neighborhood of $E = 0$.

We can compute the intervals for which no solution exists more explicitly for a periodic array of δ-potentials. For this, we let the potential barrier width $b \to 0$ and the barrier height $|V_0| \to \infty$ such that the product bV_0 remains bounded, i.e., $bV_0 \to \varepsilon \neq 0$. Then $V_L(x)$ converges (in some sense) to $\varepsilon \sum_{m \in \mathbb{Z}} \delta(x - ma)$,

$$\gamma b = \sqrt{\frac{2m(qV_0 - E)b^2}{\hbar^2}} \to 0, \quad \cosh(\gamma b) \to 1$$

and

$$\frac{(\gamma^2 - \alpha^2)b}{2} \frac{\sinh(\gamma b)}{\gamma b} = \frac{m(qV_0 - 2E)b}{\hbar^2} \frac{\sinh(\gamma b)}{\gamma b} \to \frac{mq\varepsilon}{\hbar^2}.$$

Thus, (1.22) becomes in the limit

$$f(\alpha a) = Q \frac{\sin(\alpha a)}{\alpha a} + \cos(\alpha a) = \cos(ka), \tag{1.23}$$

where $Q = mq\varepsilon a/\hbar^2$. Figure 1.7 illustrates the function $f = f(\alpha a)$. In regions where $|f(\alpha a)| \leq 1$, there exists at least one solution $k \in B$ of (1.23); in regions with $|f(\alpha a)| > 1$, no solution k exists. Every connected subset of $[0, \infty) \setminus R(E)$, where $R(E) = \{E_0 \geq 0 : \text{there exists } k \in \mathbb{R} \text{ such that } E(k) = E_0\}$, is an energy gap. $\square$

An energy gap separates two energy bands. The nearest energy band below the energy gap (if it is unique) is called the *valence band*, the nearest energy band above the energy gap is termed the *conduction band* (see Fig. 1.8).

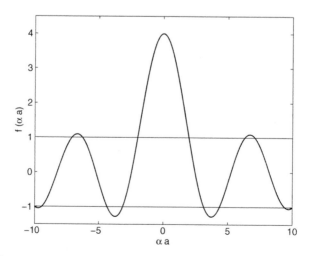

Fig. 1.7 The function $f = f(\alpha a)$ of (1.23)

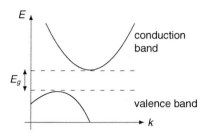

Fig. 1.8 Schematic band structure with energy gap E_g

Now we are able to state the definition of a semiconductor: It is a solid with an energy gap whose value is positive and smaller than a few electron volt (up to about 3 or 4 eV). In Table 1.1 the values of the energy gaps for some common semiconductor materials are collected.

The band structure of real crystals in three space dimensions is much more complicated than the one-dimensional situation of the Kronig–Penney model. Indeed, electrons traveling in different directions encounter different potential patterns, generated by the lattice atoms, and therefore, the $E(k)$ diagram is a function of the three-dimensional wave vector k. In physics textbooks, usually a projection of the full $E(k)$ diagram is shown. As an example, Fig. 1.9 shows the schematic band structures of silicon and gallium arsenide. In place of the positive and negative k axes of the one-dimensional case, two different crystal directions are shown, namely the $k = (0,0,1)^\top$ direction along the $+k$ axis (called the Δ line) and the $k = (1,1,1)^\top$ direction along the $-k$ axis (called the Λ line). The point $k = (0,0,0)$ is termed the Γ point. The points at the boundary of the Brillouin zone in the Λ and Δ directions are called L and X points, respectively (see Fig. 1.10; [9, 23]).

Table 1.1 Energy gaps of selected semiconductors (from [6, Appendix C]; the value for $Al_{0.3}Ga_{0.7}As$ is taken from [22])

Material	Symbol	Energy gap in eV
Indium arsenide	InAs	0.356
Germanium	Ge	0.661
Silicon	Si	1.124
Gallium arsenide	GaAs	1.424
Aluminum gallium arsenide	$Al_{0.3}Ga_{0.7}As$	1.80
Aluminum arsenide	AlAs	2.239
Gallium phosphide	GaP	2.272
Cadmium sulfur	CdS	2.514
Gallium nitride	GaN	3.44

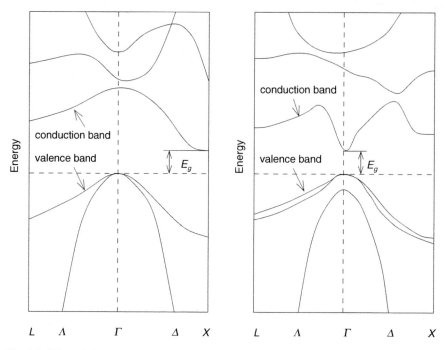

Fig. 1.9 Schematic band structure of silicon (*left*) and gallium arsenide (*right*) (see [12, Figs. 3.7 and 3.9])

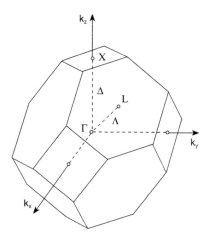

Fig. 1.10 Brillouin zone of semiconductors like silicon, germanium, gallium arsenide, etc.

1.4 The Semi-Classical Picture

In this section, we assume that the transport of electrons in a semiconductor can be described by the time-dependent one-particle Schrödinger equation

$$i\hbar\partial_t\psi = -\frac{\hbar^2}{2m}\Delta\psi - q(V_L(x) + V(x))\psi, \quad x \in \mathbb{R}^3,$$

where $V_L(x)$ is the (periodic) potential due to the interactions of the valence electrons with the lattice ions and $V(x)$ is a slowly varying (nonperiodic) potential modeling, for instance, the external applied potential. The solution of this equation is very difficult such that approximate models need to be used. One possibility is a semi-classical treatment which describes the carrier dynamics in the potential $V(x)$ by Newton's laws without explicitly treating the crystal potential $V_L(x)$. The influence of $V_L(x)$ is indirectly taken into account by the use of the energy band structure in the description of the velocity and the mass of the carrier ensemble.

The lattice potential is assumed to be spatially rapidly oscillating in macroscopic scale, with typical length λ, whereas the external potential varies over comparatively long distances, with typical length scale L. Defining the dimensionless parameters $h_0 = \hbar T \tau^2 / m\lambda^2$ and $\varepsilon = \lambda/L$, where T is the macroscopic and τ the microscopic time scale, the above Schrödinger equation can be written in dimensionless variables as

$$ih_0\partial_t\psi = -\frac{h_0^2}{2}\Delta\psi - \frac{h_0^2}{\varepsilon^2}V_L\left(\frac{x}{\varepsilon}\right)\psi - V(x)\psi, \quad x \in \mathbb{R}^3. \tag{1.24}$$

We refer to [24, 25] for details about the scaling. Here, we have used the same symbols for the physical and dimensionless variables. If the quantum wave energy is small compared to the microscopic kinetic energy, i.e., $h_0 \ll 1$, the so-called Vlasov equation (see Sect. 3.2) is obtained in the classical limit $h_0 \to 0$. This limit, without the additional periodic potential, was rigorously performed in [26, 27]. The homogenization limit $\varepsilon \to 0$ was analyzed in [25, 28]. The limit equation obtained by Poupaud and Ringhofer [25] is a Schrödinger equation with the Hamiltonian $-(h_0^2/2)\operatorname{div}(M\nabla\psi) + V(x)\psi$, where the matrix M is the Hessian of the energy band, $M = \mathrm{d}^2 E_n(0)/\mathrm{d}k^2$, and $E_n(0)$ is the strict minimum of E_n. This matrix is also called the *effective mass tensor* (cf. the discussion at the end of Sect. 1.5). The combined classical and homogenization limit is called the *semi-classical limit*. Here, the limits $h_0 \to 0$ and $\varepsilon \to 0$ are performed in such a way that the resulting semi-classical equation still contains quantum mechanical effects. Bechouche et al. [24] proved that the unscaled semi-classical equations of motion are, in the limit $\varepsilon = h_0 \to 0$,

$$\hbar\dot{x} = \nabla_k E_n(k), \quad \hbar\dot{k} = q\nabla_x V, \tag{1.25}$$

where x is the position of the electron at time t, k is the pseudo-wave vector introduced in Sect. 1.3, and the dot denotes differentiation with respect to time. For related semi-classical limits, we also refer to [25, 29–31] and references therein. The idea of the proof in [24, 25] is to formulate the Schrödinger equation with the so-called Wigner function (see Sect. 11.1) and to perform the limit in the Wigner equation, leading to a semi-classical Vlasov equation.

In the following, we will derive the semi-classical equations of motion (1.25) and the effective mass tensor in a more heuristic way. First, we motivate the left equation in (1.25). We assume that the electrons remain for all time in the same energy band, i.e., band crossings are not allowed (for band crossings, see [31]). Then we may omit the index n in $\psi_{n,k}$. Let ψ_k be a solution of the stationary Schrödinger equation

$$-\frac{\hbar^2}{2m}\Delta\psi_k - q(V_L(x) + V(x))\psi_k = E_n(k)\psi_k \quad \text{in } D \qquad (1.26)$$

with boundary conditions $\psi_k(x+y) = e^{ik \cdot y}\psi(x)$ for $x \in D$, $y \in L$, where D is the primitive cell of the lattice and L the Bravais lattice (see Sect. 1.1). Recall that the momentum is represented quantum mechanically by the operator $P\psi = -i\hbar\nabla_x\psi$, and its expectation value of a quantum system in the (normalized) state ψ_k is given by

$$\langle P \rangle_k = \int_D \overline{\psi}_k P \psi_k \, dx = \frac{\hbar}{i} \int_D \overline{\psi}_k \nabla_x \psi_k \, dx.$$

Then we define the mean velocity of this state by

$$v_n(k) = \frac{\langle P \rangle_k}{m}.$$

In the semi-classical setting, we introduce a "trajectory" of the quantum system corresponding to the mean velocity by $\dot{x} = v_n(k)$. In this interpretation, x and k are functions of time. Employing (1.26), we can relate the mean velocity to the energy band.

Lemma 1.9. *The semi-classical trajectory with mean velocity $v_n(k)$ is given by*

$$\dot{x} = v_n(k) = \frac{1}{\hbar}\nabla_k E_n(k), \quad t > 0.$$

Proof. By Bloch's Theorem 1.5, ψ_k can be written as

$$\psi_k(x) = e^{ik \cdot x} u_k(x). \qquad (1.27)$$

Differentiating (1.26) with respect to k and using (1.27) gives

$$\begin{aligned}
(\nabla_k E_n)\psi_k &= -\frac{\hbar^2}{2m}\Delta_x\left(e^{ik \cdot x}\nabla_k u_k + ix\psi_k\right) \\
&\quad - (qV_L + qV + E_n)\left(e^{ik \cdot x}\nabla_k u_k + ix\psi_k\right) \\
&= \left(-\frac{\hbar^2}{2m}\Delta_x - (qV_L + qV + E_n)\right)\left(e^{ik \cdot x}\nabla_k u_k\right) - \frac{i\hbar^2}{m}\nabla_x\psi_k \\
&\quad + ix\left(-\frac{\hbar^2}{2m}\Delta_x - (qV_L + qV + E_n)\right)\psi_k.
\end{aligned}$$

Observing that the last term vanishes in view of (1.26), multiplication of the above equation with $\overline{\psi}_k$ and integration over D yields

$$\nabla_k E_n \int_D |\psi_k|^2 \, dx + \frac{i\hbar^2}{m} \int_D \overline{\psi}_k \nabla_x \psi_k \, dx$$

$$= \int_D \overline{\psi}_k \left(-\frac{\hbar^2}{2m} \Delta_x - (qV_L + qV + E_n) \right) \left(e^{ik \cdot x} \nabla_k u_k \right) \, dx$$

$$= \int_D e^{ik \cdot x} \nabla_k u_k \left(-\frac{\hbar^2}{2m} \Delta_x - (qV_L + qV + E_n) \right) \overline{\psi}_k \, dx = 0,$$

where we have used integration by parts and again (1.26). The boundary integral in the integration-by-parts formula vanishes since u_k is periodic on D. Thus, if ψ_k is normalized,

$$\nabla_k E_n = \frac{\hbar^2}{im} \int_D \overline{\psi}_k \nabla_x \psi_k \, dx = \frac{\hbar}{m} \langle P \rangle_k = \hbar v_n(k).$$

This shows the lemma. □

The second equation in (1.25) is more difficult to justify (see [2, p. 220] or [9, p. 39]). If we suppose that the total energy, consisting of the band energy $E_n(k)$ and the potential energy $-qV(x)$, is constant along the trajectories $x = x(t), k = k(t)$, its derivative with respect to time vanishes,

$$0 = \frac{d}{dt}(E_n(k) - qV(x)) = \nabla_k E_n(k) \cdot \dot{k} - q\nabla_x V(x) \cdot \dot{x} = v_n(k) \cdot (\hbar \dot{k} - q\nabla_x V(x)). \tag{1.28}$$

This identity is satisfied if $\hbar \dot{k} - q\nabla_x V = 0$, which is the second equation in (1.25). Clearly, this equation is not necessary for the energy to be conserved since (1.28) only shows that $\hbar \dot{k} - q\nabla_x V$ is perpendicular to the velocity $v_n(k)$.

Another motivation comes from the fact that the *momentum operator* $P = -i\hbar \nabla_x$, acting on a plane wave $\psi(x) = e^{ik \cdot x}$, yields $P\psi = \hbar k \psi$ such that $\hbar k$ may be interpreted as a momentum value. Strictly speaking, this interpretation does not apply to the motion in a periodic potential since the momentum operator, acting on $\psi_k(x) = e^{ik \cdot x} u_k(x)$, gives

$$P\psi_k(x) = -i\hbar \nabla_x (e^{ik \cdot x} u_k) = \hbar k \psi_k + e^{ik \cdot x} \nabla_x u_k,$$

which is generally not equal to a multiple of ψ_k. However, one may see $\hbar k$ as a natural extension of the momentum in the case of a periodic potential. In order to emphasize this similarity, the expression $p = \hbar k$ is called the *crystal momentum*. Notice, however, that $\hbar k$ is *not* the momentum of the Bloch electron since the rate of change of its momentum only includes the external field $q\nabla_x V$ and not the periodic field of the lattice (also see the discussion in [2, p. 139, 219]). Now, assuming that Newton's law $\dot{p} = F$ is valid also for the crystal momentum $p = \hbar k$, we arrive to the second equation in (1.25) if the force F is given by the electric field $q\nabla_x V$.

The mean velocity v_n is defined by $\langle P \rangle_k = m v_n$. Employing the crystal momentum p instead of the physical momentum $\langle P \rangle_k$, we may define $p = m^* v_n$, where m^* is another mass. In the case of a free-electron motion (see Example 1.2), the mass m^* is the rest mass of the electron since $E(k) = \hbar^2 |k|^2 / 2m$ yields $\hbar k = p = m^* v_n = m^* \nabla_k E / \hbar = m^* \hbar k / m$ and hence $m^* = m$. What is the meaning of m^* in the case of a periodic potential? We differentiate the momentum $p = m^* v_n$ with respect to time and employ the first equation in (1.25),

$$\dot{p} = m^* \dot{v}_n = \frac{m^*}{\hbar} \frac{d^2 E_n}{dk^2} k = \frac{m^*}{\hbar^2} \frac{d^2 E_n}{dk^2} p,$$

which shows that

$$(m^*)^{-1} = \frac{1}{\hbar^2} \frac{d^2 E_n}{dk^2}. \tag{1.29}$$

This equation is considered as a definition of the *effective mass* m^*. The right-hand side of this definition is the Hessian matrix of E_n, so the symbol $(m^*)^{-1}$ is a 3×3 matrix.

The effective mass has the advantage that under some conditions, the behavior of the electrons in a crystal can be described similarly as that of a free-electron gas. In order to see this, we evaluate the Hessian of E_n near a local minimum (of the conduction band), i.e., $\nabla_k E_n(k_0) = 0$. Then $d^2 E_n(k_0)/dk^2$ is a symmetric positive definite matrix which can be diagonalized and the diagonal matrix has positive entries. We assume that the coordinates are chosen such that $d^2 E_n(k_0)/dk^2$ is already diagonal,

$$\frac{1}{\hbar^2} \frac{d^2 E_n}{dk^2}(k_0) = \begin{pmatrix} 1/m_1^* & 0 & 0 \\ 0 & 1/m_2^* & 0 \\ 0 & 0 & 1/m_3^* \end{pmatrix}.$$

Assume that the energy values are shifted in such a way that $E_n(k_0) = 0$. (This is possible by fixing a reference point for the energy.) Let us further assume that already $E_n(0) = 0$, otherwise define $\tilde{E}_n(k) = E_n(k + k_0)$. If the function $k \mapsto E_n(k)$ is smooth, Taylor's formula implies that

$$E_n(k) = E_n(0) + \nabla_k E_n(0) \cdot k + \frac{1}{2} k^\top \left(\frac{d^2 E_n}{dk^2}(0) \right) k + \mathcal{O}(|k|^3)$$

$$= \frac{\hbar^2}{2} \left(\frac{k_1^2}{m_1^*} + \frac{k_2^2}{m_2^*} + \frac{k_3^2}{m_3^*} \right) + \mathcal{O}(|k|^3) \quad \text{for } k \to 0,$$

where $k = (k_1, k_2, k_3)^\top$ and $\mathcal{O}(|k|^3)$ denote terms of order $|k|^3$. If the effective masses are equal in all directions, i.e., $m^* = m_1^* = m_2^* = m_3^*$, we can write, neglecting higher-order terms,

$$E_n(k) = \frac{\hbar^2}{2m^*} |k|^2. \tag{1.30}$$

This relation is valid for wave vectors k sufficiently close to a local band minimum (of the conduction band). The scalar m^* is called here the *isotropic effective mass*. Comparing this expression with the dispersion relation of a free-electron gas,

$$E(k) = \frac{\hbar^2}{2m}|k|^2,$$

we infer that the energy of an electron near a band minimum (of an isotropic semiconductor) equals the energy of a free-electron in a vacuum where the electron rest mass m is replaced by the effective mass m^*.

Expression (1.30) is referred to as the *parabolic band approximation* and usually, the range of wave vectors is extended to the whole space, $k \in \mathbb{R}^3$ (see the scaling argument in Sect. 1.6). This simple model is appropriate for low applied fields for which the carriers are close to the conduction band minimum. For high applied fields, however, the higher-order terms in the above Taylor expansion cannot be ignored. In order to account for nonparabolic effects, often the *nonparabolic band approximation* in the sense of Kane is used:

$$E_n(1 + \alpha E_n) = \frac{\hbar^2}{2m^*}|k|^2, \tag{1.31}$$

where m^* is determined from (1.29) at the conduction band minimum at $k = 0$, the nonparabolicity parameter α is given by

$$\alpha = \frac{1}{E_g}\left(1 - \frac{m^*}{m}\right)^2,$$

and E_g is the band gap (see [6, Sect. 2.1] or [10, (1.40)]). In Table 1.2 some values for α are shown. Formula (1.31) can be obtained from approximate solutions of the Schrödinger equation (1.18) derived by the so-called $k \cdot p$ theory (see Sect. 1.5).

When we consider the effective mass definition (1.29) near a maximum (of the valence band), we find that the Hessian of E_n is negative definite. This would lead to a negative effective mass. In order to obtain a positive mass, we may also change the sign for the group velocity v_n since this is consistent with the definition $p = m^* v_n$. A reversed sign in the velocity means that the particles, under the influence of an electric field, travel in the opposite direction compared to electrons. This is the case if the particles have a positive charge. Employing a positive charge leads again to a positive effective mass. The corresponding (pseudo-) particles are called *holes* (or *defect electrons*). Physically, a hole is a vacant orbital in a valence band. Thus,

Table 1.2 Values of the nonparabolicity parameter α for some semiconductors (from [9, Table 1.1]; the value for $Al_{0.3}Ga_{0.7}As$ is taken from [32])

Material	Si	Ge	GaAs	$Al_{0.3}Ga_{0.7}As$	InAs
α in $(eV)^{-1}$	0.5	0.65	0.64	0.72	2.73

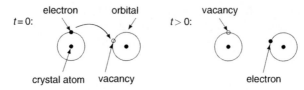

Fig. 1.11 Motion of a valence band electron to a neighboring vacant orbital or, equivalently, of a hole in the inverse direction

the current flow in a semiconductor crystal comes from two sources: the flow of electrons in the conduction band and the flow of holes in the valence band. It is a convention to consider the motion of the valence band vacancies rather than the electrons moving from one vacant orbital to the next (see Fig. 1.11).

We summarize: Close to the bottom $k = 0$ of the conduction band in an isotropic semiconductor, the band energy becomes

$$E_n(k) = E_c + \frac{\hbar^2}{2m_e^*}|k|^2, \tag{1.32}$$

whereas near the top $k = 0$ of the valence band we have

$$E_n(k) = E_v - \frac{\hbar^2}{2m_h^*}|k|^2, \tag{1.33}$$

where E_c is the energy at the conduction band minimum, E_v the energy at the valence band maximum, m_e^* the effective electron mass, and m_h^* the effective hole mass. Clearly, the energy gap E_g is given by $E_g = E_c - E_v$ (see Fig. 1.12).

For semiconductors possessing an energy band with ellipsoidal isoenergetic surfaces, we find that

$$E_n(k) = \frac{\hbar^2}{2}\left(\frac{|k_\ell|^2}{m_\ell^*} + \frac{|k_t|^2}{m_t^*}\right), \quad k = (k_\ell, k_t)^\top \in \mathbb{R}^3, \; k_\ell \in \mathbb{R}, \; k_t \in \mathbb{R}^2,$$

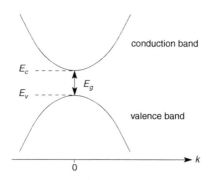

Fig. 1.12 Schematic conduction and valence bands near the extrema at $k = 0$

Table 1.3 Relative effective electron masses for some semiconductor materials (from [9, Table 1.1]). The electron rest mass is denoted by m_e, and m_ℓ^* and m_t^* refer to the longitudinal and transversal effective masses, respectively

Material	m_ℓ^*/m_e	m_e^*/m_e	m_t^*/m_e
Si	0.98	–	0.19
Ge	1.64	–	0.082
GaP	1.12	–	0.22
GaAs	–	0.067	–
InAs	–	0.023	–
InP	–	0.080	–
ZnS	–	0.28	–

where m_ℓ^* and m_t^* denote the longitudinal and transversal effective masses, respectively, and the equation describes conduction bands at the L point and along the Δ line (see Sect. 1.1). Some values for the effective masses of common cubic semiconductors can be found in Table 1.3.

1.5 The $k \cdot p$ Method

In the previous section we have seen that the mean velocity and the effective mass of the electrons in a semiconductor can be computed in the semi-classical picture from the energy band structure; see formulas (1.25) and (1.29). In this section we describe the $k \cdot p$ method which allows us to derive an approximation of the energy band $E_n(k)$ close to the bottom of the conduction band or close to the top of the valance band. The main assumption of this method is that the energy at $k = 0$ is known. Then $E_n(k)$ close to the Γ-point $k = 0$ can be computed using time-independent perturbation theory. We proceed in the following as in [3, Sect. 8.7] and [12, Sect. 4.1]. For other methods of calculating the band structure, see, e.g., [2, Chaps. 8, 9, 10, 11].

The starting point is the Schrödinger equation (1.18) for the functions $u_{n,k}$ of the Bloch function $\psi_{n,k} = e^{ik \cdot x} u_{n,k}$, here written in the form

$$(H_0 + \varepsilon H_1)u_{n,k} = E_n(k)u_{n,k},\tag{1.34}$$

where

$$H_0 = -\frac{\hbar^2}{2m}\Delta_x - qV_L(x)$$

is the single-electron Hamiltonian and

$$\varepsilon H_1 = -\frac{i\hbar^2}{m}k \cdot \nabla_x + \frac{\hbar^2}{2m}|k|^2$$

is considered to be a perturbation of H_0. The parameter ε measures the magnitude of $|k|$. Defining the quantum momentum operator $P = -i\hbar\nabla_x$, we can formally write

$$\varepsilon H_1 = \frac{\hbar}{m}k \cdot P + \frac{\hbar^2}{2m}|k|^2,$$

which explains the name of the $k \cdot p$ method (usually written as $k \cdot p$ instead of $k \cdot P$).

We assume that $|k|$ is "small" compared to a reference wave vector. Notice that for $k = 0$, the operator $H_0 + \varepsilon H_1$ reduces to H_0. Furthermore, we suppose that the solutions of the eigenvalue problem

$$H_0 u_n^{(0)} = E_n^{(0)} u_n^{(0)} \quad \text{in } D, \tag{1.35}$$

together with periodic boundary conditions, are known. Here, D is the primitive cell and $E_n^{(0)} = E_n(0)$. Since the operator H_0 is symmetric and real, the eigenfunctions $u_n^{(0)}$ and the eigenvalues $E_n^{(0)}$ are real. We show the following result.

Theorem 1.10 (Effective mass). *Let the solutions $(u_n^{(0)})$ to (1.35) form a non-degenerate orthonormal basis of $L^2(D;\mathbb{C})$ (i.e., all eigenspaces are one-dimensional). Then, up to second order in ε,*

$$E_n(k) = E_n^{(0)} + \frac{\hbar^2}{2}k^\top(m^*)^{-1}k, \tag{1.36}$$

where the matrix $(m^*)^{-1}$ consists of the elements $1/m_{j\ell}^*$ with

$$\frac{m}{m_{j\ell}^*} = \delta_{j\ell} - \frac{2\hbar^2}{m}\sum_{q \neq n}\frac{P_{qnj}P_{nq\ell}}{E_q^{(0)} - E_n^{(0)}}, \quad \text{where} \quad P_{qnj} = \int_D u_q^{(0)}\frac{\partial u_n^{(0)}}{\partial x_j}\,dx. \tag{1.37}$$

Notice that the one-dimensionality of the eigenspaces implies that $E_q^{(0)} \neq E_n^{(0)}$ for all $q \neq n$ and so (1.37) is well defined.

Proof. We apply a perturbation method to (1.34) (see [3, Sect. 4.1] or [12, Appendix C.1.1]). For this, we develop

$$u_{n,k} = u_n^{(0)} + \varepsilon u_n^{(1)} + \varepsilon^2 u_n^{(2)} + \cdots, \quad E_n(k) = E_n^{(0)} + \varepsilon E_n^{(1)} + \varepsilon^2 E_n^{(2)} + \cdots.$$

Inserting these expressions into (1.34) and equating terms with the same order of ε leads to

$$\varepsilon^0 : \quad H_0 u_n^{(0)} = E_n^{(0)} u_n^{(0)}, \tag{1.38}$$

$$\varepsilon^1 : \quad H_0 u_n^{(1)} + H_1 u_n^{(0)} = E_n^{(0)} u_n^{(1)} + E_n^{(1)} u_n^{(0)}, \tag{1.39}$$

$$\varepsilon^2 : \quad H_0 u_n^{(2)} + H_1 u_n^{(1)} = E_n^{(0)} u_n^{(2)} + E_n^{(1)} u_n^{(1)} + E_n^{(2)} u_n^{(0)}. \tag{1.40}$$

The zeroth-order equation (1.38) is clearly the same as (1.35). In order to derive the first-order correction, we multiply (1.39) by $u_q^{(0)}$ and integrate over D. Then, observing that $(u_q^{(0)}, u_n^{(0)}) = \int_D u_q^{(0)} u_n^{(0)} \, dx = \delta_{qn}$, we obtain

$$\left(u_q^{(0)}, H_0 u_n^{(1)}\right) + \left(u_q^{(0)}, H_1 u_n^{(0)}\right) = E_n^{(0)} \left(u_q^{(0)}, u_n^{(1)}\right) + E_n^{(1)} \delta_{qn}.$$

Integrating by parts twice (or employing the symmetry of H_0), it follows that

$$\left(E_q^{(0)} - E_n^{(0)}\right) \left(u_q^{(0)}, u_n^{(1)}\right) + \left(u_q^{(0)}, H_1 u_n^{(0)}\right) = E_n^{(1)} \delta_{qn}.$$

For $q = n$, this gives an expression for $E_n^{(1)}$ only depending on $u_n^{(0)}$:

$$E_n^{(1)} = \left(u_n^{(0)}, H_1 u_n^{(0)}\right). \tag{1.41}$$

For $q \neq n$ we have

$$\left(u_q^{(0)}, u_n^{(1)}\right) = \frac{\left(u_q^{(0)}, H_1 u_n^{(0)}\right)}{E_q^{(0)} - E_n^{(0)}}. \tag{1.42}$$

This is well defined since $E_q^{(0)} \neq E_n^{(0)}$ for all $q \neq n$. The sequence $(u_q^{(0)})$ is an orthonormal basis of $L^2(D; \mathbb{C})$, so we can develop $u_n^{(1)}$ in this basis,

$$u_n^{(1)} = \sum_q \left(u_q^{(0)}, u_n^{(1)}\right) u_q^{(0)}. \tag{1.43}$$

In the sum we need an expression for the term $(u_n^{(0)}, u_n^{(1)})$. This term is not determinable from the above calculation. If we want to have asymptotic normalization of $u_{n,k}$, $\|u_{n,k}\| = 1 + \mathcal{O}(\varepsilon^2)$ as $\varepsilon \to 0$, we must choose $(u_n^{(0)}, u_n^{(1)}) = 0$. In view of (1.42), the expression (1.43) becomes

$$u_n^{(1)} = \sum_{q \neq n} \frac{\left(u_q^{(0)}, H_1 u_n^{(0)}\right)}{E_q^{(0)} - E_n^{(0)}} u_q^{(0)}.$$

Thus, up to first order, the eigenfunctions are given by

$$u_n^{(0)} + \varepsilon u_n^{(1)} = u_n^{(0)} + \sum_{q \neq n} \frac{\left(u_q^{(0)}, \varepsilon H_1 u_n^{(0)}\right)}{E_q^{(0)} - E_n^{(0)}} u_q^{(0)},$$

and the eigenvalues are $E_n^{(0)} + \varepsilon E_n^{(1)} = E_n^{(0)} + (u_n^{(0)}, \varepsilon H_1 u_n^{(0)})$. Notice that these corrections only depend on the unperturbed eigenfunctions $u_n^{(0)}$ which are assumed to be known.

In order to derive the second-order correction, we multiply (1.40) by $u_q^{(0)}$ and integrate over D:

$$\left(u_q^{(0)}, H_0 u_n^{(2)}\right) + \left(u_q^{(0)}, H_1 u_n^{(1)}\right) = E_n^{(0)}\left(u_q^{(0)}, u_n^{(2)}\right) + E_n^{(1)}\left(u_q^{(0)}, u_n^{(1)}\right) + E_n^{(2)}\delta_{qn}.$$

As above, we obtain

$$\left(E_q^{(0)} - E_n^{(0)}\right)\left(u_q^{(0)}, u_n^{(2)}\right) + \left(u_q^{(0)}, H_1 u_n^{(1)}\right) = E_n^{(1)}\left(u_q^{(0)}, u_n^{(1)}\right) + E_n^{(2)}\delta_{qn}.$$

Using (1.41) and (1.43), the case $q = n$ yields

$$E_n^{(2)} = \left(u_n^{(0)}, H_1 u_n^{(1)}\right) - E_n^{(1)}\left(u_n^{(0)}, u_n^{(1)}\right)$$

$$= \sum_q \left(u_q^{(0)}, u_n^{(1)}\right)\left(u_n^{(0)}, H_1 u_q^{(0)}\right) - \left(u_n^{(0)}, H_1 u_n^{(0)}\right)\left(u_n^{(0)}, u_n^{(1)}\right)$$

$$= \sum_{q \neq n} \left(u_n^{(0)}, H_1 u_q^{(0)}\right)\left(u_q^{(0)}, u_n^{(1)}\right) = \sum_{q \neq n}\left(u_n^{(0)}, H_1 u_q^{(0)}\right)\frac{\left(u_q^{(0)}, H_1 u_n^{(0)}\right)}{E_q^{(0)} - E_n^{(0)}}.$$

In the last equation we have employed (1.42). Thus, the second-order correction to the eigenvalues is

$$E_n^{(0)} + \varepsilon E_n^{(1)} + \varepsilon^2 E_n^{(2)} = E_n^{(0)} + \left(u_n^{(0)}, \varepsilon H_1 u_n^{(0)}\right)$$

$$+ \sum_{q \neq n}\frac{\left(u_n^{(0)}, \varepsilon H_1 u_q^{(0)}\right)\left(u_q^{(0)}, \varepsilon H_1 u_n^{(0)}\right)}{E_q^{(0)} - E_n^{(0)}}.$$

It remains to compute the scalar products. We write

$$\left(u_n^{(0)}, \varepsilon H_1 u_q^{(0)}\right) = -\frac{i\hbar^2}{m}k \cdot \left(u_n^{(0)}, \nabla u_q^{(0)}\right) + \frac{\hbar^2}{2m}|k|^2\left(u_n^{(0)}, u_q^{(0)}\right)$$

$$= -\frac{i\hbar^2}{m}k \cdot P_{nq} + \frac{\hbar^2}{2m}|k|^2\delta_{nq},$$

where

$$P_{nq} = \int_D u_n^{(0)}\nabla u_q^{(0)}\, dx.$$

The periodicity of $u_n^{(0)}$ on D gives

$$P_{nn} = \frac{1}{2}\int_D \nabla\left(\left(u_n^{(0)}\right)^2\right)\, dx = 0,$$

and therefore,

$$\left(u_n^{(0)}, \varepsilon H_1 u_q^{(0)} \right) = \begin{cases} \dfrac{\hbar^2}{2m} |k|^2 & \text{for } n = q, \\[2ex] -\dfrac{i\hbar^2}{m} k \cdot P_{nq} & \text{for } n \neq q. \end{cases}$$

This shows that $E_n(k)$ is, up to second order in ε,

$$\begin{aligned} E_n(k) &= E_n^{(0)} + \varepsilon E_n^{(1)} + \varepsilon^2 E_n^{(2)} \\ &= E_n^{(0)} + \frac{\hbar^2}{2m}|k|^2 - \frac{\hbar^4}{m^2} \sum_{q \neq n} \frac{(k \cdot P_{nq})(k \cdot P_{qn})}{E_q^{(0)} - E_n^{(0)}} \\ &= E_n^{(0)} + \frac{\hbar^2}{2m}|k|^2 - \frac{\hbar^4}{m^2} \sum_{q \neq n} \sum_{j,\ell} k_j k_\ell \frac{P_{nqj} P_{qn\ell}}{E_q^{(0)} - E_n^{(0)}} = E_n^{(0)} + \frac{\hbar^2}{2} \sum_{j,\ell} \frac{k_j k_\ell}{m_{j\ell}^*} \end{aligned}$$
(1.44)

which proves the theorem. □

Equation (1.44) shows that the first-order correction of the energy simply yields the free-electron mass. The second-order correction is needed to obtain an effective mass which is different from the free-electron mass. This is the reason why we computed the corrections up to second order.

In most semiconductors, the bottom of the conduction band is nondegenerate at $k = 0$ (the energy bands do not cross; see Fig. 1.9), and Theorem 1.10 is applicable. However, the valence band maximum of semiconductors with diamond and zinc blende structure is degenerate [11, p. 148], and hence, the above result does not hold. Mathematically, we have in such a situation several eigenfunctions with the same eigenvalue, for instance for $n \neq q$,

$$H_0 u_n^{(0)} = E_n^{(0)} u_n^{(0)} \quad \text{and} \quad H_0 u_q^{(0)} = E_q^{(0)} u_q^{(0)}, \quad \text{but} \quad E_n^{(0)} = E_q^{(0)}.$$

Then, the expression (1.37) may not be defined. It is still possible to derive a formula similar to (1.36) in the degenerate case by applying degenerate perturbation theory. The idea is to find a linear combination

$$\widetilde{u}_n^{(0)} = \sum_{\alpha=1}^{A} c_\alpha u_{n,\alpha}^{(0)}$$

of the eigenfunctions $u_{n,\alpha}^{(0)}$ with the same energy $E_n^{(0)}$ such that the nominator in the first-order correction

$$\widetilde{u}_n^{(1)} = \sum_{q \neq n} \frac{\left(\widetilde{u}_q^{(0)}, H_1 \widetilde{u}_n^{(0)} \right)}{E_q^{(0)} - E_n^{(0)}} \widetilde{u}_q^{(0)}$$

vanishes if $E_q^{(0)} = E_n^{(0)}$. Thus, the problem is to find coefficients c_α such that $(\widetilde{u}_q^{(0)}, H_1 \widetilde{u}_n^{(0)}) = 0$. It can be shown [12, Sect. 4.1.4] that the energies $E_n(k)$ are, up to second order, the eigenvalues of the matrix $H_{n,k} \in \mathbb{R}^{A \times A}$ with elements

$$(H_{n,k})_{\mu\nu} = E_n^{(0)}\delta_{\mu\nu} - \frac{i\hbar^2}{m}\sum_j k_j \widetilde{P}_{nnj}^{\mu\nu} + \frac{\hbar^2}{2}\sum_{j,\ell}\frac{k_j k_\ell}{(m^*)_{j\ell}^{\mu\nu}}, \qquad (1.45)$$

where the coefficients $\widetilde{P}_{nqj}^{\mu\nu}$ are defined similarly as in Theorem 1.10 and

$$\frac{m}{(m^*)_{j\ell}^{\mu\nu}} = \delta_{j\ell}\delta_{\mu\nu} - \frac{2\hbar^2}{m}\sum_{q\neq n}\sum_{\alpha=1}^{A}\frac{\widetilde{P}_{nqj}^{\mu\alpha}\widetilde{P}_{qn\ell}^{\nu\alpha}}{E_q^{(0)} - E_n^{(0)}}.$$

Analogous results as above can be derived for holes in the valence band. In this case, the energy $E_n(k)$ can be approximately written as

$$E_n(k) = E_n^c(0) - \frac{\hbar^2}{2}k^\top (m_h^*)^{-1}k,$$

where $E_n^c(0)$ is the top energy of the valence band and m_h^* is the effective mass tensor for the holes, similarly defined as above. In this case, $\widetilde{P}_{nnj}^{\mu\nu} = 0$ for all j, such that the linear term in k in (1.45) vanishes.

In our arguments, we have neglected the magnetic dipole moment associated with the electron spin. This magnetic moment removes partially the degeneracy at the valence band maximum through its interaction with the magnetic dipole moment which is associated with the electron orbit. For details about this spin–orbit interaction, we refer to [12, Sect. 4.2.4].

The above derivation of the effective mass is valid for a Hamiltonian which takes into account the periodic lattice potential but not an external macroscopic potential. A formula for the effective mass for Hamiltonians including external potentials was rigorously derived by Poupaud and Ringhofer [25] and Allaire and Piatnitski [28]. Both works rely on Bloch wave regularity and the assumption of simple eigenvalues of the corresponding Hamiltonian. Recently, Ben Abdallah and Barletti [33] have reconsidered the effective mass approximation by employing an envelope function decomposition inspired by Burt [34] (also see the work of Kohn and Luttinger [35]) and performed the homogenization limit $\varepsilon \to 0$ in (1.24).

1.6 Semiconductor Statistics

In this section we will answer the question how many electrons and holes are in a semiconductor of finite size which is in thermal equilibrium (i.e., no current flow). Let $f(E)$ be the occupation density of the quantum state of energy E. We can interpret $f(E)$ as the mean number of electrons in a quantum state of energy $E = E_n(k)$. Then the number of electrons equals the sum of all $f(E_n(k))$ over (n,k), where n is the band number and k the pseudo-wave vector:

$$N^* = \sum_n N_n^* = 2\sum_n\sum_{k\in B} f(E_n(k)). \qquad (1.46)$$

The factor 2 takes into account the two possible states of the spin of an electron, and B is the Brillouin zone. This formula leads to two questions:

1. How can the sum over many k be computed practically?
2. How does the function f depend on the energy?

To answer the first question, we recall that k can take the values

$$k = \sum_{j=1}^{3} \frac{n_j}{N_j} a_j^*, \quad n_j \in \mathbb{Z}, \quad -\frac{N_j}{2} \leq n_j \leq \frac{N_j}{2},$$

where a_j^* is the primitive vector of the reciprocal lattice and N_j is the number of primitive cells in the jth spatial direction (see (1.17)). In one space dimension, we can simplify this expression. Let the crystal be given by a chain of $M+1$ atoms with distance h. Then the length of the chain is $L = Mh$. A primitive cell is given by $D = [-h/2, h/2]$ and, since $\mathrm{vol}(D) = 2\pi/\mathrm{vol}(B)$, by Lemma 1.1, the Brillouin zone equals $B = [-\pi/h, \pi/h]$. Thus, the wave vector k can take one of the discrete values

$$k_j = \frac{2\pi j}{L} = \frac{2\pi j}{Mh}, \quad -\frac{M}{2} \leq j \leq \frac{M}{2}.$$

Since $L = Mh$ is the chain length, each state occupies $2\pi/L$ in the wave-vector space. The number of states between k and $k + \triangle k$ is $L \triangle k/2\pi$. In d dimensions, this generalizes to $L^d \triangle k/(2\pi)^d$. To be more precise, we have to take into account the spin of the electrons. Then, the number of states divided by the volume in the k-space equals $2\mathrm{vol}(\Omega)/(2\pi)^d$, where $\mathrm{vol}(\Omega) = L^d$ denotes the volume of the semiconductor. We call this number the *density of states* in k-space,

$$N(k) = \frac{2\mathrm{vol}(\Omega)}{(2\pi)^d}. \tag{1.47}$$

Now we come back to the first question. Typically, $L = 1\,\mu\mathrm{m} = 10^{-6}\,\mathrm{m}$ and $h = 10^{-10}\,\mathrm{m}$, so $M = 10^4$. Therefore, one may consider k to be approximately continuous. To simplify the presentation, we assume in the following that M is even. Then the sum

$$\sum_{j=-M/2}^{M/2-1} g(k_j),$$

where g is some function, transforms into the integral

$$\sum_{j=-M/2}^{M/2-1} g(k_j) = \sum_{j=-M/2}^{M/2-1} g\left(\frac{2\pi j}{Mh}\right) \approx \int_{-M/2}^{M/2} g\left(\frac{2\pi j}{Mh}\right) \mathrm{d}j = \frac{L}{2\pi} \int_{-\pi/h}^{\pi/h} g(k)\,\mathrm{d}k.$$

In d-space dimensions, the factor $L/2\pi$ becomes $(L/2\pi)^d = \mathrm{vol}(\Omega)/(2\pi)^d$. Therefore, we can write

$$\frac{\text{vol}(\Omega)}{(2\pi)^d} \int_B g(k)\,dk \quad \text{instead of} \quad \sum_{k \in B} g(k). \tag{1.48}$$

The advantage of the integral formulation is that integrals can usually be more easily computed than sums. In the *continuum limit* $M \to \infty$ and $h \to 0$ such that Mh stays finite, we can extend the Brillouin zone $B = [-\pi/h, \pi/h]^d$ to $\mathbb{R}^d$ and write

$$\frac{\text{vol}(\Omega)}{(2\pi)^d} \int_{\mathbb{R}^d} g(k)\,dk.$$

Notice that we have considered here a general d-dimensional situation. The reason is that it is possible to confine carriers in one (or two) space dimensions, i.e., the carriers are confined in the x-y plane (or in the x-direction) and are free to move in the z-direction (or in the y-z plane). Such structures can be constructed with so-called semiconductor heterostructures and are called quantum wires or quantum wells, respectively (see, e.g., [9, Sect. 1.5.2]).

We summarize: The number of electrons in the nth band of a semiconductor crystal Ω with finite size, which is in thermal equilibrium, is (with the above approximations) given by

$$N_n^* = \frac{2\text{vol}(\Omega)}{(2\pi)^d} \int_B f(E_n(k))\,dk. \tag{1.49}$$

Now we answer the second question. We observe that electrons are fermions, i.e., particles with half-integral spin, satisfying the following properties:

1. Electrons cannot be distinguished from each other.
2. The Pauli exclusion principle holds, i.e., each quantum state can be occupied by not more than two electrons with opposite spins.

We arrange M electrons into ℓ energy bands each of which has g_n quantum states, $n = 1, \ldots, \ell$. Suppose that m_n electrons are occupying quantum states in the nth band, where $m_n \leq g_n$. Recall that, by the Pauli exclusion principle, each electron is occupying exactly one quantum state. The occupation probability $f(E)$ at energy E is equal to the number of occupied states m_n divided by the number of states g_n, i.e., we have to compute the quotient m_n/g_n. To this end, we are looking for the most probable configuration $(m_1, \ldots, m_\ell)$ which is obtained by maximizing the number of different configurations in all bands under the condition that the total number of electrons $\sum_n m_n$ and the total energy E are conserved.

First, we derive a formula for the number of different configurations. The number Q_n of different arrangements in the nth band equals the number of all possible configurations,

$$g_n(g_n - 1) \cdots (g_n - m_n + 1),$$

divided by the number of all possible permutations of the m_n electrons (since they are indistinguishable), $m_n!$, hence

$$Q_n = \frac{g_n(g_n-1)\cdots(g_n-m_n+1)}{m_n!} = \frac{g_n!}{(g_n-m_n)!m_n!}.$$

The total number of configurations reads in the limit $\ell \to \infty$

$$Q(m_1,m_2,\ldots) = \prod_{n=1}^{\infty} Q_n = \prod_{n=1}^{\infty} \frac{g_n!}{(g_n-m_n)!m_n!}.$$

In order to manipulate this function, it is convenient to consider

$$\log Q(m_1,m_2,\ldots) = \sum_{n=1}^{\infty} \left(\log g_n! - \log(g_n-m_n)! - \log m_n! \right).$$

Moreover, since computations with the factorial are cumbersome, we approximate $\log Q$ by employing Stirling's formula

$$n! \sim \frac{n^n}{e^n} \quad \text{or} \quad \log n! \sim n\log n - n \quad (n \to \infty),$$

where the notation $a(n) \sim b(n)$ for $n \to \infty$ means that $\lim_{n\to\infty} a(n)/b(n) = 1$. Then, approximately,

$$\log Q(m_1,m_2,\ldots) = \sum_{n=1}^{\infty} \left(\log g_n! - (g_n-m_n)\log(g_n-m_n) - m_n\log m_n + g_n \right).$$

This function is related to the thermodynamic entropy (see [3, p. 268]). The most probable configuration of $(m_1,m_2,\ldots)$ is that one which maximizes $\log Q$, under the constraints that the particle number and the energy are conserved,

$$\max_{m_j} \log Q(m_1,m_2,\ldots) \quad \text{subject to} \quad \sum_{n=1}^{\infty} m_n = M \text{ and } \sum_{n=1}^{\infty} m_n E_n = E.$$

In other words, the occupation probability $f(E_j)$ is equal to the maximizer of the above constrained extremal problem divided by the number of states g_j. The solution of this problem is given in the following lemma which answers the second question stated at the beginning of this section.

Lemma 1.11 (Fermi–Dirac distribution). *The mean number of electrons in a quantum state of energy E is given by the* Fermi–Dirac distribution function

$$f(E) = \frac{1}{1 + e^{(E-q\mu)/k_B T}}, \tag{1.50}$$

where k_B is the Boltzmann constant.

The two parameters T and μ are Lagrange multipliers coming from the constrained extremal problem. Thermodynamics shows that T can be interpreted as the *temperature* of the system and μ as the *chemical potential* [3, Chap. 5]. The meaning will become more transparent below (see Remark 1.12).

Proof. We solve the constrained extremal problem with Lagrange multipliers, i.e., we introduce

$$F(m_1, m_2, \ldots; \lambda_1, \lambda_2) = \log Q + \lambda_1 \left(\sum_{n=1}^{\infty} m_n - M \right) + \lambda_2 \left(\sum_{n=1}^{\infty} m_n E_n - E \right).$$

A necessary condition is

$$0 = \frac{\partial F}{\partial m_j} = \log(g_j - m_j) + 1 - (\log m_j + 1) + \lambda_1 + \lambda_2 E_j$$

$$= \log \left(\frac{g_j}{m_j} - 1 \right) + \lambda_1 + \lambda_2 E_j.$$

Solving for m_j / g_j yields

$$\frac{m_j}{g_j} = \frac{1}{1 + e^{-\lambda_1 - \lambda_2 E_j}}.$$

Defining the temperature T and chemical potential μ by

$$\lambda_1 = \frac{q\mu}{k_B T} \quad \text{and} \quad \lambda_2 = -\frac{1}{k_B T},$$

we obtain

$$\frac{m_j}{g_j} = \frac{1}{1 + e^{(E_j - q\mu)/k_B T}} = f(E_j).$$

Since the left-hand side is the mean number of electrons in a quantum state of energy E_j, the lemma is shown. $\square$

Remark 1.12 (Fermi–Dirac and Maxwell–Boltzmann distributions). The properties of the Fermi–Dirac distribution can be understood as follows (also see [3, p. 298ff.]). At zero temperature, this function becomes

$$f(E) = \begin{cases} 1 & \text{for } E < q\mu, \\ 0 & \text{for } E > q\mu, \end{cases} \quad \text{and} \quad f(q\mu) = \frac{1}{2}$$

(see Fig. 1.13). This means that all states which have an energy smaller than the chemical potential are occupied, and all states with an energy larger than $q\mu$ are empty. Physically, this behavior comes from the Pauli principle according to which two electrons must not occupy the same quantum state. At zero temperature, the states with lowest energy are filled first. The energy of the state filled by the last particle is equal to the chemical potential $q\mu$. This number is also called the *Fermi energy* E_F. For nonzero temperature, there is a positive probability that some energy states above $q\mu$ will be occupied, i.e., some particles jump to higher energy levels due to thermal excitation.

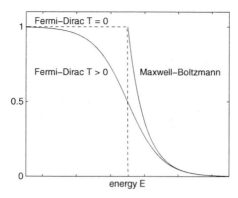

Fig. 1.13 The Fermi–Dirac distribution at zero and nonzero temperatures and the Maxwell–Boltzmann approximation

Strictly speaking, the Fermi energy is defined as $E_F = q\mu$ only if $T = 0$. By abuse of notation, we will also employ this terminology for $T > 0$ (e.g., like in [23, Sect. 7.2]).

For energies much larger than the Fermi energy in the sense of $E - E_F \gg k_B T$, we can approximate the Fermi–Dirac distribution by the *Maxwell–Boltzmann distribution*

$$f(E) = e^{-(E-E_F)/k_B T} \tag{1.51}$$

since $1/(1+e^x) \sim e^{-x}$ as $x \to \infty$ (Fig. 1.13). Semiconductors whose electron distribution can be described by this distribution are called *nondegenerate*. Semiconductor materials in which the Fermi–Dirac distribution has to be used (for instance, in the case of high doping) are termed *degenerate*. □

The electron density in a given band $E_j(k)$ is determined by the number of electrons (1.49), $N^* = N_j^*$, divided by the volume of the semiconductor domain:

$$n = \frac{N^*}{\text{vol}(\Omega)} = \frac{2}{(2\pi)^d} \int_B f(E_j(k)) \, dk, \quad \text{where } f(E) = \frac{1}{1+e^{(E-q\mu)/k_B T}}. \tag{1.52}$$

We wish to formulate the integral not in the k-space but in the energy space. For this, we introduce the *Dirac delta distribution* δ as that functional which associates the value $g(0)$ with an appropriate function g, i.e., $\delta(g) = g(0)$. This is also written as $\langle \delta, g \rangle = g(0)$ or as the *symbolic integral*

$$\int_{\mathbb{R}} \delta(x) g(x) \, dx = g(0). \tag{1.53}$$

We recall that this notation has to be considered with care: The symbol δ is not a function but a functional and (1.53) is not an integral but a symbolic representation, which is useful for the following computations.

With the Dirac distribution, the expression (1.52) for the electron density can be reformulated. We obtain from (1.53)

$$n = \frac{2}{(2\pi)^d} \int_B \int_{\mathbb{R}} \delta(E - E_j(k)) f(E) \, dE \, dk$$

$$= \int_{\mathbb{R}} \left(\frac{2}{(2\pi)^d} \int_B \delta(E - E_j(k)) \, dk \right) f(E) \, dE.$$

Thus, we can write

$$n = \int_{\mathbb{R}} N_j(E) f(E) \, dE,$$

where the integral

$$N_j(E) = \frac{2}{(2\pi)^d} \int_B \delta(E - E_j(k)) \, dk \tag{1.54}$$

is called the *density of states* of the jth band of energy E. In the physical literature, sometimes the notation DOS instead of N_j is used. The quantity $N_j(E) \triangle E$ is
approximately the number of quantum states $\triangle N^*$ between E and $E + \triangle E$. Thus,
$N_j(E)$ is approximately $\triangle N^* / \triangle E$ or, in the infinitesimal sense, $N_j(E) = dN^* / dE$.
Notice that the density of states in k-space is constant, see (1.47), but the density of
states in energy space (1.54) generally is not.

The integral (1.54) can be formulated more rigorously by applying the *coarea
formula*, which is a curvilinear generalization of Fubini's theorem [36]. The formula
reads as follows. Let $B \subset \mathbb{R}^d$ be an appropriate domain, $g : B \to \mathbb{R}$ be continuous,
and $E : B \to \mathbb{R}$ be continuously differentiable such that $1 / |\nabla_k E_j(k)|$ is integrable.
Then

$$\int_B g(k) \, dk = \int_{\mathbb{R}} \int_{E^{-1}(e)} g(k) \frac{dS_e(k)}{|\nabla_k E(k)|} \, de, \tag{1.55}$$

where $E^{-1}(e) = \{k \in B : E_j(k) = e\}$ is the level set of energy e and dS_e is a surface
element. Formally, by (1.53), this gives

$$N_j(E) = \frac{2}{(2\pi)^d} \int_{\mathbb{R}} \int_{\{E_j(k)=e\}} \delta(E - E_j(k)) \frac{dS_e(k)}{|\nabla_k E_j(k)|} \, de$$

$$= \frac{2}{(2\pi)^d} \int_{\mathbb{R}} \int_{\{E_j(k)=e\}} \delta(E - e) \frac{dS_e}{|\nabla_k E_j|} \, de$$

$$= \frac{2}{(2\pi)^d} \int_{\{E_j(k)=E\}} \frac{dS_E}{|\nabla_k E_j|}. \tag{1.56}$$

The density of states is thus written as a surface integral over the isoenergy surface
$E_j^{-1}(E)$.

We summarize these results in the following proposition.

Proposition 1.13 (Electron density). *The electron density n in a given band $E_j(k)$
reads as*

$$n = \int_{\mathbb{R}} N_j(E) f(E) \, dE,$$

*where the density of states $N_j(E)$ at energy E is defined in (1.54) or (1.56) and the
Fermi–Dirac distribution function $f(E)$ is given in (1.50).*

In a similar way, we can compute the density of holes in the jth band. Taking into account that the mean number of holes in a quantum state of energy E equals the mean number of *empty* states of energy E, $1 - f(E)$, we have

$$p = \int_{\mathbb{R}} N_j(E)(1 - f(E)) \, dE.$$

For the electron or hole density in the conduction or valence band, respectively, we write

$$n = \int_{\mathbb{R}} N_c(E) f(E) \, dE, \quad p = \int_{\mathbb{R}} N_v(E)(1 - f(E)) \, dE, \qquad (1.57)$$

where $N_c(E)$ and $N_v(E)$ denote the density of states in the conduction band $E_c(k)$ and valance band $E_v(k)$, respectively.

In the following, we derive more explicit formulas for the particle densities in the parabolic band approximation (1.32) and (1.33) with the extended Brillouin zone $B = \mathbb{R}^d$. We first compute the density of states.

Lemma 1.14 (Density of states for parabolic bands). *In the parabolic band approximation $E(k) = E_0 + \hbar^2 |k|^2 / 2m^*$ we obtain for $E \geq E_0$:*

$$N(E) = \frac{m^*}{\pi \hbar^2} \frac{\sqrt{2m^*(E - E_0)}}{\pi \hbar} \quad \text{for three-dimensional carriers,}$$

$$N(E) = \frac{m^*}{\pi \hbar^2} \quad \text{for two-dimensional carriers,}$$

$$N(E) = \frac{m^*}{\pi \hbar^2} \frac{\hbar}{\sqrt{2m^*(E - E_0)}} \quad \text{for one-dimensional carriers.}$$

For $E < E_0$, we have $N(E) = 0$ in all three cases.

Proof. We consider first the three-dimensional case. We start from (1.54), use spherical coordinates (ρ, θ, ϕ), and substitute $z = \hbar^2 \rho^2 / 2m^*$ to obtain

$$N(E) = \frac{2}{(2\pi)^3} \int_{\mathbb{R}^3} \delta\left(E - E_0 - \frac{\hbar^2}{2m^*}|k|^2\right) dk$$

$$= \frac{1}{4\pi^3} \int_0^{2\pi} \int_0^{\pi} \int_0^{\infty} \delta\left(E - E_0 - \frac{\hbar^2}{2m^*}\rho^2\right) \rho^2 \sin\theta \, d\rho \, d\theta \, d\phi$$

$$= \frac{m^*}{\pi^2 \hbar^2} \frac{\sqrt{2m^*}}{\hbar} \int_0^{\infty} \delta(E - E_0 - z) \sqrt{z} \, dz.$$

Introducing the Heaviside function H by $H(x) = 0$ for $x < 0$ and $H(x) = 1$ for $x \geq 0$ and using (1.53), we obtain

$$N(E) = \frac{m^*}{\pi \hbar^2} \frac{\sqrt{2m^*}}{\pi \hbar} \int_{\mathbb{R}} \delta(E - E_0 - z) \sqrt{z} H(z) \, dz$$

$$= \frac{m^*}{\pi \hbar^2} \frac{\sqrt{2m^*}}{\pi \hbar} \sqrt{E - E_0} H(E - E_0).$$

In the two-dimensional case, we start again from (1.54) and use polar coordinates (ρ, ϕ) and the substitution $z = \hbar^2 \rho^2 / 2m^*$,

$$
\begin{aligned}
N(E) &= \frac{2}{(2\pi)^2} \int_0^{2\pi} \int_0^\infty \delta\left(E - E_0 - \frac{\hbar^2}{2m^*}\rho^2\right) \rho \, d\rho \, d\phi \\
&= \frac{m^*}{\pi\hbar^2} \int_0^\infty \delta(E - E_0 - z) \, dz = \frac{m^*}{\pi\hbar^2} H(E - E_0).
\end{aligned}
$$

Finally, in the one-dimensional case,

$$
\begin{aligned}
N(E) &= \frac{2}{2\pi} \int_{\mathbb{R}} \delta\left(E - E_0 - \frac{\hbar^2}{2m^*}k^2\right) dk = \frac{1}{\pi} \frac{\sqrt{2m^*}}{2\hbar} \int_0^\infty \delta(E - E_0 - z) \frac{dz}{\sqrt{z}} \\
&= \frac{m^*}{\pi\hbar^2} \frac{\hbar}{\sqrt{2m^*(E - E_0)}} H(E - E_0),
\end{aligned}
$$

showing the lemma. □

Remark 1.15 (Density of states for nonparabolic bands). In the nonparabolic band approximation (1.31),

$$
E(1 + \alpha E) = \frac{\hbar^2}{2m^*}|k|^2, \quad \alpha > 0,
$$

the density of states becomes

$$
N(E) = \frac{m^*}{\pi\hbar^2} \frac{\sqrt{2m^* E}}{\pi\hbar} \sqrt{1 + \alpha E}\,(1 + 2\alpha E) \quad \text{in the three-dimensional case,}
$$

$$
N(E) = \frac{m^*}{\pi\hbar^2}(1 + 2\alpha E) \quad \text{in the two-dimensional case}
$$

(see [9, Problem 1.4]). □

For the particle densities in thermal equilibrium and in the parabolic band approximation, the following result holds.

Lemma 1.16 (Electron and hole densities for parabolic bands). *Let the conduction and valence bands be given by the parabolic band approximations (1.32) and (1.33). Then the three-dimensional particle densities are*

$$
n = N_c F_{1/2}\left(\frac{q\mu - E_c}{k_B T}\right), \quad p = N_v F_{1/2}\left(\frac{E_v - q\mu}{k_B T}\right),
$$

where

$$
N_c = 2\left(\frac{m_e^* k_B T}{2\pi\hbar^2}\right)^{3/2}, \quad N_v = 2\left(\frac{m_h^* k_B T}{2\pi\hbar^2}\right)^{3/2} \tag{1.58}
$$

are the effective densities of states *and*

$$F_{1/2}(z) = \frac{2}{\sqrt{\pi}} \int_0^\infty \frac{\sqrt{x}\,\mathrm{d}x}{1+e^{x-z}}, \quad z \in \mathbb{R}, \tag{1.59}$$

is the Fermi integral *(of index 1/2)* [37, 38]. *Furthermore,* m_e^* *and* m_h^* *denote the (isotropic) effective mass of the electrons and holes, respectively.*

Proof. From (1.57), Lemma 1.14, and the substitution $x = (E - E_c)/k_BT$ we obtain

$$
\begin{aligned}
n &= \frac{m_e^*}{\pi\hbar^2}\frac{\sqrt{2m_e^*}}{\pi\hbar} \int_{E_c}^\infty \frac{\sqrt{E-E_c}}{1+e^{(E-q\mu)/k_BT}}\,\mathrm{d}E \\
&= \frac{4}{\sqrt{\pi}}\left(\frac{m_e^* k_BT}{2\pi\hbar^2}\right)^{3/2} \int_0^\infty \frac{\sqrt{x}\,\mathrm{d}x}{1+e^{x-(q\mu-E_c)/k_BT}} = N_c F_{1/2}\left(\frac{q\mu - E_c}{k_BT}\right).
\end{aligned}
$$

In a similar way,

$$
\begin{aligned}
p &= \frac{m_h^*}{\pi\hbar^2}\frac{\sqrt{2m_h^*}}{\pi\hbar} \int_{-\infty}^{E_v} \frac{\sqrt{E_v-E}}{1+e^{-(E-q\mu)/k_BT}}\,\mathrm{d}E \\
&= \frac{4}{\sqrt{\pi}}\left(\frac{m_h^* k_BT}{2\pi\hbar^2}\right)^{3/2} \int_{-\infty}^0 \frac{\sqrt{x}\,\mathrm{d}x}{1+e^{x-(E_v-q\mu)/k_BT}} = N_v F_{1/2}\left(\frac{E_v - q\mu}{k_BT}\right).
\end{aligned}
$$

This proves the lemma. □

Next, we compute the particle densities for some particular cases.

Lemma 1.17 (Two-dimensional electron density). *The electron density in a quantum well equals*

$$n = \frac{m_e^* k_BT}{\pi\hbar^2} \log\left(1 + e^{(q\mu-E_c)/k_BT}\right).$$

Proof. In a quantum well, electrons are confined in one direction. Therefore, using the density of states function for two-dimensional carriers (see Lemma 1.14),

$$
\begin{aligned}
n &= \frac{m_e^*}{\pi\hbar^2} \int_{E_c}^\infty \frac{\mathrm{d}E}{1+e^{(E-q\mu)/k_BT}} = \frac{m_e^* k_BT}{\pi\hbar^2}\left[-\log(1+e^{-(E-q\mu)/k_BT})\right]_{E_c}^\infty \\
&= \frac{m_e^* k_BT}{\pi\hbar^2}\log(1+e^{-(E_c-q\mu)/k_BT}),
\end{aligned}
$$

ending the proof. □

Lemma 1.18 (Electron and hole densities for parabolic bands). *The three-dimensional electron and hole densities in the parabolic band and Maxwell–Boltzmann approximation* $|q\mu - E_c|,\ |q\mu - E_v| \ll k_BT$ *are*

$$n = N_c \exp\left(\frac{q\mu - E_c}{k_BT}\right), \quad p = N_v \exp\left(\frac{E_v - q\mu}{k_BT}\right),$$

where N_c *and* N_v *are the effective densities of states defined in (1.58).*

Proof. For $z \to -\infty$ we can approximate

$$F_{1/2}(z) = e^z \frac{2}{\sqrt{\pi}} \int_0^\infty \frac{\sqrt{x}\,dx}{e^z + e^x} \sim e^z \frac{2}{\sqrt{\pi}} \int_0^\infty \frac{\sqrt{x}\,dx}{e^x} = e^z \frac{2}{\sqrt{\pi}} \Gamma\left(\frac{3}{2}\right) = e^z,$$

where $\Gamma(x)$ is the Gamma function,

$$\Gamma(x) = \int_0^\infty y^{x-1} e^{-y}\,dy, \tag{1.60}$$

with the properties $\Gamma(\frac{1}{2}) = \sqrt{\pi}$ and $\Gamma(x+1) = x\Gamma(x)$ for $x > 0$. Thus, the result follows from Lemma 1.16. □

Finally, we discuss two notions needed in the subsequent chapters, the intrinsic density and doping of semiconductors.

A pure semiconductor with no impurities is called an intrinsic semiconductor. In this case, electrons in the conduction band can only come from valence band levels leaving a vacancy (hole) behind them (see Fig. 1.11). Therefore, the number of electrons in the conduction band is equal to the number of holes in the valence band,

$$n = p = n_i.$$

The quantity n_i is called the *intrinsic density*. It can be computed in the nondegenerate parabolic band case from Lemma 1.18:

$$n_i = \sqrt{np} = \sqrt{N_c N_v} \exp\left(\frac{E_v - E_c}{2k_B T}\right) = \sqrt{N_c N_v} \exp\left(-\frac{E_g}{2k_B T}\right), \tag{1.61}$$

where $E_g = E_c - E_v$ is the energy gap. This allows us to determine the Fermi energy $E_F = q\mu$ of an intrinsic semiconductor:

$$E_F = E_c + k_B T \log \frac{n}{N_c} = E_c + k_B T \log \frac{n_i}{N_c} = E_c - \frac{E_g}{2} + \frac{k_B T}{2} \log \frac{N_v}{N_c}$$
$$= \frac{1}{2}(E_c + E_v) + \frac{3}{4} k_B T \log \frac{m_h^*}{m_e^*}.$$

This asserts that at zero temperature, the Fermi energy lies precisely in the middle of the energy gap. Furthermore, since $\log(m_h^*/m_e^*)$ is of order one, the correction is only of order $k_B T$ for nonzero temperature. In most semiconductors at room temperature, the energy gap is much larger than $k_B T \approx 0.026\,\text{eV}$. This shows that the nondegeneracy assumptions

$$E - E_F \geq E_c - E_F = \frac{E_g}{2} - \frac{3}{4} k_B T \log \frac{m_h^*}{m_e^*} \gg k_B T,$$

$$E_F - E \geq E_F - E_v = \frac{E_g}{2} - \frac{3}{4} k_B T \log \frac{m_h^*}{m_e^*} \gg k_B T$$

are satisfied and that the result is consistent with our assumptions.

The intrinsic density is too small to result in a significant conductivity for nonzero temperature. For instance, in silicon we have $n_i \approx 6.93 \cdot 10^9\,\mathrm{cm}^{-3}$ compared to N_c and N_v being of the order of $10^{19}\,\mathrm{cm}^{-3}$. Replacing some atoms in the semiconductor crystal by atoms which provide free-electrons in the conduction band or free holes in the valence band allows one to increase the conductivity. Such a process is called the *doping* of a semiconductor. Impurities are called *donors* if they supply additional electrons to the conduction band and *acceptors* if they supply additional holes to (i.e., capture electrons from) the valence band. A semiconductor which is doped with donors is termed an *n-type semiconductor*, and a semiconductor doped with acceptors is called a *p-type semiconductor*. For instance, when we dope a germanium crystal, whose atoms each have four valence electrons, with arsenic, which has five valence electrons per atom, each arsenic atom provides one additional electron. These additional electrons are only weakly bound to the arsenic atom. Indeed, the binding energy is about 0.013 eV [2, Table 28.2] which is smaller than the thermal energy $k_B T \approx 0.026$ eV at room temperature. More generally, denoting by E_d and E_a the energies of a donor electron and an acceptor hole, respectively, then $E_c - E_d$ and $E_a - E_v$ are small compared to $k_B T$ (see Fig. 1.14). This means that the additional carriers contribute at room temperature to the electron and hole density and increase the conductivity of the semiconductor.

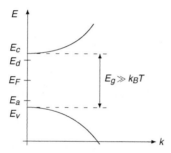

Fig. 1.14 Illustration of the energy gap E_g and the donor and acceptor energies E_d and E_a. The Fermi energy E_F approximately lies in the middle of the energy gap for moderate temperatures

Let $N_d(x)$, $N_a(x)$ denote the densities of the donor and acceptor impurities, respectively. Then the *doping profile* or *doping concentration* is $C(x) = N_a(x) - N_d(x)$ and the total space charge is given by

$$\rho = -qn + qp + qN_a(x) - qN_d(x) = -q(n - p - C(x)).$$

References

1. R. Adler, A. Smith, and R. Longini. *Introduction to Semiconductor Physics*. John Wiley & Sons, New York, 1964.
2. N. Ashcroft and N. Mermin. *Solid State Physics*. Sanners College, Philadelphia, 1976.
3. K. Brennan. *The Physics of Semiconductors*. Cambridge University Press, Cambridge, 1999.

4. J. Davies. *The Physics of Low-Dimensional Semiconductors*. Cambridge University Press, Cambridge, 1998.
5. C. Kittel. *Introduction to Solid State Physics*. 7th edition, John Wiley & Sons, New York, 1996.
6. K. Seeger. *Semiconductor Physics. An Introduction*. Springer, Berlin, 2004.
7. H. Grahn. *Introduction to Semiconductor Physics*. World Scientific, Singapore, 1999.
8. J. Inkson. *Many-Body Theory of Solids*. Plenum Press, New York, 1984.
9. M. Lundstrom. *Fundamentals of Carrier Transport*. 2nd edition, Cambridge University Press, Cambridge, 2000.
10. D. Neamen. *Semiconductor Physics and Devices*. 3rd edition, McGraw Hill, New York, 2002.
11. U. Rössler. *Solid State Theory. An Introduction*. Springer, Berlin, 2004.
12. W. Wenckebach. *Essentials of Semiconductor Physics*. John Wiley & Sons, Chichester, 1999.
13. P. Yu and M. Cardona. *Fundamentals of Semiconductors*. Springer, Berlin, 2001.
14. P. Markowich, C. Ringhofer, and C. Schmeiser. *Semiconductor Equations*. Springer, Vienna, 1990.
15. F. Bloch. Über die Quantenmechanik der Elektronen in Kristallgittern. *Z. Phys.* 52 (1928), 555–600.
16. M. Reed and B. Simon. *Methods of Modern Mathematical Physics IV: Analysis of Operators*. Academic Press, New York, 1978.
17. C. Wilcox. Theory of Bloch waves. *J. d'Analyse Math.* 33 (1978), 146–167.
18. H. Ibach and H. Lüth. *Solid-State Physics*. Springer, Berlin, 1995.
19. D. Arnold. *Functional Analysis*. Lecture Notes, Penn State University, USA, 1997.
20. P. Degond. Mathematical modelling of microelectronics semiconductor devices. In: *Some Current Topics on Nonlinear Conservation Laws*, Studies Adv. Math. 15, 77–110. Amer. Math. Soc., Providence, 2000.
21. R. Kronig and W. Penney. Quantum mechanics of electrons in crystal lattices. *Proc. Royal Soc. (London)* A 130 (1931), 499–513.
22. K. Jackson and W. Schröter (eds.). *Handbook of Semiconductor Technology*, Vol. 1. Wiley-VCH, Weinheim, 2000.
23. M. Grundmann. *The Physics of Semiconductors*. Springer, Berlin, 2006.
24. P. Bechouche, N. Mauser, and F. Poupaud. Semiclassical limit for the Schrödinger-Poisson equation in a crystal. *Commun. Pure Appl. Math.* 54 (2001), 851–890.
25. F. Poupaud and C. Ringhofer. Semi-classical limits in a crystal with exterior potentials and effective mass theorems. *Commun. Part. Diff. Eqs.* 21 (1996), 1897–1918.
26. P.-L. Lions and T. Paul. Sur les mesures de Wigner. *Rev. Mat. Iberoamer.* 9 (1993), 553–618.
27. P. Markowich and N. Mauser. The classical limit of a self-consistent quantum-Vlasov equation in 3D. *Math. Models Meth. Appl. Sci.* 3 (1993), 109–124.
28. G. Allaire and A. Piatnitski. Homogenization of the Schrödinger equation and effective mass theorems. *Commun. Math. Phys.* 258 (2005), 1–22.
29. P. Gérard. Mesures semi-classiques et ondes de Bloch. *Séminaires sur les Equations aux Dérivées Partielles*, 1990–1991, Exp. No. XVI, Ecole Polytechnique, Palaiseau, France, 1991.
30. P. Markowich, N. Mauser, and F. Poupaud. A Wigner-function approach to (semi) classical limits: electrons in a periodic potential. *J. Math. Phys.* 35 (1994), 1066–1094.
31. G. Panati, H. Spohn, and S. Teufel. Effective dynamics for Bloch electrons: Peierls substitution and beyond. *Commun. Math. Phys.* 242 (2003), 547–578.
32. F. Malcher, G. Lommer, and U. Rössler. Electron states in GaAs/Ga$_{1-x}$Al$_x$As heterostructures: nonparabolicity and spin-splitting. *Superlattices Microstruct.* 2 (1986), 267–272.
33. N. Ben Abdallah and L. Barletti. Work in preparation, 2008.
34. M. Burt. The justification for applying the effective-mass approximation to microstructures. *J. Phys.: Condens. Matter* 4 (1992), 6651–6690.
35. W. Kohn and J. Luttinger. Theory of donor states in silicon. *Phys. Rev.* 98 (1955), 915–922.
36. H. Federer. *Geometric Measure Theory*. Springer, New York, 1969.
37. J. Blakemore. *Semiconductor Statistics*. Pergamon Press, Oxford, 1962.
38. V. Bonch-Bruevich and S. Kalashnikov. *Halbleiterphysik*. VEB Deutscher Verlag der Wissenschaften, Berlin, 1982.

Chapter 2
Derivation of Macroscopic Equations

In this chapter we explain the strategy how to derive macroscopic models from the semiconductor Boltzmann equation by the so-called moment method. The macroscopic models contain only averaged physical quantities. A main assumption of the derivation of the transport equations is that scattering of the electrons is (in some sense) dominant. Then, depending on the considered timescale, hydrodynamic or diffusive fluid-type models are derived.

2.1 The Boltzmann Equation and its Scalings

In the previous chapter we have considered an electron ensemble being in thermal equilibrium (no current flow). Here, we wish to find an evolution equation for an electron ensemble not being in equilibrium. Since the number of conduction electrons in a semiconductor crystal is usually very large, we employ, as in Sect. 1.6, a statistical approach.

Let $f(x,k,t)$ be the *distribution function* of an electron ensemble, where $x \in \mathbb{R}^3$ denotes the spatial variable, $k \in B$ the pseudo-wave vector (see Sect. 1.3), and $t > 0$ the time. More precisely, $f(x,k,t)$ is the ratio of the number of occupied quantum states in the infinitesimal volume element $dx\,dk$ in the conduction band to the total number of states in $dx\,dk$ in the conduction band. The function $f(x,k,t)$ is often referred to as the *occupation number* of the state k in the point x. In this sense, $0 \le f(x,k,t) \le 1$ for all (x,k,t) holds.

We can define in a similar way the distribution function $f_h(x,k,t)$ of a hole ensemble, where $f_h(x,k,t)$ is the ratio of the number of occupied states in $dx\,dk$ in the valence band to the total number of states in $dx\,dk$ in the valence band. However, for simplicity, we consider in this chapter only the transport of electrons.

From the distribution function, we can derive macroscopic quantities, like the particle density, velocity, and energy, which can be measured experimentally. Macroscopic models then describe the evolution of these quantities. The number N^* of

Jüngel, A.: *Derivation of Macroscopic Equations*. Lect. Notes Phys. **773**, 45–54 (2009)
DOI 10.1007/978-3-540-89526-8_2   © Springer-Verlag Berlin Heidelberg 2009

electrons in the conduction band of a semiconductor crystal $\Omega \subset \mathbb{R}^3$ is, according to (1.49), given by

$$N^*(x,t) = \frac{\text{vol}(\Omega)}{4\pi^3} \int_B f(x,k,t)\,dk.$$

Thus, the *electron density* n, i.e., the number of electrons per unit volume, equals

$$n(x,t) = \frac{N^*(x,t)}{\text{vol}(\Omega)} = \frac{1}{4\pi^3} \int_B f(x,k,t)\,dk.$$

Introducing the notation

$$\langle g \rangle = \frac{1}{4\pi^3} \int_B g(k)\,dk,$$

we can write the above expression compactly as $n = \langle f \rangle$. The *mean velocity* $u(x,t)$ and the *mean energy density* $(ne)(x,t)$ are defined, respectively, by

$$u = \frac{1}{n}\langle vf \rangle, \quad ne = \langle Ef \rangle,$$

where $v = \nabla_k E / \hbar$ is the semi-classical mean velocity and E the conduction-band energy (see Lemma 1.9).

In the parabolic band approximation $E(k) = \hbar^2 |k|^2 / 2m^*$, where m^* is the effective electron mass, we have $v(k) = \hbar k / m^*$ and thus,

$$nu = \frac{1}{m^*}\langle pf \rangle, \quad ne = \frac{1}{m^*}\left\langle \frac{1}{2}|p|^2 f \right\rangle,$$

where $p = \hbar k$ is the crystal momentum (see Sect. 1.4). We call the expression $n = \langle f \rangle$ the *zeroth moment*, $\langle pf \rangle$ the *first moment*, and $\langle \frac{1}{2}|p|^2 f \rangle$ the *second moment* associated with f. Furthermore, 1, p, and $\frac{1}{2}|p|^2$ are termed *weight functions*.

Now, we derive an evolution equation for the distribution function. We assume that f is constant along the trajectory $(x(t),k(t))$. We refer to Sect. 3.1 for some explanations of this hypothesis. Then, along the trajectories,

$$0 = \frac{df}{dt} = \partial_t f + \dot{x} \cdot \nabla_x f + \dot{k} \cdot \nabla_k f.$$

The semi-classical equations (1.25) yield

$$\partial_t f + v(k) \cdot \nabla_x f + \frac{q}{\hbar} \nabla_x V(x) \cdot \nabla_k f = 0, \quad x \in \mathbb{R}^3, \ k \in B, \ t > 0.$$

This is a partial differential equation which is complemented by an initial condition for $f(x,k,0)$. In Sect. 3.2 we give a more precise derivation of this equation.

Scattering allows particles to jump to another trajectory. We assume that the rate of change df/dt due to convective or electric effects and the rate of change $Q(f)$ due to collisions balance, $df/dt = Q(f)$. Then

$$\partial_t f + v(k) \cdot \nabla_x f + \frac{q}{\hbar} \nabla_x V(x) \cdot \nabla_k f = Q(f), \quad x \in \mathbb{R}^3, \, k \in B, \, t > 0. \qquad (2.1)$$

This is the *semiconductor Boltzmann equation*. We call $Q(f)$ a collision operator. Usually, it is a nonlocal and nonlinear operator in f. Since we will discuss the scattering mechanism in more detail in Chap. 4, we will only use a very simple example in Sects. 2.3 and 2.4.

In order to identify small parameters and to perform the asymptotic limits leading to the macroscopic models, we scale the Boltzmann equation (2.1). We define a characteristic length λ (for instance, the device diameter) and the characteristic velocity $u = \sqrt{k_B T_L / m^*}$, where T_L is the lattice temperature. The velocity u corresponds to a particle with kinetic energy of the order $k_B T_L$. Furthermore, we define the characteristic pseudo-wave vector $m^* u / \hbar$, the characteristic electric potential $U_T = k_B T_L / q$, which is also referred to as the *thermal voltage*, and a characteristic time τ, which will be specified below. A second timescale is given by the time τ_c between two consecutive collisions. The distance $\lambda_c = u \tau_c$, which a particle travels between two consecutive collisions, is called the *mean free path*. This defines the dimensionless variables

$$x = \lambda x_s, \quad t = \tau t_s, \quad k = \frac{m^* u}{\hbar} k_s$$

and the dimensionless functions

$$V = U_T V_s, \quad Q(f) = \frac{1}{\tau_c} Q_s(f).$$

The distribution function is a number and therefore, it is already dimensionless.

We derive macroscopic models by assuming that scattering is dominant in the sense that the mean free path is much smaller than the device diameter, i.e., $\alpha = \lambda_c / \lambda \ll 1$. This means that a particle will undergo many collisions along its way through the device. The number α is called the *Knudsen number*.

The constant τ is still unspecified. We consider two scalings. First, we set $\tau = \tau_c / \alpha$. Since $\alpha \ll 1$, this means that we consider a timescale much larger than the collision time. A computation shows that the Boltzmann equation (2.1) becomes, after omitting the index "s" for the dimensionless variables and parameters,

$$\alpha \partial_t f + \alpha \left(v(k) \cdot \nabla_x f + \nabla_x V \cdot \nabla_k f \right) = Q(f). \qquad (2.2)$$

This is the Boltzmann equation in the *hydrodynamic scaling*. Next, we set $\tau = \tau_c / \alpha^2$. In this scaling, which we call the *diffusion scaling*, the characteristic time is even larger than that from the hydrodynamic scaling. Here, the Boltzmann equation can be written as

$$\alpha^2 \partial_t f + \alpha \left(v(k) \cdot \nabla_x f + \nabla_x V \cdot \nabla_k f \right) = Q(f). \qquad (2.3)$$

The choice of the scaling – hydrodynamic or diffusive – depends on the equilibrium states associated with the collision operators. Before we can make precise the structure of the collision operator, which will be used in the derivation, we need the notion of the *Maxwellian* which is introduced in the following section.

2.2 Maxwellians

In thermal equilibrium, the distribution function is equal to the Fermi–Dirac distribution (1.50) or to the Maxwell–Boltzmann distribution (1.51),

$$f_{\text{eq}} = e^{-(E-E_F)/k_B T_L},$$

where E_F is the Fermi energy and T_L the (constant) lattice temperature. The energy E is the sum of the potential energy $-qV$ due to the doping potential and the band energy $E_n(k)$, $E = -qV + E_n$. Under the assumption that the range of the velocity v is the whole space $\mathbb{R}^3$, we will show that f_{eq} is a solution of the Boltzmann equation (2.1) if and only if the Fermi energy E_F is constant.

Since in equilibrium nothing changes with time, f_{eq} does not depend on time and the net collision rate is zero, $Q(f_{\text{eq}}) = 0$. A computation shows that the equilibrium distribution f_{eq} satisfies

$$v(k) \cdot \nabla_x f_{\text{eq}} + \frac{q}{\hbar} \nabla_x V \cdot \nabla_k f_{\text{eq}} = v(k) \cdot \nabla_x E_F \frac{f_{\text{eq}}}{k_B T_L} = 0,$$

if E_F is constant. On the other hand, if f_{eq} solves the Boltzmann equation, the above computation shows that $v(k) \cdot \nabla_x E_F = 0$ for all $k \in B$. This implies that $\nabla_x E_F = 0$, and E_F is constant.

The equilibrium distribution

$$f_{\text{eq}} = N(x)e^{-E_n(k)/k_B T_L} \quad \text{with} \quad N(x) = e^{(qV(x)+E_F)/k_B T_L} \tag{2.4}$$

with constant E_F is called a *Maxwellian*.

In order to derive the thermal equilibrium state, we can also argue thermodynamically. In thermal equilibrium, the *entropy* of the system with a fixed number of particles should be maximal. This maximum-entropy argument was already used in statistical mechanics by Jaynes in 1957 [1]. In the following, we generalize this idea by allowing not only a fixed number of electrons but also a fixed momentum or energy. (In Sect. 1.6 we have already used a similar argument.) Depending on the fixed quantities, we obtain different Maxwellians. To make this statement precise, we consider for simplicity Maxwell–Boltzmann statistics and parabolic bands $E_n(k) = \hbar^2 |k|^2 / 2m^*$. We refer to Chap. 8 for more general situations. The kinetic entropy is defined by

$$S(f) = -\int_{\mathbb{R}^3 \times B} f\left(k_B T_L \log f - k_B T_L + E(k)\right) dx \, dk.$$

More precisely, $-S$ is the free energy of the system. Furthermore, we introduce the weight functions $\kappa_0(k) = 1$, $\kappa_1(k) = \hbar k/m^*$, and $\kappa_2(k) = \hbar^2|k|^2/2m^*$. We abbreviate as in Sect. 2.1 $\langle g \rangle = \int_{\mathbb{R}^3} g(k)\,dk/4\pi^3$ for any function $g(k)$ and we call the integrals

$$m_0 = \langle \kappa_0 f \rangle = \langle f \rangle,$$

$$m_1 = \langle \kappa_1 f \rangle = \left\langle \frac{\hbar k}{m^*} f \right\rangle,$$

$$m_2 = \langle \kappa_2 f \rangle = \left\langle \frac{\hbar^2|k|^2}{2m^*} f \right\rangle$$

the zeroth, first, and second moment of f, respectively. They have a physical interpretation: m_0 is the particle density, m_1 the momentum density, and m_2 the energy density.

The precise maximization problem reads as follows: For given $f(x,k,t)$ and moments $m_i = \langle \kappa_i f \rangle$, find $M[f]$ such that

$$S(M[f]) = \max\{S(g) : \langle \kappa_i g \rangle = m_i \text{ for } i = 0,1,2\}.$$

We will prove in Sect. 8.1 that, if this constrained extremal problem is solvable, its formal solution is given by

$$M[f] = \exp\left(\kappa \cdot \lambda - \frac{\hbar^2|k|^2}{2k_B T_L m^*} \right),$$

where $\kappa = (\kappa_i)$ and $\lambda = (\lambda_i)$ are the Lagrange multipliers arising from the constrained maximization problem and depending on x and t through the distribution function f. We call $M[f]$ the *Maxwellian* of f.

Example 2.1 (Maxwellians). We consider some examples.

1. *Classical Maxwellian:* Let $\kappa_0(k) = 1$, i.e., we solve the maximization problem for a given particle density or for a fixed number of electrons. Then, the Maxwellian equals $M[f] = \exp(\lambda_0 - \hbar^2|k|^2/2k_B T_L m^*)$. The Lagrange multiplier λ_0 can be eliminated by observing that the electron density is given by $n = m_0 = \int_{\mathbb{R}^3} M[f]\,dk/4\pi^3$. Indeed, we have, with spherical coordinates,

$$n = \frac{e^{\lambda_0}}{4\pi^3} \int_{\mathbb{R}^3} e^{-\hbar^2|k|^2/2k_B T_L m^*} \, dk = \frac{e^{\lambda_0}}{\pi^2} \int_0^\infty e^{-\hbar^2\rho^2/2k_B T_L m^*} \rho^2 \, d\rho = N_c(T_L)e^{\lambda_0},$$

where

$$N_c(T_L) = 2\left(\frac{m^* k_B T_L}{2\pi\hbar^2} \right)^{3/2}$$

is the effective density of state of the conduction band introduced in (1.58). Hence, the Maxwellian can be written as

$$M[f] = \frac{n}{N_c(T_L)} \exp\left(-\frac{\hbar^2|k|^2}{2k_B T_L m^*} \right). \tag{2.5}$$

This is the Maxwellian (2.4) with $N = n/N_c(T_L)$ and $E_n(k) = \hbar^2|k|^2/2m^*$.

2. *Heated Maxwellian:* Let $\kappa = (1, \hbar^2|k|^2/2m^*)$. Then the Maxwellian reads as

$$M[f] = \exp\left(\lambda_0 + \frac{\hbar^2|k|^2}{2m^*}\lambda_1 - \frac{\hbar^2|k|^2}{2k_BT_Lm^*}\right).$$

Introducing the *electron temperature* T by $1/k_BTm^* = -\lambda_1 + 1/k_BT_Lm^*$, we can write $M[f] = \exp(\lambda_0 - \hbar^2|k|^2/2k_BTm^*)$ and similarly as above, we obtain

$$M[f] = \frac{n}{N_c(T)}\exp\left(-\frac{\hbar^2|k|^2}{2k_BTm^*}\right). \tag{2.6}$$

Notice that compared to (2.5), the constant lattice temperature T_L is replaced by the particle temperature T. The function (2.6) is referred to as the *heated Maxwellian*.

3. *Shifted Maxwellian:* Let $\kappa = (1, \hbar k/m^*, \hbar^2|k|^2/2m^*)$. Then it follows that

$$M[f] = \exp\left(\lambda_0 + \frac{\hbar k}{m^*}\cdot\lambda_1 + \frac{\hbar^2|k|^2}{2m^*}\lambda_2 - \frac{\hbar^2|k|^2}{2k_BT_Lm^*}\right),$$

and introducing the electron temperature as above by $1/k_BTm^* = -\lambda_2 + 1/k_BT_Lm^*$ and the mean velocity $u = k_BT\lambda_1/m^*$, the above formula becomes

$$M[f] = \exp\left(\lambda_0 + \frac{m^*|u|^2}{2k_BT} - \frac{|\hbar k - m^*u|^2}{2k_BTm^*}\right). \tag{2.7}$$

A calculation analogous to that of the previous examples shows that the electron density equals $n = N_c(T)e^{\lambda_0 - m^*|u|^2/2k_BT}$. Hence, we can write the Maxwellian as

$$M[f] = \frac{n}{N_c(T)}\exp\left(-\frac{|\hbar k - m^*u|^2}{2k_BTm^*}\right). \tag{2.8}$$

This expression is called the (heated) *shifted Maxwellian*. The variable u has the meaning of a mean velocity since $\langle(\hbar k/m^*)M[f]\rangle = nu$ is the momentum. $\square$

2.3 Hydrodynamic Models

We consider now the scaled Boltzmann equation in the hydrodynamic scaling (2.2). To simplify the presentation, we make the following assumptions (see Chap. 9 for more general results):

1. The (scaled) energy band is parabolic, i.e., $v(k) = k$.
2. The collision operator is of relaxation-time type, i.e., $Q(f) = (M[f] - f)/\tau$, where $M[f]$ is the Maxwellian of f given the moments corresponding to the weight functions $\kappa_0(k) = 1$, $\kappa_1(k) = k$, and $\kappa_2(k) = |k|^2/2$.

If the distribution function is homogeneous in space and there are no electric forces, the Boltzmann equation becomes

$$\partial_t f = \frac{1}{\tau}(M[f] - f).$$

Thus, τ can be interpreted as the typical (scaled) time in which the distribution function converges to the Maxwellian. Thus, the scaled Boltzmann equation reads as

$$\alpha \partial_t f_\alpha + \alpha(k \cdot \nabla_x f_\alpha + \nabla_x V \cdot \nabla_k f_\alpha) = \frac{1}{\tau}(M[f_\alpha] - f_\alpha). \qquad (2.9)$$

The main idea for the derivation of macroscopic models is to multiply the Boltzmann equation by the weight functions $\kappa_i(k)$ and to integrate the resulting equation over the wave-vector space. This leads to the so-called *moment equations*,

$$\partial_t \langle \kappa_i f_\alpha \rangle + \mathrm{div}_x \langle \kappa_i k f_\alpha \rangle - \nabla_x V \cdot \langle \nabla_k \kappa_i f_\alpha \rangle = 0, \quad i = 0, 1, 2. \qquad (2.10)$$

The right-hand side vanishes since the moments of $M[f_\alpha]$ and f_α coincide by definition. Notice that we have integrated by parts in the last term on the left-hand side. The moments $m_i = \langle \kappa_i f_\alpha \rangle$ are the variables of the model. They are the electron density $n = m_0$, the momentum or current density $nu = m_1$, and the energy density $ne = m_2$.

The integrals on the left-hand side can be expressed in terms of the moments except the integrals $\langle \kappa_1 k f_\alpha \rangle = \langle k \otimes k f \rangle$ and $\langle \kappa_2 k f_\alpha \rangle = \langle \frac{1}{2} k |k|^2 f_\alpha \rangle$. (The notation $k \otimes k$ signifies the matrix with elements $k_i k_j$ for $i, j = 1, 2, 3$.) This is a conceptional problem. Given any set of weight functions, there is at least one integral with a higher-order moment which cannot be expressed in terms of the lower-order moments. This is called the *closure problem*. An additional condition to close the equations is necessary. Sensible closure conditions are motivated by investigation of the asymptotic limit $\alpha \to 0$ which consists of two steps.

In the first step, we perform the (formal) limit $\alpha \to 0$ in the Boltzmann equation (2.9). This gives $Q(f) = 0$, where $f = \lim_{\alpha \to 0} f_\alpha$, and hence $f = M[f]$. For the weight functions defined above, $M[f]$ is given by (2.8) in scaled form, i.e., $M[f] = \frac{1}{2}(2\pi/T)^{3/2} n \exp(-|k - u|^2/2T)$. In the second step we perform the limit $\alpha \to 0$ in the moment equations (2.10):

$$\partial_t \langle \kappa_i M[f] \rangle + \mathrm{div}_x \langle \kappa_i k M[f] \rangle - \nabla_x V \cdot \langle \nabla_k \kappa_i M[f] \rangle = 0. \qquad (2.11)$$

As $M[f]$ is known, we can compute the integrals.

Since the moments of the Maxwellian are, by definition, the same as the moments of f, we obtain $\langle M[f] \rangle = n$, $\langle k M[f] \rangle = nu$, and $\langle \frac{1}{2} |k|^2 M[f] \rangle = ne$. Furthermore, a computation shows that

$$\langle k \otimes k M[f] \rangle = P + nu \otimes u,$$

where the *stress tensor* P is defined by $P = \langle (k - u) \otimes (k - u) M[f] \rangle$, and

$$\left\langle \frac{1}{2}|k|^2 kM[f] \right\rangle = q + (P + ne\,\mathrm{Id})u,$$

where $q = \langle \frac{1}{2}(k-u)|k-u|^2 M[f] \rangle$ is the *heat flux* and Id is the identity matrix. Given the above Maxwellian, the heat flux vanishes since the integrand is odd in $k - u$. Moreover, $P = nT\,\mathrm{Id}$ and $ne = \frac{3}{2}nT$. We refer to Sect. 9.1 for the detailed computations.

Thus, we can reformulate (2.11) for $\kappa_0(k) = 1$, $\kappa_1(k) = k$, and $\kappa_2(k) = \frac{1}{2}|k|^2$ as

$$\partial_t n + \mathrm{div}\,(nu) = 0, \tag{2.12}$$
$$\partial_t (nu) + \mathrm{div}\,(nu \otimes u) + \nabla(nT) - n\nabla V = 0, \tag{2.13}$$
$$\partial_t (ne) + \mathrm{div}\,(nu(T+e)) - nu \cdot \nabla V = 0. \tag{2.14}$$

These equations are referred to as the *hydrodynamic model*. In the absence of the electric field $-\nabla V$, the above equations express the conservation of mass, momentum, and energy, respectively. We refer to Chap. 9 for a discussion of this result.

Notice that we obtain a hierarchy of hydrodynamic models by choosing different sets of weight functions or, equivalently, moments. For instance, if we only choose $\kappa_0(k) = 1$ and $\kappa_1(k) = k$, this leads to (2.12) and (2.13) with $T = \mathrm{const.}$, since the corresponding Maxwellian has constant temperature. This model hierarchy will be studied in more detail in Chap. 9.

Remark 2.2. The above derivation essentially consists of the formulation of the moment equations (2.10) and the replacement of f_α by the corresponding Maxwellian. This replacement was motivated by the hydrodynamic limit $\alpha \to 0$. Concerning the collision operator, we have only needed the following properties: The kernel consists exactly of the corresponding Maxwellian and the averages $\langle \kappa_i Q(f) \rangle$ are known. In the particular example of the relaxation-time operator, the averages vanish, which expresses conservation of mass, momentum, and energy. More general scattering integrals are considered in Chap. 9.   □

2.4 Diffusion Models

Diffusion models are derived from the scaled Boltzmann equation (2.3),

$$\alpha^2 \partial_t f_\alpha + \alpha\,(k \cdot \nabla_x f_\alpha + \nabla_x V \cdot \nabla_k f_\alpha) = Q(f_\alpha).$$

As in the previous section, we have assumed the parabolic band approximation and we employ the relaxation-time operator $Q(f) = (M[f] - f)/\tau$, where $M[f]$ is the Maxwellian of f, given the moments corresponding to weight functions of *even* order in k.

The derivation is based on three steps. The first step is as in the previous section. Performing the formal limit $\alpha \to 0$ in the above Boltzmann equation, we obtain $Q(f) = 0$, where $f = \lim_{\alpha \to 0} f_\alpha$, and hence $f = M[f]$.

For the second step we introduce the *Chapman–Enskog expansion* $f_\alpha = M[f_\alpha] + \alpha g_\alpha$ (in fact, this equation defines g_α), which includes the first-order correction αg_α. This correction is needed for the following reason. In the previous section, we have defined the current density as the integral $J = -nu = -\langle kM[f]\rangle$. In the present situation, the Maxwellian is even in k (since the weight functions are assumed to be even in k) and thus, the integral of $kM[f]$ vanishes, $J = 0$. In order to obtain a nonvanishing current density, we need to introduce the first-order correction αg_α.

Inserting the Chapman–Enskog expansion into the Boltzmann equation gives, after division by α,

$$\alpha \partial_t (M[f_\alpha] + \alpha g_\alpha) + (k \cdot \nabla_x M[f_\alpha] + \nabla_x V \cdot \nabla_k M[f_\alpha])$$
$$+ \alpha (k \cdot \nabla_x g_\alpha + \nabla_x V \cdot \nabla_k g_\alpha) = \frac{1}{\alpha} Q(M[f_\alpha]) + Q(g_\alpha) = Q(g_\alpha).$$

Performing the limit $\alpha \to 0$ then yields

$$Q(g) = k \cdot \nabla_x M[f] + \nabla_x V \cdot \nabla_k M[f],$$

where $g = \lim_{\alpha \to 0} g_\alpha$. As $M[f]$ is given, this is an operator equation for g. The relaxation-time operator allows us to solve this equation explicitly,

$$g = M[f] - \tau (k \cdot \nabla_x M[f] + \nabla_x V \cdot \nabla_k M[f]). \tag{2.15}$$

The third step is concerned with the limit $\alpha \to 0$ in the moment equations, reading as

$$\partial_t \langle \kappa_i M[f_\alpha]\rangle + \alpha \partial_t \langle \kappa_i g_\alpha\rangle + \frac{1}{\alpha} (\text{div}_x \langle \kappa_i kM[f_\alpha]\rangle + \nabla_x V \cdot \langle \nabla_k \kappa_i M[f_\alpha]\rangle)$$
$$+ \text{div}_x \langle \kappa_i k g_\alpha\rangle + \nabla_x V \cdot \langle \nabla_k \kappa_i g_\alpha\rangle = 0.$$

The integrals with the factor $1/\alpha$ vanish since the functions κ_i are assumed to be even and thus, $\kappa_i kM[f_\alpha]$ and $\nabla_k \kappa_i M[f_\alpha]$ are odd in k. Then the limit $\alpha \to 0$ leads to

$$\partial_t \langle \kappa_i M[f]\rangle + \text{div}_x \langle \kappa_i k g\rangle + \nabla_x V \cdot \langle \nabla_k \kappa_i g\rangle = 0. \tag{2.16}$$

As the function g is given by (2.15), this determines the evolution equation. The integral $J_i = -\langle \kappa_i k g\rangle$ is the flux corresponding to the moment $\langle \kappa_i M[f]\rangle$. Inserting g into the definition of J_i yields the expression

$$J_i = -\tau \langle \kappa_i k \otimes k \nabla_x M[f]\rangle - \tau \langle \kappa_i k \otimes \nabla_k M[f]\rangle \nabla_x V. \tag{2.17}$$

The first term shows that (2.16) is a diffusion equation since it contains the expression $\text{div}_x \langle \kappa_i k \otimes k \nabla_x M[f]\rangle$ with second-order derivatives. The second term expresses the convection due to the electric field $-\nabla_x V$.

The structure of the equations becomes clearer when we specify the moments. We present in the following only a simple example and refer to Chaps. 5, 6, 7, and 8 for more general situations.

Example 2.3. Let $\kappa_0(k) = 1$. Then, by (2.5), the Maxwellian reads as $M[f] = \frac{1}{2}(2\pi)^{3/2} n e^{-|k|^2/2}$ and (2.16) becomes, since $n = \langle M[f] \rangle$,

$$\partial_t n - \operatorname{div} J_n = 0. \tag{2.18}$$

It remains to compute the particle flux,

$$
\begin{aligned}
J_n &= \tau \langle k \otimes k \nabla_x M[f] + k \nabla_x V \cdot \nabla_k M[f] \rangle \\
&= \frac{\tau}{2}(2\pi)^{3/2} \langle k \otimes k e^{-|k|^2/2} \rangle (\nabla_x n - n \nabla_x V).
\end{aligned}
$$

As $\langle k \otimes k e^{-|k|^2/2} \rangle = 2(2\pi)^{-3/2} \operatorname{Id}$, we obtain

$$J_n = \tau(\nabla_x n - n \nabla_x V). \tag{2.19}$$

Equations (2.18) and (2.19) are referred to as the *drift-diffusion model*. It will be studied in more detail in Chap. 5. □

We remark that different choices of even-weight functions lead to a hierarchy of diffusive models, which will be discussed in detail in Chap. 8.

Example 2.4. The above derivation is based on the moment equations and on the diffusive limit $\alpha \to 0$, but compared to the derivation of the hydrodynamic models, first-order effects of the expansion $f_\alpha = M[f_\alpha] + \alpha g_\alpha$ are additionally taken into account. Moreover, the derivation requires more knowledge on the collision operator. Like in the previous section, we have assumed that the kernel consists of the corresponding Maxwellian and that the integrals $\langle \kappa_i Q(f) \rangle$ are known. In addition, we need to solve the operator equation $Q(g) = h$, where $h = (k \cdot \nabla_x + \nabla_x V \cdot \nabla_k) M[f]$. This equation was easy to solve for relaxation-time operators $Q(f)$, but it requires some mathematical theory for more general collision operators. We come back to this point in Sect. 5.1. □

Reference

1. E. Jaynes. Information theory and statistical mechanics. *Phys. Rev.* 106 (1957), 620–630.

Part II
Microscopic Semi-Classical Models

The fundamental evolution equations for classical charged particle flow are Newton's laws. When the number of particles is large, a statistical description is recommended, modeling the behavior of the particle ensemble by a probability density or distribution function. We derive kinetic equations describing the evolution of the distribution function in the phase space (or, more precisely, the position-wave vector space). First, we consider models without collision mechanisms, such as the Liouville and Vlasov equations. Then, we allow for scattering events leading to the Boltzmann equation.

Chapter 3
Collisionless Models

In this chapter, we consider only long-range interactions, like Coulomb forces, leading to the semi-classical Liouville or Vlasov equations. Models including short-range interactions are studied in Chap. 4.

3.1 The Liouville Equation

We first analyze the classical motion of M particles with mass m moving in a vacuum under the action of a force. The particles are described as classical particles, i.e., we associate the position vector $x_i \in \mathbb{R}^3$ and the velocity vector $v_i \in \mathbb{R}^3$ with the ith particle of the ensemble. Quantum mechanical effects are incorporated later in such a way that we obtain a semi-classical description of the electron ensemble in a semiconductor. The trajectories $(x_i(t), v_i(t))$ of the particles satisfy Newton's laws in the $6M$-dimensional ensemble position–velocity phase space

$$\dot{x} = v, \quad m\dot{v} = F, \quad t > 0, \tag{3.1}$$

with initial conditions

$$x(0) = x_0, \quad v(0) = v_0, \tag{3.2}$$

where $x = (x_1, \ldots, x_M)$, $v = (v_1, \ldots, v_M)$, the dot denotes differentiation with respect to time, and $F = (F_1, \ldots, F_M)$ is a force. It can, for instance, be given by an electric field acting on the electron ensemble,

$$F_i = -qE(x,t), \quad i = 1, \ldots, M.$$

We assume that the forces are independent of the velocity.

In semiconductors, the number of electrons M is typically very large (at least $M > 10^4$) and therefore, the numerical solution of (3.1) and (3.2) is very expensive. Since we are rather interested in the behavior of the particle *ensemble* instead of the behavior of the *individual* electrons, it seems reasonable to use a statistical

Jüngel, A.: *Collisionless Models*. Lect. Notes Phys. **773**, 57–54 (2009)
DOI 10.1007/978-3-540-89526-8_3

description. Then we are not prescribing the initial conditions (3.2) for each particle but the probability density $f_I(x,v)$ of the initial position and velocity of the particles. The integral

$$\int_\Omega f_I(x,v)\,dx\,dv$$

represents the expected number of particles at time $t=0$ in the subset Ω of the (x,v) space.

Let $f(x,v,t)$ be the *probability density* or *distribution function* of the ensemble at time t. We wish to derive an evolution equation for f. Under some condition on the dynamical system, the distribution function is constant along the trajectory $(x(t),v(t))$:

$$f(x(t),v(t),t) = f_I(x_0,v_0), \quad t > 0 \tag{3.3}$$

(also see Sect. 2.1). To specify this condition, let x_0, $v_0 \in \mathbb{R}^{3M}$ and $X = (X_1,\dots,X_M)$, $U = (U_1,\dots,U_M) : \mathbb{R}^{6M} \to \mathbb{R}^{3M}$ be two functions. Furthermore, let Φ_t be the flow map of the system of differential equations

$$\dot{x} = X(x,v), \quad \dot{v} = U(x,v), \quad t > 0, \quad x(0) = x_0, \quad v(0) = v_0, \tag{3.4}$$

i.e., Φ_t is defined as $\Phi_t(x_0,v_0) = (x(t),v(t))$, where (x,v) is the solution of (3.4) (which is assumed to exist). We define the distribution function, for given f_I, as

$$\int_{\Phi_t(\Omega)} f(\xi,\eta,t)\,d\xi\,d\eta = \int_\Omega f_I(x,v)\,dx\,dv \tag{3.5}$$

for all $\Omega \subset \mathbb{R}^{6M}$, where $\Phi_t(\Omega)$ is the image of Ω under Φ_t, i.e., the set of all points $(\xi,\eta) \in \mathbb{R}^{6M}$ which can be written as $(\xi,\eta) = \Phi_t(x,v)$ for some $(x,v) \in \mathbb{R}^{6M}$.

Theorem 3.1 (Liouville). *If*

$$\sum_{i=1}^{M} \left(\frac{\partial X_i}{\partial x_i} + \frac{\partial U_i}{\partial v_i} \right)(x(t),v(t)) = 0 \quad \text{for all } t > 0, \tag{3.6}$$

then (3.3) *holds.*

Proof. Let $t > 0$ be fixed. A transformation of variables in (3.5) gives

$$\int_\Omega f(\Phi_t(x,v),t)\,|\det D\Phi_t(x,v)|\,dx\,dv = \int_\Omega f_I(x,v)\,dx\,dv,$$

where $D\Phi_t$ is the Jacobian of Φ_t. We show that $\det D\Phi_t(x,v) = 1$ for all $(x,v) \in \Omega$. To prove this relation, we observe that the flow map satisfies the equation $\partial_t \Phi_t(x,v) = (X(\Phi_t(x,v)), U(\Phi_t(x,v)))^\top$. Differentiation with respect to (x,v) yields, by the chain rule,

$$\partial_t D\Phi_t = (DX \circ \Phi_t, DU \circ \Phi_t)^\top D\Phi_t.$$

As the derivative of the determinant of a matrix B is given by $\partial_t \det B = \text{Tr}(B^{-1}\partial_t B) \det B$ [1], where Tr is the trace of a matrix, we have

$$\partial_t(\det D\Phi_t) = \text{Tr}\left((D\Phi_t)^{-1}\partial_t(D\Phi_t)\right)\det D\Phi_t$$
$$= \text{Tr}\left((D\Phi_t)^{-1}(DX \circ \Phi_t, DU \circ \Phi_t)^\top D\Phi_t\right)\det D\Phi_t$$
$$= \text{Tr}(DX \circ \Phi_t, DU \circ \Phi_t)\det D\Phi_t.$$

Assumption (3.6) translates to

$$\text{Tr}\left(DX(\Phi_t(x,v)), DU(\Phi_t(x,v))\right)^\top = \text{Tr}(DX, DU)^\top(x(t), v(t)) = 0.$$

Hence, $\det D\Phi_t(x,v)$ is constant for all $t > 0$, and since $\det D\Phi_0(x,v)$ is the unit matrix, we obtain $\det D\Phi_t(x,v) = 1$ which shows the theorem. $\square$

If the right-hand sides of the differential equations in (3.4) are given by Newton's laws (3.1), the hypothesis of the Liouville Theorem (3.1) is satisfied since we have assumed that the forces do not depend on the velocity. Thus, differentiating (3.3) with respect to t gives the differential equation

$$0 = \frac{d}{dt}f(x(t), v(t), t) = \partial_t f + \dot{x} \cdot \nabla_x f + \dot{v} \cdot \nabla_v f,$$

and employing Newton's laws (3.1) leads to the *Liouville equation*

$$\partial_t f + v \cdot \nabla_x f + \frac{1}{m}F \cdot \nabla_v f = 0, \quad (x,v) \in \mathbb{R}^{6M}, \ t > 0. \tag{3.7}$$

It is supplemented with the initial condition

$$f(x, v, 0) = f_I(x, v), \quad (x, v) \in \mathbb{R}^{6M}.$$

In semiconductors, the electron ensemble cannot be described classically and the above argument is not valid. In Sect. 1.4, however, we have motivated that the motion of the electrons can be modeled semi-classically by the equations

$$\hbar\dot{x}_i = \nabla_{k_i}E_n(k_i), \quad \hbar\dot{k}_i = q\nabla_{x_i}V, \quad i = 1, \ldots, M, \tag{3.8}$$

where E_n is the energy of the nth band depending on the pseudo-wave vector k_i and $V(x,t)$ is the electric potential (cf. (1.25)). As above, we introduce the vectors $x = (x_1, \ldots, x_M)$ and $k = (k_1, \ldots, k_M)$. The distribution function f depends now on (x, k, t) rather than on (x, v, t). We assume that the electrons of the ensemble stay in the same energy band so that we can drop the index n. We claim that Liouville's theorem can be applied to this semi-classical picture. Indeed, the classical momentum $p_i = mv_i$ of the ith particle translates into the crystal momentum $p_i = \hbar k_i$ (see Sect. 1.4), and the hypothesis (3.6) becomes

$$\sum_{i=1}^{M}\left(\frac{\partial\dot{x}_i}{\partial x_i} + \frac{\partial\dot{k}_i}{\partial k_i}\right) = 0.$$

This condition is satisfied since $\dot{x}_i$ depends on k but not on x, and $\dot{k}_i$ depends on x but not on k. Taking into account (3.8), the Liouville equation reads as follows (also see Sect. 2.1):

$$0 = \frac{\mathrm{d}}{\mathrm{d}t} f(x(t), k(t), t) = \partial_t f + \dot{x} \cdot \nabla_x f + \dot{k} \cdot \nabla_k f$$

$$= \partial_t f + \frac{1}{\hbar} \nabla_k E \cdot \nabla_x f + \frac{q}{\hbar} \nabla_x V \cdot \nabla_k f, \qquad (3.9)$$

to be solved for $x \in \mathbb{R}^{3M}$, $k \in B^M$, and $t > 0$, where B is the Brillouin zone introduced in Sect. 1.1. Equation (3.9) is referred to as the *semi-classical Liouville equation*. The products are scalar products in $\mathbb{R}^{6M}$, i.e., they are defined as

$$\nabla_k E \cdot \nabla_x f = \sum_{i=1}^{M} \nabla_{k_i} E \cdot \nabla_{x_i} f, \quad \nabla_x V \cdot \nabla_k f = \sum_{i=1}^{M} \nabla_{x_i} V \cdot \nabla_{k_i} f,$$

and the dot on the right-hand sides denotes the usual scalar product in $\mathbb{R}^3$. We have to complement the Liouville equation by initial and boundary conditions since B is a bounded set of $\mathbb{R}^3$. The initial condition is given by

$$f(x, k, 0) = f_I(x, k), \quad (x, k) \in \mathbb{R}^{3M} \times B^M. \qquad (3.10)$$

Often, periodic boundary conditions

$$f(x, k_1, \ldots, k_i, \ldots, k_M, t) = f(x, k_1, \ldots, -k_i, \ldots, k_M, t), \quad k_i \in \partial B, \qquad (3.11)$$

for $i = 1, \ldots, M$, are chosen (see, e.g., [2, formula (1.2.49)]). This formulation makes sense since B is point symmetric with respect to the origin, i.e., $k_i \in B$ if and only if $-k_i \in B$.

The Liouville equation possesses some important properties. We recall the definition of the electron-ensemble density (see Sect. 2.1)

$$n(x, t) = \frac{1}{(4\pi^3)^M} \int_{B^M} f(x, k, t) \, \mathrm{d}k.$$

The electron-ensemble current density is defined as the first moment of the distribution function,

$$J_n(x, t) = -\frac{q}{(4\pi^3)^M} \int_{B^M} f(x, k, t) v(k) \, \mathrm{d}k,$$

where $v(k) = \nabla_k E / \hbar$ is the mean velocity.

Proposition 3.2 (Properties of the Liouville equation). *Let $f(x, k, t)$ be the solution of the Liouville initial boundary-value problem (3.9), (3.10), and (3.11). Then the following properties hold:*

1. *Positivity preservation: If $f_I \geq 0$ in $\mathbb{R}^{3M} \times B^M$ then $f(x, k, t) \geq 0$ for all $x \in \mathbb{R}^{3M}$, $k \in B^M$, and $t > 0$.*

2. Conservation law: It holds

$$\partial_t n - \frac{1}{q} \mathrm{div}_x J_n = 0, \quad x \in \mathbb{R}^{3M}, \, t > 0,$$

with initial condition $n(x,0) = (4\pi^3)^{-M} \int_{B^M} f_I \, dk, \, x \in \mathbb{R}^{3M}$.
3. Conservation of mass: The number of particles is conserved,

$$\int_{R^{3M}} n(x,t) \, dx = \frac{1}{(4\pi^3)^M} \int_{\mathbb{R}^{3M}} \int_{B^M} f_I(x,k) \, dk \, dx, \quad t > 0.$$

Proof. The first property follows immediately from (3.3). By formally integrating the Liouville equation (3.9) over $k \in B^M$, we obtain from the divergence theorem

$$\partial_t n = \frac{1}{(4\pi^3)^M} \int_{B^M} \partial_t f \, dk$$

$$= -\frac{1}{(4\pi^3)^M} \int_{B^M} \mathrm{div}_x(vf) \, dk - \frac{1}{(4\pi^3)^M} \frac{q}{\hbar} \int_{B^M} \mathrm{div}_k(\nabla_x V f) \, dk = \frac{1}{q} \mathrm{div}_x J_n.$$

The last property follows from the second one after integrating over the spatial variable $x \in \mathbb{R}^{3M}$. □

We notice that in the parabolic band approximation $E(k_i) = \hbar^2 |k_i|^2 / 2m^*$ (see (1.30)), where m^* denotes the effective electron mass, the semi-classical Liouville equation (3.9) reduces to the classical Liouville equation (3.7) with $m = m^*$ since $v = \nabla_k E / \hbar = \hbar k / m^*$ and $\nabla_k f = (\hbar/m^*) \nabla_v f$.

The Liouville equation can be extended to describe the motion of an electron ensemble under the influence of a magnetic field with induction vector B_{ind}. The magnetic force is given by $qv \times B_{\mathrm{ind}}$, and the total force is the Lorentz force $F = q(\nabla_x V + v \times B_{\mathrm{ind}})$. Then the semi-classical Liouville equation can be written as

$$\partial_t f + v \cdot \nabla_x f + \frac{q}{\hbar} (\nabla_x V + v \times B_{\mathrm{ind}}) \cdot \nabla_k f = 0, \quad t > 0,$$

where $v = (v_1(k_1), \ldots, v_M(k_M))$, together with the initial and boundary conditions (3.10) and (3.11). The magnetic field B_{ind} may be given or coupled to the electric field $-q\nabla_x V$ and the current J_n through the Maxwell equations. For mathematical results on the corresponding *Vlasov–Maxwell system*, we refer to [3–5].

3.2 The Vlasov Equation

The main disadvantage of the semi-classical Liouville initial boundary-value problem (3.9), (3.10), and (3.11) is that it has to be solved in a very high-dimensional phase space. Modeling the moderate number of $M = 10^4$ electrons, the dimension becomes $6 \cdot 10^4$ which is prohibitive for numerical simulations. In this section, we will derive an equation which acts in a six-dimensional phase space, by replacing the

electron-ensemble electric potential by an effective single-particle potential, similar
to the Hartree–Fock approximation of Sect. 1.3. The idea of the derivation is first to
assume a certain structure of the electric force, then to integrate the Liouville equation in sub-phase spaces, and finally, to carry out the formal limit $M \to \infty$, where M
is the number of particles. We proceed similarly as in [2].

Let an ensemble of M electrons be given and denote by $x = (x_1, \ldots, x_M)$ and $k = (k_1, \ldots, k_M)$ the position and wave vector coordinates of the particles, respectively.
We impose the following assumptions:

1. The motion is governed by an external electric field E_{ext} and by two-particle
   (long-range) interaction forces E_{int},

$$F_i(x,t) = -qE_{\text{ext}}(x_i,t) - \frac{q}{4\pi^3} \sum_{j=1,\, j\neq i}^{M} E_{\text{int}}(x_i,x_j), \quad i = 1, \ldots, M,$$

   where the interaction forces are assumed to be anti-symmetric,

$$E_{\text{int}}(x_i,x_j) = -E_{\text{int}}(x_j,x_i) \quad \text{for all } i, j. \tag{3.12}$$

2. The pointwise limit $E_0 = \lim_{M\to\infty} M E_{\text{int}}$ exists, i.e., the interaction force is of
   order $1/M$.
3. The initial density is independent of the numbering of the particles,

$$f_I(x_1,\ldots,x_M,k_1,\ldots,k_M) = f_I\left(x_{\pi(1)},\ldots,x_{\pi(M)},k_{\pi(1)},\ldots,k_{\pi(M)}\right) \tag{3.13}$$

   for all $x_i \in \mathbb{R}^3$, $k_i \in B$, $i = 1, \ldots, M$, and for all permutations π of $\{1,\ldots,M\}$.
4. The subensemble initial density

$$f_I^{(a)}(x_1,\ldots,x_a,k_1,\ldots,k_a) = \frac{1}{(4\pi^3)^{M-a}} \int_{(\mathbb{R}^3 \times B)^{M-a}} f_I \, dx_{a+1} \cdots dx_M \, dk_{a+1} \cdots dk_M$$

   can be factorized,

$$f_I^{(a)} = \prod_{i=1}^{a} F_I(x_i,k_i), \tag{3.14}$$

   for all $a = 1, \ldots, M-1$, where F_I is a given function.

We discuss these assumptions. We have excluded in the first condition velocity-
dependent forces, which exclude magnetic fields. The first assumption, which is
crucial for the derivation of the Vlasov equation, means that the force field F_i exerted on the ith electron is given by the sum of an external electric field acting on
the ith electron and of the sum of $M-1$ two-particle interaction forces of order $1/M$
between the ith electron and all other electrons. The interaction force E_{int} is independent of the electron index which express the fact that the electrons are indistinguishable. The action–reaction law implies that the force exerted by the jth electron
on the ith electron equals the negative force of the ith electron on the jth electron,
i.e., E_{int} is anti-symmetric. This property and the third assumption imply that also
$f(x,k,t)$ is independent of the numbering of the particles for all $t > 0$. Finally, the
fourth hypothesis means that the electrons of a subensemble with a-particles move

independently of each other *initially*. Intuitively, this hypothesis is reasonable if a is small compared to M. A discussion of the validity of this condition, which is called the *initial chaos assumption*, can be found in [6, Chap. 2.3].

We derive first an equation for the distribution function $f^{(a)}$ of a subensemble consisting of $a < M$ electrons,

$$f^{(a)}(x_1,\ldots,x_a,k_1,\ldots,k_a,t) = \frac{1}{(4\pi^3)^{M-a}} \int_{(\mathbb{R}^3 \times B)^{M-a}} f(x,k,t)\,dx_{(a+1)}\,dk_{(a+1)},$$

(3.15)

where $dx_{(a+1)} = dx_{a+1}\cdots dx_M$ and $dk_{(a+1)} = dk_{a+1}\cdots dk_M$.

Lemma 3.3 (BBGKY hierarchy). *Let f be a solution of the semi-classical Liouville initial-value problem (3.9), (3.10), and (3.10) and let $1 \leq a \leq M-1$. We suppose that the first and third assumptions from above hold. Then $f^{(a)}$, defined in (3.15), solves*

$$0 = \partial_t f^{(a)} + \sum_{j=1}^{a} v(k_j) \cdot \nabla_{x_j} f^{(a)} - \frac{q}{\hbar} \sum_{j=1}^{a} E_{\text{ext}}(x_j,t) \cdot \nabla_{k_j} f^{(a)}$$

$$- \frac{q}{\hbar} \sum_{j,\ell=1}^{a} E_{\text{int}}(x_j,x_\ell) \cdot \nabla_{k_j} f^{(a)}$$

(3.16)

$$- (M-a)\frac{q}{4\pi^3 \hbar} \sum_{j=1}^{a} \text{div}_{k_j} \int_{\mathbb{R}^3 \times B} E_{\text{int}}(x_j,x_*) f_*^{(a+1)}\,dx_*\,dk_*,$$

with initial conditions

$$f^{(a)}(x_1,\ldots,x_a,k_1,\ldots,k_a,0) = f_I^{(a)}(x_1,\ldots,x_a,k_1,\ldots,k_a),$$

(3.17)

for $(x_1,\ldots,x_a) \in \mathbb{R}^{3a}$, $(k_1,\ldots,k_a) \in B^a$, where

$$f_*^{(a+1)} = f^{(a+1)}(x_1,\ldots,x_a,x_*,k_1,\ldots,k_a,k_*,t).$$

The system of equations (3.16) is called the *BBGKY hierarchy*, from Bogoliubov [7], Born and Green [8], Kirkwood [9], and Yvon [10]. It describes the evolution of the a-particle distribution function $f^{(a)}$ in terms of the $(a+1)$-particle distribution function $f^{(a+1)}$.

Proof. We integrate the semi-classical Liouville equation

$$\partial_t f + \sum_{j=1}^{M} v(k_j) \cdot \nabla_{x_j} f - \frac{q}{\hbar} \sum_{j=1}^{M} E_{\text{ext}}(x_j,t) \cdot \nabla_{k_j} f - \frac{q}{4\pi^3 \hbar} \sum_{j,\ell=1}^{M} E_{\text{int}}(x_j,x_\ell) \cdot \nabla_{k_j} f = 0$$

(3.18)

with respect to $x_{a+1},\ldots,x_M,k_{a+1},\ldots,k_M$ in order to obtain an equation for $f^{(a)}$. We reformulate the corresponding integrals term by term.

The first integral coming from the first term on the left-hand side of (3.18) equals $\partial_t f^{(a)}$ after the integration. For the integral of the second term we compute, using the divergence theorem,

$$\frac{1}{(4\pi^3)^{M-a}} \sum_{j=1}^{M} \int_{(\mathbb{R}^3 \times B)^{M-a}} v(k_j) \cdot \nabla_{x_j} f \, dx_{(a+1)} \, dk_{(a+1)}$$

$$= \frac{1}{(4\pi^3)^{M-a}} \sum_{j=1}^{a} v(k_j) \cdot \nabla_{x_j} \int_{(\mathbb{R}^3 \times B)^{M-a}} f \, dx_{(a+1)} \, dk_{(a+1)}$$

$$+ \frac{1}{(4\pi^3)^{M-a}} \sum_{j=a+1}^{M} \int_{(\mathbb{R}^3 \times B)^{M-a}} \text{div}_{x_j} (v(k_j) f) \, dx_{(a+1)} \, dk_{(a+1)}$$

$$= \sum_{j=1}^{a} v(k_j) \cdot \nabla_{x_j} f^{(a)}.$$

The integral of the third term in (3.18) can be treated in a similar way, leading to

$$\frac{1}{(4\pi^3)^{M-a}} \sum_{j=1}^{M} \int_{(\mathbb{R}^3 \times B)^{M-a}} \text{div}_{k_j} (E_{\text{ext}}(x_j, t) f) \, dx_{(a+1)} \, dk_{(a+1)}$$

$$= \sum_{j=1}^{a} \text{div}_{k_j} \left(E_{\text{ext}}(x_j, t) f^{(a)} \right).$$

The integral of the last term on the left-hand side of (3.18) becomes

$$\frac{1}{(4\pi^3)^{M-a}} \sum_{j,\ell=1}^{M} \int_{(\mathbb{R}^3 \times B)^{M-a}} \text{div}_{k_j} (E_{\text{int}}(x_j, x_\ell) f) \, dx_{(a+1)} \, dk_{(a+1)}$$

$$= \sum_{j,\ell=1}^{a} E_{\text{int}}(x_j, x_\ell) \cdot \nabla_{k_j} f^{(a)}$$

$$+ \frac{1}{(4\pi^3)^{M-a}} \sum_{j=a+1}^{M} \sum_{\ell=1}^{M} \int_{(\mathbb{R}^3 \times B)^{M-a}} \text{div}_{k_j} (E_{\text{int}}(x_j, x_\ell) f) \, dx_{(a+1)} \, dk_{(a+1)}$$

$$+ \frac{1}{(4\pi^3)^{M-a}} \sum_{j=1}^{a} \sum_{\ell=a+1}^{M} \int_{(\mathbb{R}^3 \times B)^{M-a}} \text{div}_{k_j} (E_{\text{int}}(x_j, x_\ell) f) \, dx_{(a+1)} \, dk_{(a+1)}.$$

The last integral on the right-hand side of the above equation vanishes by the divergence theorem. For the first integral, we use the anti-symmetry of the external field (3.12) and the third assumption (3.13). Indeed, it is possible to renumber the particles such that the last integral equals

$$\frac{1}{(4\pi^3)^{M-a}} \sum_{j=1}^{a} (M-a) \int_{(\mathbb{R}^3 \times B)^{M-a}} \text{div}_{k_j} (E_{\text{int}}(x_j, x_{a+1}) f) \, dx_{(a+1)} \, dk_{(a+1)}$$

$$= \frac{M-a}{4\pi^3} \sum_{j=1}^{a} \text{div}_{k_j} \int_{\mathbb{R}^3 \times B} E_{\text{int}}(x_j, x_*) f_*^{(a+1)} \, dx_* \, dk_*,$$

Thus, integration of (3.18) yields the system of equations (3.16). $\square$

The BBGKY hierarchy does not simplify the Liouville equation. Indeed, in order to find $f^{(1)}$, we need to know $f^{(2)}$ and so on. In order to find $f^{(M-1)}$, the knowledge about $f^{(M)}$ is necessary, which is the solution of the Liouville equation. In view of the initial chaos assumption (3.14), we may expect that if $M \to \infty$, this property also holds for positive time. Therefore, we perform the limit $M \to \infty$ in (3.16) to find the evolution of a one-particle distribution function.

Let $f_M^{(a)}$ be a solution of (3.16) and (3.17). We assume that $f_M^{(a)}$ converges point-wise to some function $f^{(a)}$ as $M \to \infty$ (similar for its derivatives). By the second assumption, the internal field E_{int} is of order $1/M$ and vanishes in the formal limit $M \to \infty$. Moreover, $(M-a)E_{\text{int}}$ converges to E_0. Thus, the BBGKY hierarchy becomes in the formal limit of infinitely many particles

$$0 = \partial_t f^{(a)} + \sum_{j=1}^{a} v(k_j) \cdot \nabla_{x_j} f^{(a)} - \frac{q}{\hbar} \sum_{j=1}^{a} E_{\text{ext}}(x_j,t) \cdot \nabla_{k_j} f^{(a)}$$

$$- \frac{q}{4\pi^3 \hbar} \sum_{j=1}^{a} \operatorname{div}_{k_j} \int_{\mathbb{R}^3 \times B} E_0(x_j,x_*) f_*^{(a+1)} \, dx_* \, dk_*, \tag{3.19}$$

$$f^{(a)}(x_1,\ldots,x_a,k_1,\ldots,k_a,0) = f_I^{(a)}(x_1,\ldots,x_a,k_1,\ldots,k_a), \tag{3.20}$$

where $(x_1,\ldots,x_a) \in \mathbb{R}^{3a}$ and $(k_1,\ldots,k_a) \in B^a$. We claim that a one-particle distribution function contains all the dynamics of this many-particle problem under the initial chaos assumption.

Theorem 3.4 (Semi-classical Vlasov equation). *Let the assumptions on page 62 hold and let f^* be a solution of the one-particle semi-classical Vlasov equation*

$$\partial_t f^* + v(k) \cdot \nabla_x f^* - \frac{q}{\hbar} E_{\text{eff}} \cdot \nabla_k f^* = 0, \quad x \in \mathbb{R}^3, \, k \in B, \, t > 0, \tag{3.21}$$

$$f^*(x,k,0) = f_I^*(x,k), \quad x \in \mathbb{R}^3, \, k \in B,$$

with periodic boundary conditions

$$f^*(x,k,t) = f^*(x,-k,t), \quad x \in \mathbb{R}^3, \, k \in B, \, t > 0,$$

where

$$E_{\text{eff}}(x,t) = E_{\text{ext}}(x,t) + \int_{\mathbb{R}^3} n(x_*,t) E_0(x,x_*) \, dx_* \tag{3.22}$$

is the effective field and

$$n(x,t) = \int_{\mathbb{R}^3} f^*(x,k,t) \frac{dk}{4\pi^3}$$

represents the electron density. Then the functions

$$f^{(a)}(x_1,\ldots,x_a,k_1,\ldots,k_a,t) = \prod_{i=1}^{a} f^*(x_i,k_i,t) \tag{3.23}$$

are a solution of the limit BBGKY hierarchy (3.19) and (3.20).

Proof. We multiply the Vlasov equation (3.21), evaluated at (x_i, k_i), by

$$Q_i = \prod_{j=1, j \neq i}^{a} f^*(x_j, k_j, t)$$

and sum over $i = 1, \ldots, a$. Then, the time derivative in (3.21) becomes, for $f_i^* = f^*(x_i, k_i, t)$,

$$\sum_{i=1}^{a} Q_i \partial_t f_i^* = \sum_{i=1}^{a} \prod_{j \neq i} f_j^* \partial_t f_i^* = \partial_t \prod_{i=1}^{a} f_i^* = \partial_t f^{(a)}.$$

In a similar way, we compute

$$\sum_{i=1}^{a} Q_i v(k_i) \cdot \nabla_{x_i} f_i^* = \sum_{i=1}^{a} v(k_i) \cdot \nabla_{x_i} f^{(a)},$$

$$\sum_{i=1}^{a} Q_i E_{\text{ext}}(x_i, t) \cdot \nabla_{k_i} f_i^* = \sum_{i=1}^{a} E_{\text{ext}}(x_i, t) \cdot \nabla_{k_i} f^{(a)}.$$

The expression involving the limiting internal field E_0 can be reformulated as

$$\sum_{i=1}^{a} Q_i \left(\int_{\mathbb{R}^3 \times B} f^*(x_*, k_*, t) E_0(x_i, x_*) \, dx_* \, dk_* \right) \cdot \nabla_{k_i} f_i^*$$

$$= \sum_{i=1}^{a} \operatorname{div}_{k_i} \int_{\mathbb{R}^3 \times B} \prod_{j=1}^{a} f^*(x_j, k_j, t) f^*(x_*, k_*, t) E_0(x_i, x_*) \, dx_* \, dk_*$$

$$= \sum_{i=1}^{a} \operatorname{div}_{k_i} \int_{\mathbb{R}^3 \times B} f_*^{(a+1)} E_0(x_i, x_*) \, dx_* \, dk_*.$$

Putting together the above expressions, we see that the ansatz (3.23) indeed solves (3.19) and (3.20). $\quad\square$

Each solution of the semi-classical Vlasov equation provides a solution of the limiting semi-classical Liouville equation under the initial chaos assumption. Thus, the solution of the many-particle problem is reduced to the solution of a one-particle problem. In this sense, the Vlasov equation is derived from the Liouville equation. Figure 3.1 illustrates the steps of the derivation.

The Vlasov equation has the form of a Liouville equation for a single particle with the force field $-qE_{\text{eff}}$. Many-particle effects are taken into account through the effective field E_{eff} which depends on the particle density and hence, on the distribution function f^*. Therefore, (3.21) is a nonlinear equation with a nonlocal quadratic

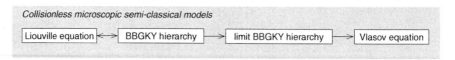

Fig. 3.1 Illustration of the derivation of the Vlasov equation

nonlinearity. It describes the macroscopic motion of the many-particle system with weak long-range forces. It does not provide a description of strong short-range interactions such as scattering of particles, which are considered in Chap. 4.

The classical Vlasov equation can be directly derived from the Newton equations $\dot{x}_i = v_i$, $\dot{v}_i = F_i/m$, $i = 1,\ldots,M$, for identical particles of mass m which are initially equally separated. For this, let the interaction force F_i be given by $F_i = \sum_{j\neq i} F(x_i - x_j)/M$. The measure

$$\mu_M(x,v,t) = \frac{1}{M} \sum_{i=1}^{M} \delta(x - x_i(t)) \otimes \delta(v - v_i(t)),$$

where (x_i, v_i) solves the Newton equations and the symbol $\otimes$ denotes the tensor product, can be shown to be a solution of the Vlasov equation in the sense of distributions. The question is whether the weak limit of μ_M as $M \to \infty$ solves the Vlasov equation or not. The first positive answers were given by Neunzert and Wick [11], Dobrushin [12], and Braun and Hepp [13]. A physical derivation can be found in [14]. Further results were obtained in, for instance, [15–17]. The result of Braun and Hepp [13] is based on the boundedness of the forces F_i, thus excluding forces of the type $F(x) \leq c|x|^{-\alpha}$ with $\alpha > 0$. The recent work of Hauray and Jabin [18] allows for such forces with $\alpha < 1$. Unfortunately, the case $\alpha = 1$ cannot be treated, which would cover the case of the Vlasov–Poisson system in three space dimensions.

The Vlasov equation was derived also in other contexts. For instance, it was deduced from the Liouville equation for plasmas permeated by a uniform ambient magnetic field in [19]. A derivation from the many-particle Schrödinger equation with pair interaction in the classical limit is presented in [20].

The first proof of the global-in-time existence of smooth solutions of the Vlasov–Poisson system was found by Pfaffelmoser in 1989 [21]. Previous results [22, 23] were concerned with symmetric or small initial data. Later, other and simpler proofs were given by Lions and Perthame [24], Schaeffer [25], and Horst [26] improving the result of [25].

As for the Liouville equation, the quantity $f^*(x,k,t)$ can be interpreted as the probability density of a particle to be in the state (x,k) at time t. Indeed, from the trajectory equations

$$\dot{x} = v(k), \quad \dot{k} = -\frac{q}{\hbar} E_0, \quad t > 0,$$

and the semi-classical Vlasov equation, we obtain

$$0 = \partial_t f^* + v(k) \cdot \nabla_x f^* - \frac{q}{\hbar} E_{\text{eff}} \cdot \nabla_k f^* = \frac{d}{dt} f^*(x(t),k(t),t), \quad t > 0, \tag{3.24}$$

from which we conclude that f^* is constant along the trajectories,

$$f^*(x(t),k(t),t) = f^*(x(0),k(0),0), \quad t > 0.$$

In particular, if the number density is nonnegative initially, it remains nonnegative for all times.

Finally, we wish to formulate the Vlasov equation in the case of *Coulomb forces*, which are the most important long-range forces between two electrons. The effective internal field is given by

$$E_0(x,y) = -\frac{q}{4\pi\varepsilon_s}\frac{x-y}{|x-y|^3}, \quad x,y \in \mathbb{R}^3, \ x \neq y, \tag{3.25}$$

where ε_s denotes the semiconductor permittivity, which is a material constant. We assume that the external field is generated by doping atoms in the semiconductor crystal of charge $+q$,

$$E_{\text{ext}}(x,t) = \frac{q}{4\pi\varepsilon_s}\int_{\mathbb{R}^3} C(y)\frac{x-y}{|x-y|^3}\,dy, \tag{3.26}$$

where $C(x)$ is the doping concentration (see Sect. 1.6).

Proposition 3.5 (Semi-classical Vlasov–Poisson system). *In the case of the Coulomb forces* (3.25) *and* (3.26), *the semi-classical Vlasov equation can be written as the* Vlasov–Poisson system

$$\partial_t f^* + v(k) \cdot \nabla_x f^* + \frac{q}{\hbar}\nabla_x V \cdot \nabla_k f^* = 0, \tag{3.27}$$

$$\varepsilon_s \Delta V = q(n-C), \quad x \in \mathbb{R}^3, \ k \in B, \ t > 0, \tag{3.28}$$

with periodic boundary conditions for f^* on ∂B and the initial condition $f^*(\cdot,\cdot,0) = f_I^*$.

If the semiconductor structure consists of several materials, the permittivity is space dependent and we have to replace the *Poisson equation* (3.28) by $\text{div}\,(\varepsilon_s \nabla V) = q(n-C)$.

Proof. It is well known from potential theory that the function

$$\phi(x) = -\frac{1}{4\pi}\int_{\mathbb{R}^3}\frac{g(y)}{|x-y|}\,dy, \quad x \in \mathbb{R}^3,$$

solves the Poisson equation $\Delta\phi = g$ in $\mathbb{R}^3$ under some regularity assumptions on g. Differentiation gives the formulas

$$g(x) = \Delta\phi(x) = \frac{1}{4\pi}\text{div}_x\int_{\mathbb{R}^3} g(y)\frac{x-y}{|x-y|^3}\,dy,$$

$$0 = \text{curl}\,\nabla\phi(x) = \frac{1}{4\pi}\text{curl}_x\int_{\mathbb{R}^3} g(y)\frac{x-y}{|x-y|^3}\,dy.$$

This shows that $\operatorname{div} E_{\text{ext}} = qC/\varepsilon_s$, $\operatorname{curl} E_{\text{ext}} = 0$, and

$$\operatorname{div} E_{\text{eff}}(x,t) = \operatorname{div} E_{\text{ext}}(x,t) + \operatorname{div} \int_{\mathbb{R}^3} n(x_*,t) E_0(x,x_*) \, dx_*$$
$$= \frac{q}{\varepsilon_s} (C(x) - n(x,t)), \tag{3.29}$$

$$\operatorname{curl} E_{\text{eff}}(x,t) = \operatorname{curl} E_{\text{ext}}(x,t) + \int_{\mathbb{R}^3} n(x_*,t) \operatorname{curl} E_0(x,x_*) \, dx_* = 0.$$

Since E_{eff} is vortex free, there exists a potential V such that $E_{\text{eff}} = -\nabla V$. Thus, by (3.29),

$$\varepsilon_s \Delta V = -\varepsilon_s \operatorname{div} E_{\text{eff}} = q(n - C),$$

and the proposition follows. $\quad\square$

References

1. M. Golberg. The derivative of a determinant. *The Amer. Math. Monthly* 79 (1972), 1124–1126.
2. P. Markowich, C. Ringhofer, and C. Schmeiser. *Semiconductor Equations.* Springer, Vienna, 1990.
3. Y. Guo. The Vlasov-Maxwell-Boltzmann system near Maxwellians. *Invent. Math.* 153 (2003), 593–630.
4. J. Schaeffer. Steady states of the Vlasov-Maxwell system. *Quart. Appl. Math.* 63 (2005), 619–643.
5. R. Strain. The Vlasov-Maxwell-Boltzmann system in the whole space. *Commun. Math. Phys.* 268 (2006), 543–567.
6. C. Cercignani, R. Illner, and M. Pulvirenti. *The Mathematical Theory of Dilute Gases.* Springer, New York, 1994.
7. N. Bogoliubov. Problems of a dynamical theory in statistical physics. In: J. de Boer and U. Uhlenbeck (eds.), *Studies in Statistical Mechanics*, Vol. I, 1–118. North-Holland, Amsterdam, 1962.
8. M. Born and H. Green. *A General Kinetic Theory of Fluids.* Cambridge University Press, Cambridge, 1949.
9. J. Kirkwood. Statistical mechanical theory of transport processes I. General theory. *J. Chem. Phys.* 14 (1946), 180–201. Errata: 14 (1946), 347.
10. J. Yvon. *La théorie statistique des fluides.* Hermann, Paris, 1935.
11. H. Neunzert and J. Wick. Theoretische und numerische Ergebnisse zur nichtlinearen Vlasov-Gleichung. *Lecture Notes Math.* 267, 159–185. Springer, Berlin, 1972.
12. R. Dobrushin. Vlasov equations. *Funktional. Anal. i Prilozhen* 13 (1979), 48–58.
13. W. Braun and K. Hepp. The Vlasov dynamics and its fluctuations in the $1/N$ limit of interacting classical particles. *Commun. Math. Phys.* 56 (1977), 101–2113.
14. R. Wilson, W. Futterman, and J. Yue. Derivation of the Vlasov equation. *Phys. Rev. A* 3 (1971), 453–470.
15. J. Batt, N-particle approximation to the Vlasov-Poisson system. *Nonlin. Anal.* 47 (2001), 1445–1456.
16. H. Spohn. *Large Scale Dynamics of Interacting Particles.* Springer, Berlin, 1991.
17. S. Wollman. On the approximation of the Vlasov-Poisson system by particles methods. *SIAM J. Numer. Anal.* 37 (2000), 1369–1398.

18. M. Hauray and P.-E. Jabin. *N*-particles approximation of the Vlasov equations with singular potential. *Arch. Rat. Mech. Anal.* 183 (2007), 489–524.
19. T. Passot. From kinetic to fluid descriptions of plasmas. In: *Topics in Kinetic Theory*, Fields Inst. Commun. 46, 213–242. Amer. Math. Soc., Providence, 2005.
20. C. Bardos, F. Golse, A. Gottlieb, and N. Mauser. On the derivation of nonlinear Schrödinger and Vlasov equations. In: *Dispersive Transport Equations and Multiscale Models*, IMA Math. Anal. 136, 1–23. Springer, New York, 2004.
21. K. Pfaffelmoser. Global classical solutions of the Vlasov-Poisson system in three dimensions for general initial data. *J. Diff. Eqs.* 95 (1992), 281–303.
22. C. Bardos and P. Degond. Global existence for the Vlasov-Poisson equation in three space variables with small initial data. *Ann. Inst. H. Poincaré, Anal. non linéaire* 2 (1985), 101–118.
23. J. Batt. Global symmetric solutions of the initial value problem of stellar dynamics. *J. Diff. Eqs.* 25 (1977), 342–364.
24. P.-L. Lions and B. Perthame. Propagation of moments and regularity for the three-dimensional Vlasov-Poisson system. *Invent. Math.* 105 (1991), 415–430.
25. J. Schaeffer. Global existence of smooth solutions to the Vlasov-Poisson system in three dimensions. *Commun. Part. Diff. Eqs.* 16 (1991), 1313–1335.
26. E. Horst. On the asymptotic growth of the solutions of the Vlasov-Poisson system. *Math. Meth. Appl. Sci.* 16 (1993), 75–86.

Chapter 4
Scattering Models

The Vlasov (or Liouville) equation of the previous chapter does not take into account short-range particle interactions, like collisions of the particles with other particles or with the crystal lattice. In this chapter, we extend the Vlasov equation to include scattering mechanisms which leads to the Boltzmann equation. We present only a phenomenological derivation. For rigorous results, we refer to [1, Sect. 1.5.3] and [2, Chap. 4].

4.1 The Boltzmann Equation

The Vlasov equation along trajectories

$$\frac{\mathrm{d}f}{\mathrm{d}t} = 0$$

states that the probability density f (of occupation of states) does not change in time (see Sect. 3.1). Scattering allows particles to jump to another trajectory. Our main assumption is that the rate of change of f due to convection and the effective field, $\mathrm{d}f/\mathrm{d}t$, and the rate of change of f due to collisions, $Q(f)$, balance:

$$\frac{\mathrm{d}f}{\mathrm{d}t} = Q(f).$$

Clearly, this equation has to be understood along trajectories. By (3.24), this equation equals

$$\partial_t f + v(k) \cdot \nabla_x f - \frac{q}{\hbar} E_{\mathrm{eff}} \cdot \nabla_k f = Q(f), \quad x \in \mathbb{R}^3, \, k \in B, \, t > 0, \tag{4.1}$$

where the effective field E_{eff} is given by (3.22).

It remains to derive an expression for $Q(f)$. We assume that scattering of particles occurs instantaneously and only changes the crystal momentum of the particles. The

Jüngel, A.: *Scattering Models*. Lect. Notes Phys. **773**, 71–95 (2009)
DOI 10.1007/978-3-540-89526-8_4

rate $P(x, k' \to k, t)$ at which a particle at (x, t) changes its Bloch state k' into another Bloch state k due to a scattering event is proportional to

- the occupation probability $f(x, k', t)$ and
- the probability $1 - f(x, k, t)$ that the state (x, k) is not occupied.

Here, we used a statistical version of the Pauli exclusion principle which is valid for fermions and in particular for electrons. Thus,

$$P(x, k' \to k, t) = s(x, k', k) f(x, k', t)(1 - f(x, k, t)),$$

where the proportionality factor $s(x, k', k)$ is called the *transition* or *scattering rate*. The rate of change of f due to collisions is the sum of all in-scattering rates from some k' to k minus the out-scattering rate from k to some k',

$$P(x, k' \to k, t) - P(x, k \to k', t),$$

for all possible Bloch states k' in the "volume element" dk'. In the limit, the sum becomes an integral and we obtain

$$(Q(f))(x, k, t) = \int_B \left(P(x, k' \to k, t) - P(x, k \to k', t) \right) dk' \tag{4.2}$$

$$= \int_B \left(s(x, k', k) f'(1 - f) - s(x, k, k') f(1 - f') \right) dk',$$

where $f = f(x, k, t)$ and $f' = f(x, k', t)$. Equation (4.1), together with the effective-field equation

$$E_{\text{eff}}(x, t) = E_{\text{ext}}(x, t) + \int_{\mathbb{R}^3} n(\xi, t) E_0(x, \xi) d\xi, \quad n = \int_B f \frac{dk}{4\pi^3}, \tag{4.3}$$

where E_{ext} and E_0 are given functions, and the collision operator (4.2) is called the *semi-classical Boltzmann equation*. When E_{ext} and E_0 are given by the Coulomb forces (3.25) and (3.26), equations (4.1), (4.2), and (4.3) are called the *Boltzmann–Poisson system*, which can be written as (3.27) and (3.28) with f instead of f^* and with the right-hand side $Q(f)$ in (3.27). Again we impose the initial and periodic boundary conditions

$$f(x, k, t) = f(x, -k, t), \quad x \in \mathbb{R}^3, k \in \partial B, t > 0, \tag{4.4}$$

$$f(x, k, 0) = f_I(x, k), \quad x \in \mathbb{R}^3, k \in B. \tag{4.5}$$

The Boltzmann equation has two nonlinearities:

- a quadratic nonlocal nonlinearity in the position variable caused by the self-consistent field E_{eff} in (4.3) and
- another quadratic nonlocal nonlinearity in the wave vector caused by the collision integral (4.2).

These nonlinearities make the mathematical analysis of the initial boundary-value problem (4.1), (4.2), (4.3), (4.4), and (4.5) very difficult. The Boltzmann equation was first formulated by Boltzmann in 1872 for the nonequilibrium transport of dilute gases, and some properties of the solutions were shown, but no existence analysis was available [3]. The first result on the existence and uniqueness of solutions to the homogeneous gas-dynamics Boltzmann equation $\partial_t f = \tilde{Q}(f)$ was published by Carleman in 1933 under the assumption that f depends on the modulus of the velocity $|v|$ and t only [4]. For a very general class of collision operators, Arkeryd developed an L^1 theory for the homogeneous equation [5]. For the nonhomogeneous Boltzmann equation $\partial_t f + v \cdot \nabla_x f = Q(f)$, only existence results for initial data close to the equilibrium state or close to a homogeneous distribution were proved up to 1987 [2, Sect. 5.1]. The first result of global-in-time existence of so-called renormalized solutions with general initial data was then shown by Di Perna and Lions [6] using a smoothing effect of the flow term $v \cdot \nabla_x f$, known as the velocity averaging lemma [7]. We refer to [8] for a review on the solution of the Boltzmann equation and to [9, 10] for more details and references. The existence of weak solutions to the *semiconductor* Boltzmann–Poisson system was first proved by Poupaud [11]. The result was extended by Mustieles in unbounded wave-vector spaces to conclude smooth global-in-time solutions in one or two space dimensions and local-in-time solutions in three dimensions [12, 13]. The existence of smooth global solutions in three space dimensions was proved by Andréasson [14]. In all these works, the scattering rate was assumed to be smooth. Majorana et al. proved the existence of solutions of the spatially homogeneous Boltzmann equation allowing for nonsmooth scattering rates [15–17].

The Boltzmann equation can be solved numerically by direct simulation Monte Carlo methods, in view of the high computational cost of a conventional quadrature rule for the evaluation of the collision integral [18–20]. An approximation can be obtained by spherical harmonics expansions of the distribution function before performing Monte Carlo simulations [21–23]. As probabilistic methods yield low accurate results for instationary solutions and a low convergence rate, deterministic methods were also designed, for instance, particle methods [2], discrete velocity models [24], spectral approximations [25], power-series time discretizations [26], and more recently, high-order finite-difference weighted essentially non-oscillatory (WENO) solver [27–29]. Simulations of the Boltzmann–Poisson system coupled with the effective mass Schrödinger equation were performed in [30]. We refer to [31, 32] for more details and references.

In Fig. 4.1 we present a summary of the models derived in this and the previous chapter and the relations between them.

We give now some examples of collision operators. In semiconductor crystals, scattering of electrons is due to lattice defects, phonons, and other carriers. We consider only the following important collision events:

- electron–phonon scattering,
- ionized impurity scattering, and
- carrier–carrier scattering.

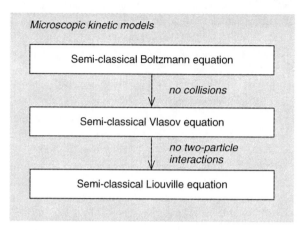

Fig. 4.1 Relations between the kinetic models

Carrier–carrier scattering includes electron–electron or electron–hole collisions. Also collisions of a carrier with a carrier ensemble (collective carrier–carrier scattering), i.e., the interaction of a carrier with oscillations in the carrier density, may occur when the carrier density is sufficiently high. Extensive treatments of scattering mechanisms in semiconductors can be found in, e.g., [33, Chap. 9], [34, Chap. 7], [35, Chap. 2], [36, Chap. 6], and in the textbooks [37–39].

Phonon scattering. At nonzero temperature, the atoms in the crystal lattice vibrate around their fixed equilibrium. These vibrations are quantized and the quantum of lattice vibrations is called a *phonon*. We can distinguish so-called *acoustic phonons* and *optical phonons*. Acoustic phonons arise from displacements of lattice atoms in the same direction such as sound waves. Optical phonons describe displacements in the wave vector and are able to interact strongly with light. Denoting by $\hbar\omega_\alpha$ the energy of a phonon with frequency ω_α, the phonon occupation number N_α is computed from Bose–Einstein statistics,

$$N_\alpha = \frac{1}{e^{\hbar\omega_\alpha/k_B T} - 1},$$

where the index α refers to either "op" for optical phonons or "ac" for acoustic phonons. Notice that Bose–Einstein statistics has to be used for indistinguishable particles not obeying the Pauli exclusion principle and therefore also for phonons (see [33, p. 307ff.] for a derivation).

An electron in the Bloch state k' with conduction-band energy $E(k')$ before the collision with a phonon with quantized frequency ω_α can change to the state k after the collision if

$$E(k') - E(k) = \pm\hbar\omega_\alpha, \tag{4.6}$$

where the plus sign refers to phonon emission and the minus sign for phonon absorption. The transition rate $s(x,k,k')$ is nonzero only if the energy conservation condition (4.6) is satisfied. Therefore,

$$s_\alpha(x,k,k') = \sigma_\alpha(x,k,k') \tag{4.7}$$
$$\times \left((1+N_\alpha)\delta(E(k') - E(k) + \hbar\omega_\alpha) + N_\alpha\delta(E(k') - E(k) - \hbar\omega_\alpha) \right),$$

where the number $\sigma_\alpha(x,k,k')$ is assumed to be symmetric in k and k' and δ is the delta distribution. The first delta distribution contributes when an energy of $\hbar\omega_\alpha$ has been absorbed, i.e., $E(k') = E(k) - \hbar\omega_\alpha$, whereas the second term contributes when an energy of $\hbar\omega_\alpha$ has been emitted, i.e., $E(k') = E(k) + \hbar\omega_\alpha$. The factors $1 + N_\alpha$ and N_α come from the eigenvalues of the so-called creation and annihilation operators [40, Appendix B]. The transition rate is known as the *Fermi golden rule* (see [33, Sect. 4.4], [35, Sect. 1.7.1], or [40, Appendix C]). It is valid over long time durations and for single-state transitions.

Generally, the phonon energy $\hbar\omega_\alpha$ can be interpreted as a function of the wave vectors k and k'. Often, it depends on the difference $k - k'$ of the wave vectors before and after a scattering event only [41, Sect. 2]. For optical (nonpolar) phonon scattering, the dependency is weak such that ω_{op} can be considered to be constant [35, Sect. 1.8]. On the other hand, the energy of acoustic phonons is rather small compared to the kinetic energy of a carrier and can be neglected near room temperature, $\hbar\omega_{ac} \approx 0$. Then (4.7) can be simplified for elastic acoustic phonon scattering to

$$s_{ac}(x,k,k') = \sigma_0\delta(E(k') - E(k)), \tag{4.8}$$

where $\sigma_0 = \sigma_{ac}(2N_{ac} + 1)$ does not depend on k or k'.

The collision operator reads according to (4.2)

$$(Q_\alpha(f))(x,k,t) = \int_B \left(s_\alpha(x,k',k)f'(1-f) - s_\alpha(x,k,k')f(1-f') \right) dk', \tag{4.9}$$

where $\alpha = ac, op$. For acoustic phonon scattering in the elastic approximation, for which (4.8) holds, we can employ the symmetry of δ to obtain

$$\int_B \left(s_{ac}(x,k',k)f'f - s_{ac}(x,k,k')ff' \right) dk' = \int_B \sigma_0\delta(E(k') - E(k))(f'f - ff')dk'$$
$$= 0,$$

and the collision operator becomes

$$(Q_{ac}(f))(x,k,t) = \int_B \sigma_0\delta(E(k') - E(k))(f' - f)dk'. \tag{4.10}$$

The above expression for s_α shows that the scattering rates can be highly nonsmooth. In fact, they may not be functions but distributions. As already mentioned in Sect. 1.6, integrals involving the delta distribution can be interpreted by employing the coarea formula (1.55). Indeed, let us as in [41] $S_\varepsilon = \{k \in B : E(k) = \varepsilon\}$ be the surface of constant energy ε, $dS_\varepsilon(k)$ the Euclidean surface element on S_ε, and $dN_\varepsilon(k) = dS_\varepsilon(k)/|\nabla_k E(k)|$ the coarea measure. Then $N(\varepsilon) = 2(2\pi)^{-3}\int_{S_\varepsilon} dN_\varepsilon$ is the density of states of energy ε (see (1.56)). The coarea formula can now be formulated as

$$\int_B g(k)\,dk = \int_{\mathbb{R}}\int_{S_\varepsilon} g(k)\,dN_\varepsilon(k)\,d\varepsilon.$$

The collision integral (4.10), for instance, then becomes

$$(Q_{\mathrm{ac}}(f))(x,k,t) = \int_{S_{E(k)}} \sigma_0 f(k')\,dN_{E(k)}(k') - f(k)\int_{S_{E(k)}} \sigma_0\,dN_{E(k)}(k')$$

and does not contain delta distributions anymore.

Finally, we remark that there are many different phonon scattering mechanisms; see, e.g., [35] for details.

Ionized impurity scattering. A doping atom in the semiconductor material donates either an electron or a hole, leaving behind an ionized charged impurity. This fixed charge may attract or repulse an electron propagating through the crystal lattice. The interaction of carriers with neutral impurities is another scattering possibility but we do not consider it here. Since the scattering is elastic, the electron energy $E(k')$ after the collision is the same as the energy $E(k)$ before the interaction, and the transition rate is consequently

$$s_{\mathrm{imp}}(x,k,k') = \sigma_{\mathrm{imp}}(x,k,k')\delta(E(k')-E(k)),$$

where σ_{imp} is symmetric in k and k' [42]. Again, this expression can be derived from Fermi's golden rule. Notice that it is possible to describe acoustic phonon scattering in the elastic approximation by the same formula in which $\sigma_{\mathrm{imp}}(x,k,k') = \sigma_0$ (see (4.8)). The symmetry of σ_{imp} and δ implies as above that the collision operator can be written as

$$(Q_{\mathrm{imp}}(f))(x,k,t) = \int_B \sigma_{\mathrm{imp}}(x,k,k')\delta(E(k')-E(k))(f'-f)\,dk'. \tag{4.11}$$

Carrier–carrier scattering. We only consider binary electron–electron interactions. Also binary electron–hole scattering or collective carrier–carrier collisions [35, Sect. 2.10.2] are possible but we do not consider these mechanisms here. The influence of electron–electron interactions on the carrier dynamics is more pronounced in degenerate semiconductors (see Sect. 1.6). The transition rate for carriers in the Bloch states k' and k_1' which collide and scatter to the states k and k_1 is given by

$$s_{\mathrm{ee}}(x,k,k',k_1,k_1') = \sigma_{\mathrm{ee}}(x,k,k',k_1,k_1')\delta(E(k')+E(k_1')-E(k)-E(k_1)),$$

since the collisions are elastic. More precisely, the above delta distribution should be understood as B-periodic in order to account for so-called umklapp processes and to preserve the periodic structure in k. We refer to [41, Sect. 2] for details. (Umklapp processes, discovered by Peierls, occur when the pseudo-wave vector of an electron or phonon interacting with other particles leaves the Brillouin zone and is brought back to this zone by adding a reciprocal lattice vector; see [43, p. 135ff.].)

The collision rate is proportional to the occupation probabilities $f' = f(x,k',t)$ and $f'_1 = f(x,k'_1,t)$ (the states k' and k'_1 are occupied) and to $1 - f = 1 - f(x,k,t)$ and $1 - f_1 = 1 - f(x,k_1,t)$ (the states k are k_1 are not occupied). Therefore, the collision operator becomes [44, (2.7)]:

$$(Q_{ee}(f))(x,k,t) = \int_{B^3} s_{ee}(x,k,k',k_1,k'_1) \tag{4.12}$$
$$\times \left(f' f'_1 (1-f)(1-f_1) - f f_1 (1-f')(1-f'_1) \right) \, dk' \, dk_1 \, dk'_1.$$

Notice that this operator has a nonlocal nonlinearity of fourth order.

Summarizing, the collision operator in the semiconductor Boltzmann equation (4.1) can be written as the sum of the collision operators considered above:

$$Q(f) = Q_{op}(f) + Q_{ac}(f) + Q_{imp}(f) + Q_{ee}(f).$$

The Boltzmann equation (4.1) is fundamental in deriving simpler macroscopic models for semiconductors (see Chap. 2). It is the basic equation in semi-classical semiconductor modeling and usually, other models are validated by numerical comparisons with the Boltzmann equation. Nevertheless, it is important to understand its limitations (here we follow [35, Sect. 37]):

- The semiconductor Boltzmann equation is a single-particle description of a many-particle charge-carrier system. Correlations between carriers are incorporated only by the effective-field approximation.
- Quantum mechanical phenomena are only modeled in a semi-classical way. Electrons are considered as particles obeying the semi-classical Newton's laws.
- Collisions are assumed to be binary and to be instantaneous in time and local in space.
- The statistical description using the probability density $f(x,k,t)$ makes only sense if the number of carriers is sufficiently large.

We can estimate the range of validity of the semi-classical approach by using Heisenberg's uncertainty principles, expressing that it is impossible to determine momentum and position at the same time and that the energy of a particle can be determined only if it stays in the same state for a certain time [35, p. 152]. Thus, the standard deviations of momentum $\triangle p$ and space $\triangle x$ cannot be both arbitrarily small and the same statement holds for the standard deviations of energy $\triangle E$ and time $\triangle t$:

$$\triangle p \triangle x \geq \hbar \quad \text{and} \quad \triangle E \triangle t \geq \hbar. \tag{4.13}$$

When we consider an electron as a particle, we can relate the kinetic energy E and the momentum p by $E = p^2/2m^*$. (This formula holds, for instance, if the energy is equal to the parabolic band approximation $E = \hbar^2 |k|^2/2m^*$ and $p = \hbar k$ is the crystal momentum.) Then, assuming that the energy spread is of the order of the thermal energy $k_B T$, we obtain

$$\triangle p = \sqrt{2m^* \triangle E} \sim \sqrt{2m^* k_B T}.$$

The sign "$\sim$" here means "of the same order as". Introducing the *de Broglie length* $\lambda_B = h/\sqrt{2m^*k_BT}$, which is the wavelength of an electron with thermal energy, the first inequality in (4.13) leads to the requirement

$$\triangle x \geq \frac{\hbar}{\triangle p} \sim \frac{\lambda_B}{2\pi}.$$

Therefore, when treating electrons as particles, they cannot be localized sharper than $\triangle x$ which is of the order of λ_B. At room temperature and with the (longitudinal) effective mass $m^* = 0.98m_e$ of silicon (see Table 1.3), we obtain the restriction $\triangle x \sim 1.2\,\mathrm{nm}$.

Taking $\triangle t$ to be the time between two consecutive collisions and again assuming that $\triangle E \sim k_BT$, the second inequality in (4.13) gives

$$\triangle t \geq \frac{\hbar}{\triangle E} \sim \frac{\hbar}{k_BT}.$$

Supposing further that the electron has the velocity v corresponding to the thermal energy, $v = p/m^* \sim \sqrt{2k_BT/m^*}$, the distance between two collisions is

$$\triangle L = v\triangle t \geq \sqrt{\frac{2k_BT}{m^*}} \frac{h}{2\pi k_BT} = \frac{\lambda_B}{\pi}.$$

Thus, the *mean free path* $\triangle L$ should be larger than the de Broglie length $\lambda_B \sim 8\,\mathrm{nm}$ in order to guarantee validity of the Boltzmann equation approach.

4.2 Properties of Collision Operators

In the following we show some properties of the collision operators described in the previous section. In particular, we derive their conservation properties and characterize their kernels. The results are needed for the derivation of macroscopic models.

Collision operator (4.2). Instead of specifying the scattering rate $s(x,k,k')$ of the collision operator, we derive a relation which is sufficient to conclude the properties. We assume that the *principle of detailed balance* [34, 45] holds, according to which the *local* scattering probabilities vanish,

$$s(x,k',k)f'_{eq}(1-f_{eq}) = s(x,k,k')f_{eq}(1-f'_{eq}) = 0.$$

Here, the prime means evaluation at k' and f_{eq} is the thermal equilibrium occupation number density given by the Fermi–Dirac distribution

$$f_{eq}(x,k) = \frac{1}{1+\exp((E(k)-E_F)/k_BT)},$$

where $E(k)$ is the band energy and E_F the Fermi energy (see Sect. 1.6 for the derivation of the equilibrium density). This shows that

$$\frac{s(x,k,k')}{s(x,k',k)} = \frac{f'_{eq}(1-f_{eq})}{f_{eq}(1-f'_{eq})} = \exp\frac{E(k)-E(k')}{k_BT}. \tag{4.14}$$

The following result is due to Poupaud [11].

Proposition 4.1 (Collision operator (4.2)**).** *Let* (4.14) *hold for some function* $E(k)$ *and let* $s(x,k,k')$ *be smooth and positive.*

(1) *For all (regular) functions* $f(x,k,t)$, *there holds*

$$\int_B (Q(f))(x,k,t)\,dk = 0 \quad for\ x \in \mathbb{R}^3,\ t > 0.$$

(2) *For all functions* $f(x,k,t) \in (0,1)$ *and nondecreasing functions* $\chi : \mathbb{R} \to \mathbb{R}$, *there holds*

$$\int_B (Q(f))(x,k,t)\chi\left(\frac{f(x,k,t)}{1-f(x,k,t)}e^{E(k)/k_BT}\right)\,dk \leq 0,$$

$$\int_B (Q(f))(x,k,t)\chi\left(\frac{1-f(x,k,t)}{f(x,k,t)}e^{-E(k)/k_BT}\right)\,dk \geq 0.$$

(3) *The kernel of* Q *only consists of Fermi–Dirac distributions, i.e.,* $Q(f) = 0$ *if and only if, for some* $-\infty \leq E_F \leq \infty$,

$$f(k) = \frac{1}{1+e^{(E(k)-E_F)/k_BT}}, \quad k \in B. \tag{4.15}$$

Proof. (1) Exchanging k and k' in the second integral of (4.2) gives

$$\int_B Q(f)\,dk = \int_{B^2} s(k',k)f'(1-f)\,dk'\,dk - \int_{B^2} s(k,k')f(1-f')\,dk'\,dk$$

$$= \int_{B^2} s(k',k)f'(1-f)\,dk'\,dk - \int_{B^2} s(k',k)f'(1-f)\,dk\,dk' = 0.$$

(2) We show only the second inequality. The proof of the first one is similar. Set

$$M(k) = e^{-E(k)/k_BT}, \quad F(k) = \frac{1-f(k)}{f(k)}M(k).$$

The function $M(k)$ is the Maxwellian (see (1.51)). With this notation, assumption (4.14) is equivalent to

$$\frac{s(k',k)}{M(k)} = \frac{s(k,k')}{M(k')}, \tag{4.16}$$

and we obtain

$$\int_B Q(f)\chi(F)\,dk = \int_{B^2} s(k,k')\left(\frac{M}{M'}f'(1-f) - f(1-f')\right)\chi(F)\,dk'\,dk$$

$$= \int_{B^2} \frac{s(k,k')}{M'}ff'(F-F')\chi(F)\,dk'\,dk, \qquad (4.17)$$

and, after exchanging k and k' and again using (4.16),

$$\int_B Q(f)\chi(F)\,dk = \int_{B^2} \frac{s(k',k)}{M}f'f(F'-F)\chi(F')\,dk\,dk'$$

$$= \int_{B^2} \frac{s(k,k')}{M'}f'f(F'-F)\chi(F')\,dk\,dk'. \qquad (4.18)$$

Adding (4.17) and (4.18) leads to

$$\int_B Q(f)\chi(F)\,dk = \frac{1}{2}\int_{B^2} \frac{s(k,k')}{M'}ff'(F-F')(\chi(F)-\chi(F'))\,dk'\,dk \geq 0, \quad (4.19)$$

since χ is nondecreasing and all the other factors are nonnegative.

(3) We see from (4.19) with $\chi(x) = x$ that $Q(f) = 0$ is equivalent to

$$ff'(F-F')^2 = 0 \quad \text{for almost all } k, k' \in B.$$

This implies $f = 0$ or $F = F'$ almost everywhere. The latter equation is equivalent to

$$\frac{1-f(k)}{f(k)}M(k) = \frac{1-f(k')}{f(k')}M(k') \quad \text{for almost all } k, k' \in B.$$

We infer that both sides are constant and denote this constant by e^{-E_F/k_BT} for some $E_F \in \mathbb{R}$. Notice that the constant is positive, except if $f = 1$. Then, solving

$$\frac{1-f(k)}{f(k)} = \frac{e^{-E_F/k_BT}}{M(k)} = e^{(E(k)-E_F)/k_BT}$$

for $f(k)$ yields (4.15). Finally, choosing $E_F = \pm\infty$ leads to the other two possibilities $f = 0$ or $f = 1$. $\quad\square$

Proposition 4.1 can be interpreted physically. Let f be a solution of the Boltzmann equations (4.1) and (4.2). Then, by the first statement of the theorem and the divergence theorem,

$$\frac{d}{dt}\int_{\mathbb{R}^3} n(x,t)\,dx = \int_{\mathbb{R}^3}\int_B \partial_t f(x,k,t)\frac{dk}{4\pi^3}\,dx \qquad (4.20)$$

$$= \int_{\mathbb{R}^3}\int_B \left(-\text{div}_x(v(k)f) + \frac{q}{\hbar}\text{div}_k(E_{\text{eff}}f) + Q(f)\right)\frac{dk}{4\pi^3}\,dx = 0.$$

This implies that the total number of electrons is conserved in time:

$$\int_{\mathbb{R}^3} n(x,t)\,dx = \int_{\mathbb{R}^3} n(x,0)\,dx \quad \text{for all } t > 0.$$

Physically, this is reasonable: collisions neither destroy nor generate particles. The second statement of the theorem is called the *H-theorem*. It means that the physical entropy of the system is increasing in time. We will explain this in detail in Remark 4.4 for a simplified situation. The third part of the theorem finally expresses the fact that the net scattering rate vanishes in thermal equilibrium or, in other words, that the Fermi–Dirac distributions span the kernel of Q.

If the scattering rate $s(k,k')$ is given by electron–phonon interactions,

$$s(k,k') = \sigma(k,k')\left((1+N)\delta(E'-E+\hbar\omega) + N\delta(E'-E-\hbar\omega)\right)$$

(see (4.7)), where σ is a symmetric function, $\hbar\omega$ is the constant phonon energy, and $N = 1/(e^{\hbar\omega} - 1)$ is the phonon occupation number, then the kernel of the corresponding collision operator may not consist of Fermi–Dirac distributions only. In fact, Majorana [46] proved that the kernel is given by all functions

$$\frac{1}{1+g(E(k))\exp(E(k)/k_BT)}, \quad \text{where } g \text{ is } \hbar\omega\text{-periodic.} \tag{4.21}$$

This result is not stable with respect to perturbations of collision mechanisms in the following sense. Consider the scattering of electrons with phonons of two different energies $\hbar\omega_1$ and $\hbar\omega_2$, corresponding to the collision operators Q_1 and Q_2, respectively. Then the kernel of $Q_1 + Q_2$ is given by all functions (4.21) such that g is both $\hbar\omega_1$- and $\hbar\omega_2$-periodic. As a consequence, the kernel is spanned by Fermi–Dirac distributions only if the quotient ω_1/ω_2 is not a rational number (also see the discussion in [47]).

The collision operator (4.2) is nonlocal and nonlinear. Therefore, its analytical treatment is quite difficult and it is reasonable to consider simplified operators. In the literature, two simplifications can be found: the low-density approximation and the relaxation-time approximation. The former model is nonlocal, but linear; the latter one is local and linear.

Low-density collision operator. First we derive the low-density approximation. We assume that the distribution function is small in the sense $0 \leq f(x,k,t) \ll 1$. Then we can approximate $1 - f(x,k,t) \approx 1$ and write, using (4.16),

$$Q(f)(k) \approx \int_B \left(s(k',k)f' - s(k,k')f\right) dk = \int_B \frac{s(k',k)}{M(k)}\left(Mf' - M'f\right) dk'.$$

This motivates the introduction of the *low-density collision operator*

$$(Q_0(f))(x,k,t) = \int_B \sigma(x,k',k)(Mf' - M'f)\,dk', \tag{4.22}$$

where the *collision cross-section* $\sigma(x,k',k) = s(x,k',k)/M(k)$ is symmetric thanks to (4.16). For this operator, a similar result like Proposition 4.1 holds.

Proposition 4.2 (Low-density collision operator). *Let $\sigma(x,k,k') > 0$ be symmetric in k and k'. Then it holds:*

(1) For all functions $f(x,k,t)$, we have

$$\int_B (Q_0(f))(x,k,t)\,dk = 0 \quad for\ x \in \mathbb{R}^3,\ t > 0.$$

(2) For all nonnegative functions $f(x,k,t)$ and nondecreasing functions $\chi : \mathbb{R} \to \mathbb{R}$,

$$\int_B (Q_0(f))(x,k,t)\chi\left(f(x,k,t)e^{E(k)/k_BT}\right)\,dk \le 0. \qquad (4.23)$$

(3) The kernel $N(Q_0)$ of Q_0 is spanned by the Maxwell–Boltzmann distribution or Maxwellian

$$M(k) = e^{-E(k)/k_BT}.$$

Proof. The proof of the first part follows immediately from the symmetry of σ. For the second part, we compute as in the proof of Proposition 4.1:

$$\int_B Q_0(f)\chi(f/M)dk = \int_{B^2} \sigma(Mf' - M'f)\chi(f/M)\,dk'\,dk$$

$$= \int_{B^2} \sigma MM'\left(\frac{f'}{M'} - \frac{f}{M}\right)\chi\left(\frac{f}{M}\right)\,dk'\,dk.$$

We add this relation to the equation in which k and k' are exchanged:

$$\int_B Q_0(f)\chi(f/M)\,dk \qquad (4.24)$$

$$= -\frac{1}{2}\int_{B^2} \sigma MM'\left(\frac{f'}{M'} - \frac{f}{M}\right)\left(\chi\left(\frac{f'}{M'}\right) - \chi\left(\frac{f}{M}\right)\right)\,dk'\,dk \le 0.$$

Finally, the third part follows from this inequality taking $\chi(x) = x$. $\quad\square$

Remark 4.3. In the parabolic band approximation

$$E(k) = \frac{\hbar^2}{2m^*}|k - k_0|^2, \quad k \in \mathbb{R}^3,$$

we can characterize the kernel $N(Q_0)$ of the low-density operator, consisting of multiples of M, by the family of functions

$$M_{n,u,T}(v) = n\left(\frac{m^*}{2\pi k_BT}\right)^{3/2}\exp\left(-\frac{m^*|v - u|^2}{2k_BT}\right), \qquad (4.25)$$

where n and T are positive. Indeed, defining $v = \hbar k/m^*$ and $u = \hbar k_0/m^*$, we obtain

$$M(v) = \exp\left(-\frac{E(k)}{k_B T}\right) = \exp\left(-\frac{m^*|v-u|^2}{2k_B T}\right).$$

The quantity n represents the density, u the velocity, and T the temperature of the particle system. The Maxwellian (4.25) is exactly the local Maxwellian used in gas dynamics [2]. It corresponds to the heated shifted Maxwellian (2.8) in the velocity variable formulation. □

Remark 4.4. Again, inequality (4.23) can be termed an H-theorem. In order to explain this notion, we neglect for the moment electric effects, i.e., we set $E_{\mathrm{eff}} = 0$ in (4.1). Let the (relative) entropy S be given by

$$S(t) = -\frac{2}{(2\pi)^3} \int_{\mathbb{R}^3} \int_B f \log\left(\frac{f}{M}\right) dk\, dx. \tag{4.26}$$

Differentiating this function and employing the Boltzmann equation (4.1) yields

$$\frac{dS}{dt} = -\frac{2}{(2\pi)^3} \int_{\mathbb{R}^3} \int_B \partial_t f \left(\log\left(\frac{f}{M}\right) + 1\right) dk\, dx$$

$$= -\frac{2}{(2\pi)^3} \int_{\mathbb{R}^3} \int_B \left(\partial_t f - v \cdot \nabla_x f \log\left(\frac{f}{M}\right) + Q_0(f) \log\left(\frac{f}{M}\right)\right) dk\, dx.$$

The last term is nonnegative, due to Proposition 4.2 (2) with $\chi(x) = \log x$. In the second term, we integrate by parts and employ again (4.1):

$$\frac{dS}{dt} \geq -\frac{2}{(2\pi)^3} \int_{\mathbb{R}^3} \int_B \left(\partial_t f - v \cdot \nabla_x f \log\left(\frac{f}{M}\right)\right) dk\, dx$$

$$= -\frac{2}{(2\pi)^3} \int_{\mathbb{R}^3} \int_B (\partial_t f + v \cdot \nabla_x f) dk\, dx$$

$$= -\frac{2}{(2\pi)^3} \int_{\mathbb{R}^3} \int_B Q_0(f) dk\, dx = 0.$$

Hence, the entropy S is nondecreasing or the negative entropy, which is often denoted by the symbol H, is nonincreasing. This fact is called the H-theorem.

If we allow for a force term, we need a slightly different argument since we have to take into account the electric energy. We assume that E_{eff} is a gradient field, i.e., there exists a potential $V(x,t)$ such that $E_{\mathrm{eff}} = -\nabla_x V$. Define the modified Maxwellian

$$M_1(x,k,t) = e^{-(E(k)-qV(x,t))/k_B T},$$

where $E(k) - qV$ also includes the electric energy component. Then the statement of Proposition 4.2 still holds, since a multiplication with a nonnegative function which depends only on x and t does not change the arguments. Therefore,

$$\int_B Q_0(f) \log\left(\frac{f}{M_1}\right) dk \leq 0.$$

Hence, differentiating the entropy (4.26) with M_1 instead of M gives, after a computation similar as above,

$$\frac{dS}{dt} = -\frac{2}{(2\pi)^3} \int_{\mathbb{R}^3} \int_B \left(\partial_t f - v \cdot \nabla_x f \log\left(\frac{f}{M_1}\right) - \frac{q}{\hbar} \nabla_x V \cdot \nabla_k f \log\left(\frac{f}{M_1}\right) \right.$$
$$\left. + Q_0(f) \log\left(\frac{f}{M_1}\right) \right) dk\,dx$$
$$\geq -\frac{2}{(2\pi)^3} \int_{\mathbb{R}^3} \int_B \left(\partial_t f + f v \cdot \nabla_x (\log f - \log M_1) \right.$$
$$\left. + \frac{q}{\hbar} f \nabla_x V \cdot \nabla_k (\log f - \log M_1) \right) dk\,dx$$
$$= \frac{2}{(2\pi)^3} \int_{\mathbb{R}^3} \int_B \left(v \cdot \nabla_x \log M_1 + \frac{q}{\hbar} \nabla_x V \cdot \nabla_k \log M_1 \right) f\,dk\,dx = 0,$$

since

$$\nabla_x \log M_1 = \frac{q}{k_B T} \nabla_x V \quad \text{and} \quad \frac{q}{\hbar} \nabla_k \log M_1 = -\frac{q}{\hbar k_B T} \nabla_k E = -\frac{qv}{k_B T}.$$

Thus, the entropy is nondecreasing. □

Relaxation-time collision operator. We rewrite the low-density collision operator as

$$Q_0(f) = \int_B \frac{s(x,k',k)}{M}(Mf' - M'f)\,dk' = \int_B \frac{s(x,k,k')}{M'}(Mf' - M'f)\,dk'$$
$$= M \int_B s(x,k,k') \frac{f'}{M'}\,dk' - f \int_B s(x,k,k')\,dk',$$

where we have employed the property (4.16) for $s(x,k',k)$. If the initial distribution is close to a multiple of the Maxwellian $M(k) = e^{-E(k)/k_B T}$, we expect that f'/M' is close to a constant (at least for small time) and one may approximate $f'/M' \approx n$, where $n > 0$. Then

$$Q_0(f) \approx (nM - f) \int_B s(x,k,k')\,dk'.$$

Introducing the *relaxation time* $\tau(x,k)$ by

$$\tau(x,k) = \left(\int_B s(x,k,k')\,dk' \right)^{-1},$$

we obtain the *relaxation-time operator*

$$(Q_\tau(f))(x,k,t) = -\frac{f(x,k,t) - n(x)M(k)}{\tau(x,k)}. \tag{4.27}$$

The quantity n is related to the equilibrium particle density f_{eq} if the Maxwellian is scaled in such a way that $\int_B M\,dk = 4\pi^3$, since then $nM = f_{eq}$ and $\int_B f_{eq}\,dk/4\pi^3 = n$.

The relaxation-time operator drives the distribution function toward the equilibrium nM. Indeed, consider the Boltzmann equation along the trajectories $(x(t), k(t))$,

$$\frac{df}{dt} = -\frac{f - nM}{\tau}, \quad t > 0, \tag{4.28}$$

where $(x(t), k(t))$ solves $\dot{x} = v(k)$, $\dot{k} = -qE_{\text{eff}}/\hbar$ with initial conditions $x(0) = x_0$, $k(0) = k_0$. For simplicity, let the relaxation time τ be constant. The above differential equation can be solved, and we see that $f - nM$ converges to zero as $e^{-t/\tau}$ if $t \to \infty$. This means that the distribution function relaxes to the equilibrium density nM along the trajectories after a time of order τ.

In Sects. 2.3 and 2.4, we have employed the relaxation-time operator in order to derive macroscopic semiconductor models. This operator is in fact widely used because of its simplicity. According to [35, Chap. 3.5], the relaxation-time approximation is appropriate under low electric fields when the scattering is elastic or isotropic.

We notice that the low-density operator equals the relaxation-time operator exactly if the collision cross-section σ only depends on the position variable and if the Maxwellian is scaled such that $\int_B M \, dk = 4\pi^3$, since

$$Q_0(f) = \int_B \sigma(x)(Mf' - M'f) \, dk'$$

$$= \sigma(x)M(k) \int_B f' \, dk' - \sigma(x)f(x,k,t) \int_B M' \, dk'$$

$$= -4\pi^3 \sigma(x) \left(f(x,k,t) - M(k)n(x,t) \right),$$

where $n(x,t) = \int_B f(x,k,t) \, dk/4\pi^3$. This corresponds to (4.27) with $\tau = 1/4\pi^3 \sigma$.

Elastic collision operator. Some scattering mechanisms can be described or approximated by elastic collisions, like ionized impurity or acoustic phonon scattering. According to (4.10) and (4.11), the corresponding collision operator can be written as

$$(Q_{\text{el}}(f))(x,k,t) = \int_B \sigma(x,k,k')\delta(E(k') - E(k))(f' - f) \, dk'.$$

We assume that the scattering rate σ is positive and symmetric,

$$\sigma(x,k,k') = \sigma(x,k',k) > 0, \quad x \in \mathbb{R}^3, \; k,k' \in B. \tag{4.29}$$

Then the following properties were proved by Ben Abdallah and Degond [41].

Proposition 4.5 (Elastic collision operator). *Let the condition (4.29) hold. Then*
(1) For all (regular) functions $f(x,k,t)$,

$$\int_B (Q_{\text{el}}(f))(x,k,t) \, dk = \int_B (Q_{\text{el}}(f))(x,k,t)E(k) \, dk = 0 \quad \text{for } x \in \mathbb{R}^3, \; t > 0.$$

(2) *The operator* $-Q_{el}$ *is symmetric and nonnegative in the sense*

$$-\int_B Q_{el}(f)f \, dk \geq 0 \quad \text{for all functions } f.$$

(3) *The kernel* $N(Q_{el})$ *of* Q_{el} *consists of all functions which depend only on the energy, being of the form* $F(x, E(k), t)$.

The first property expresses the conservation of mass and energy. The conservation of mass is already explained in (4.20). The energy conservation only holds in absence of external forces. Indeed, defining the energy density $ne = \int_B fE(k) \, dk/4\pi^3$ and integrating the Boltzmann equation (4.1) gives

$$\frac{d}{dt}\int_{\mathbb{R}^3} ne(x,t) \, dx = \int_{\mathbb{R}^3}\int_B \partial_t fE(k)\frac{dk}{4\pi^3} \, dx = -\int_{\mathbb{R}^3}\int_B \text{div}_x(vEf)\frac{dk}{4\pi^3} \, dx = 0.$$

Proof. (1) The first property follows from the symmetry of σ and δ,

$$\int_{B^2} Q_{el}(f)E^{i-1} \, dk' \, dk = \int_{B^2} \sigma(x,k,k')\delta(E'-E)f'E^{i-1} \, dk' \, dk$$

$$- \int_{B^2} \sigma(x,k,k')\delta(E-E')f'(E')^{i-1} \, dk \, dk'$$

$$= \int_{B^2} \sigma(x,k,k')\delta(E'-E)f'E^{i-1} \, dk' \, dk$$

$$- \int_{B^2} \sigma(x,k,k')\delta(E'-E)f'E^{i-1} \, dk \, dk' = 0,$$

where $E' = E(k')$ and $i = 1, 2$. We have used that $\delta(E'-E)E = \delta(E'-E)E'$.

(2) A computation similar as above gives for functions f and g

$$\int_B Q_{el}(f)g \, dk = \frac{1}{2}\int_{B^2} \sigma(x,k,k')\delta(E'-E)(f'-f)g \, dk' \, dk$$

$$+ \frac{1}{2}\int_{B^2} \sigma(x,k',k)\delta(E-E')(f-f')g' \, dk \, dk'$$

$$= -\frac{1}{2}\int_{B^2} \sigma(x,k,k')\delta(E'-E)(f'-f)(g'-g) \, dk' \, dk$$

$$= \int_B Q_{el}(g)f \, dk.$$

This shows that $-Q_{el}$ is symmetric. Moreover, it is nonnegative, since the choice $f = g$ leads to

$$-\int_B Q_{el}(f)f \, dk = \frac{1}{2}\int_{B^2} \sigma(x,k,k')\delta(E'-E)(f'-f)^2 \, dk' \, dk \geq 0. \qquad (4.30)$$

(3) Let $f \in N(Q_{el})$. Then (4.30) implies that

$$\delta(E(k')-E(k))(f(k')-f(k))^2 = 0 \quad \text{for } k, k' \in B.$$

Since the support of the delta distribution is concentrated at the origin we must have $f(k') = f(k)$ at all points $k, k' \in B$ for which $E(k') = E(k)$. Thus, f is constant on each energy surface $\{k \in B : E(k) = \varepsilon\}$ for $\varepsilon \in \mathbb{R}$ which means that f is a function of $E(k)$ only. □

Electron–electron collision operator. Electron–electron scattering can be described according to (4.12) by the nonlinear operator

$$
(Q_{ee}(f))(x,k,t) = \int_{B^3} \sigma_{ee}(x,k,k',k_1,k_1')\delta(E' + E_1' - E - E_1)
$$
$$
\times \left(f'f_1'(1-f)(1-f_1) - ff_1(1-f')(1-f_1') \right) dk'\, dk_1\, dk_1'.
$$

The scattering rate is symmetric in the following sense (for notational simplicity, we omit the dependence on x):

$$
\sigma_{ee}(k,k',k_1,k_1') = \sigma_{ee}(k',k,k_1,k_1') = \sigma_{ee}(k_1,k_1',k,k') \tag{4.31}
$$

for all $k, k', k_1, k_1' \in B$. Then the operator possesses the following properties shown in [41].

Proposition 4.6 (Electron–electron collision operator). *Let the symmetry assumption (4.31) hold. Then we have*
(1) For (regular) functions $f(x,k,t)$ it holds

$$
\int_B Q_{ee}(f)\, dk = \int_B Q_{ee}(f)E(k)\, dk = 0. \tag{4.32}
$$

(2) The kernel of Q_{ee} consists of Fermi–Dirac distributions,

$$
N(Q_{ee}) = \{f(x,k,t) : \text{there exist } \mu, T \text{ such that } f = F_{\mu,T}\},
$$

where

$$
F_{\mu,T}(k) = \frac{1}{1 + e^{(E(k) - \mu)/k_B T}}.
$$

Since $f \in N(Q_{ee})$ may depend on (x,t), so do the parameters μ and T, i.e., $f(x,k,t) = F_{\mu(x,t),T(x,t)}(k)$.

Notice that in the cases $T = 0$ or $T = \pm\infty$, the Fermi–Dirac distribution attains the limiting numbers 0, $\frac{1}{2}$, or 1. We interpret the parameter T as the *temperature* of the electrons (and then, we suppose that $T > 0$) and μ as the *chemical potential*.

Proof. First we show that for regular functions f and g,

$$
\int_B Q_{ee}(f)g\, dk = -\frac{1}{4}\int_{B^4} \sigma_{ee}(k,k',k_1,k_1')\delta(E + E_1 - E' - E_1')
$$
$$
\times (g' + g_1' - g - g_1)\left(f'f_1'(1-f)(1-f_1) - ff_1(1-f')(1-f_1') \right) d^4k,
$$

where $d^4k = dk\, dk_1\, dk'\, dk'_1$. In fact, this is a consequence of a renumbering of the variables and the symmetry property (4.31):

$$
\int_B Q_{ee}(f)g\, dk = \frac{1}{4}\int_{B^4} \sigma_{ee}(k,k',k_1,k'_1)\delta(E+E_1-E'-E'_1)
$$
$$
\times\left(f'f'_1(1-f)(1-f_1)-ff_1(1-f')(1-f'_1)\right)g\, d^4k
$$
$$
+\frac{1}{4}\int_{B^4}\sigma_{ee}(k',k,k'_1,k_1)\delta(E'+E'_1-E-E_1)
$$
$$
\times\left(ff_1(1-f')(1-f'_1)-f'f'_1(1-f)(1-f_1)\right)g'\, d^4k
$$
$$
+\frac{1}{4}\int_{B^4}\sigma_{ee}(k_1,k',k,k'_1)\delta(E+E_1-E'-E'_1)
$$
$$
\times\left(f'f'_1(1-f_1)(1-f)-f_1f(1-f')(1-f'_1)\right)g_1\, d^4k
$$
$$
+\frac{1}{4}\int_{B^4}\sigma_{ee}(k'_1,k_1,k',k)\delta(E'_1+E'-E_1-E)
$$
$$
\times\left(f_1f(1-f'_1)(1-f')-f'_1f'(1-f_1)(1-f)\right)g'_1\, d^4k
$$
$$
=\frac{1}{4}\int_{B^4}\sigma_{ee}(k,k',k_1,k'_1)\delta(E+E_1-E'-E'_1)
$$
$$
\times\left(f'f'_1(1-f)(1-f_1)-ff_1(1-f')(1-f'_1)\right)(g+g_1-g'-g'_1)d^4k.
$$

Now, taking $g(k)=1$, the first conservation property in (4.32) follows immediately. Choosing $g(k)=E(k)$, we obtain an integral over B^4 involving the product

$$
\delta(E+E_1-E'-E'_1)(E+E_1-E'-E'_1)
$$

which vanishes since the support of the delta distribution is concentrated at $E+E_1-E'-E'_1=0$. This shows (1).

We only sketch the proof of (2) since it is based on the umklapp processes whose modeling is hidden in σ_{ee} (see Sect. 4.1). Choosing $g=H(f)=\log f-\log(1-f)$, it follows that

$$
\int_B Q_{ee}(f)H(f)\, dk = -\frac{1}{4}\int_{B^4}\sigma_{ee}(k,k',k_1,k'_1)\delta(E+E_1-E'-E'_1)
$$
$$
\times\left(f'f'_1(1-f)(1-f_1)-ff_1(1-f')(1-f'_1)\right)
$$
$$
\times\left(\log\left(f'f'_1(1-f)(1-f_1)\right)-\log\left(ff_1(1-f')(1-f'_1)\right)\right)d^4k \leq 0,
$$

employing the elementary inequality $-(x-y)(\log x-\log y)\leq 0$ for all $x,y>0$. Since $(x-y)(\log x-\log y)=0$ if and only if $x=y$, we obtain for all $f\in N(Q_{ee})$,

$$
\log\left(f'f'_1(1-f)(1-f_1)\right)=\log\left(ff_1(1-f')(1-f'_1)\right),
$$

or, in terms of the function H,

$$
H(f')+H(f'_1)=H(f)+H(f_1),
$$

whenever $E(k') + E(k_1') = E(k) + E(k_1)$ and $k' + k_1' = k + k_1$ modulo B. The condition on the energy comes from the delta distribution, whereas the second condition follows from the umklapp processes. In [44] it is shown that this relation implies that $H(f(k))$ is an affine function of the energy, i.e., $H(f) = aE + b$ for some constants a and b. If $a = 0$, $H(f)$ and hence f are constant with respect to k. Therefore, we assume that $a \neq 0$. Then, introducing the variables $T = -1/a$ and $\mu = -b/a$, we obtain from

$$\log \frac{f}{1-f} = H(f) = aE + b = -\frac{E - \mu}{k_B T}$$

the equation $f = 1/(1 + e^{(E-\mu)/k_B T})$ proving the proposition. $\qquad \square$

4.3 Additional Topics

In this section we specify some boundary conditions for the Vlasov or Boltzmann equation when they are solved in a bounded spatial domain, and we detail the bipolar Boltzmann equation including generation–recombination processes.

Spatial boundary conditions. Usually, a semiconductor device is considered in a bounded position domain so that appropriate boundary conditions for the spatial variable have to be imposed. Let $\Omega \subset \mathbb{R}^3$ be a bounded domain. Its boundary $\partial\Omega$ is supposed to be decomposed into two classes of boundary segments: the union of contacts Γ_D through which carriers enter or exit and the union of insulating boundary parts Γ_N, i.e., $\partial\Omega = \Gamma_D \cup \Gamma_N$ and $\Gamma_D \cap \Gamma_N = \emptyset$.

Let $f(x, k, t)$ be a distribution function, for instance, the solution of the Vlasov or the Boltzmann equation. A boundary condition for these equations has to be imposed on the sets on which the (spatial) characteristics point into Ω. These sets are subsets of either Γ_D or Γ_N. We define the *Dirichlet inflow boundary*

$$\Gamma_D^- = \{(x, k) : x \in \Gamma_D, \ k \in B, \ v(k) \cdot \eta(x) < 0\},$$

and the *Neumann "inflow" boundary*

$$\Gamma_N^- = \{(x, k) : x \in \Gamma_N, \ k \in B, \ v(k) \cdot \eta(x) < 0\},$$

where $\eta(x)$ is the exterior normal unit vector at $x \in \partial\Omega$. At the Dirichlet inflow boundary, the particle distribution function is supposed to be known,

$$f(x, k, t) = f_D(x, k, t) \quad \text{for } (x, k) \in \Gamma_D^-, \ t > 0. \tag{4.33}$$

At the insulating boundary parts, the carriers do not enter or exit the boundary but they are reflected. Assuming elastic *specular reflection*, the distribution function on the Neumann "inflow" boundary has the property

$$f(x, k, t) = f(x, k', t), \quad \text{for } (x, k) \in \Gamma_N^-, \ t > 0, \tag{4.34}$$

where k' is such that $v(k') = v(k) - 2(v(k) \cdot \eta(x))\eta(x)$. The Vlasov or Boltzmann equation is then solved with the initial condition $f(x,k,0) = f_I(x,k)$ for $(x,k) \in \Omega \times B$ and the boundary conditions (4.33) and (4.34).

In gas dynamics, also another reflecting boundary condition is known, the *diffusive reflection*. Here, the distribution function at the Neumann boundary is equal to the corresponding Maxwellian,

$$f(x,k,t) = \rho \sigma_f(x,t)e^{-E(k)/k_B T} \quad \text{for } (x,k) \in \Gamma_N^-, \, t > 0,$$

where $\rho > 0$ is a constant and

$$\sigma_f(x,t) = \int_{v(k)\cdot\eta(x)>0} (v(k) \cdot \eta(x))f(x,k,t)\frac{dk}{4\pi^3}$$

(see [48]). A linear combination of specular and diffusive reflection is also possible,

$$f(x,k,t) = (1-\alpha)f(x,k',t) + \alpha\rho\sigma_f(x,t)e^{-E(k)/k_B T},$$

where $(x,k) \in \Gamma_N^-, \, t > 0$, and the parameter $0 \le \alpha \le 1$ is called the *accommodation constant*.

Under the inflow boundary conditions, we cannot expect to have conservation of the carrier number. Instead, by the divergence theorem, the rate of change is the difference of the incoming and outgoing fluxes,

$$\frac{d}{dt} \int_\Omega \int_B f(x,k,t)\,dk\,dx = \int_{\Gamma_D^-} (v(k) \cdot \eta)f_D\,ds - \int_{\Gamma_D^+} (v(k) \cdot \eta)f\,ds,$$

where Γ_D^+ is defined as Γ_D^- with $v(k) \cdot \eta > 0$ instead of $v(k) \cdot \eta < 0$.

If the Vlasov–Poisson or Boltzmann–Poisson problem (see Sect. 3.2 or 4.1, respectively) is studied, we need also to impose boundary conditions for the force field E_{eff}. We assume that there exists a potential V such that $E_{\text{eff}} = -\nabla_x V$ and we impose boundary conditions for V. The potential is given at the contacts and the normal component of the electric field $-\nabla_x V$ vanishes at the insulating boundary parts,

$$V = V_D \quad \text{on } \Gamma_D, \quad \nabla_x V \cdot \eta = 0 \quad \text{on } \Gamma_N, \, t > 0.$$

Bipolar model. So far we have only considered the transport of electrons in the conduction band. However, holes in the valence band also contribute to the carrier flow in the semiconductor (see Sect. 1.4). It is possible that an electron moves from the valence band to the conduction band, leaving a hole behind it in the valence band. This process is called the *generation* of an electron–hole pair (see Fig. 4.2). The electron has to overcome the energy gap, which is of the order of 1 eV. On the other hand, the thermal energy of an electron is only of the order of $k_B T \approx 0.026$ eV at room temperature. Therefore, a lot of absorption energy is necessary for such processes. The inverse process of an electron moving from the conduction to the valence band, occupying an empty state, is termed the *recombination* of an

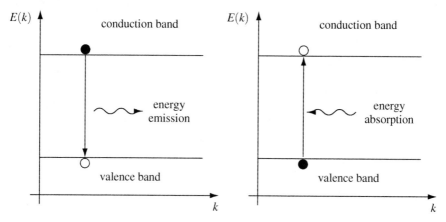

Fig. 4.2 Recombination (left) and generation (right) of an electron–hole pair

electron–hole pair. In such an event, energy is emitted. The basic mechanisms for generation–recombination processes are

- Auger/impact ionization generation–recombination,
- radiative generation–recombination,
- thermal generation–recombination.

An *Auger process* is defined as an electron–hole recombination followed by a transfer of energy to a free carrier which is excited to a state of higher energy. The inverse Auger process, i.e., the generation of an electron–hole pair, is called *impact ionization*. The energy for the pair generation comes from the collision of a high-energetic free carrier with the lattice or from electron–electron or hole–hole collisions. When an electron from the conduction band recombines with a hole from the valence band and emits a photon, we call this process *radiative* recombination. The energy of the photon is equal to the band-gap energy. Radiative generation occurs when a photon with energy larger than or equal to the gap energy is absorbed. These processes are of importance in narrow-gap semiconductors. A third source of energy comes from lattice vibrations or phonons. Thus, *thermal* recombination or generation arises from phonon emission or absorption, respectively.

The recombination–generation operator is derived, like the collision operator in Sect. 4.1, from phenomenological considerations following [49, Chap. 1.6]. The generation of an electron in the state k and a hole in the state k' is possible if both states are not occupied, and its rate is given by

$$g(x,k',k)(1-f_n)(1-f'_p),$$

where $g(x,k',k) \geq 0$ is the generation rate, $f_n = f_n(x,k,t)$ the electron distribution function and $f'_p = f_p(x,k',t)$ the hole distribution function. The rate of recombination of an electron at state k and a hole at state k' equals

$$r(x,k,k')f_n f'_p,$$

where $r(x,k,k') \geq 0$ is the recombination rate. The net rate is the difference of generation and recombination rates,

$$g(x,k',k)(1-f_n)(1-f_p') - r(x,k,k')f_nf_p'.$$

The recombination–generation operator for electrons in the conduction band is the integral over all states k':

$$(I_n(f_n,f_p))(x,k,t) = \int_B \left(g(x,k',k)(1-f_n)(1-f_p') - r(x,k,k')f_nf_p' \right) dk'. \quad (4.35)$$

In a similar way, the recombination–generation operator for holes in the valence band can be written as

$$(I_p(f_n,f_p))(x,k,t) = \int_B \left(g(x,k,k')(1-f_n')(1-f_p) - r(x,k',k)f_n'f_p \right) dk'. \quad (4.36)$$

The recombination and generation rates are related by the equation

$$r(x,k,k') = \exp\left(\frac{E_n(k) - E_p(k')}{k_B T} \right) g(x,k',k), \quad (4.37)$$

which can be derived from the principle of detailed balance like (4.14) assuming Maxwell–Boltzmann statistics. This principle holds here since recombination and generation balance in thermal equilibrium.

The operators (4.35) and (4.36) are added to the electron and hole collision operators. Then, the evolution of the distribution functions f_n and f_p is given by the system of Boltzmann equations

$$\partial_t f_n + v_n(k) \cdot \nabla_x f_n - \frac{q}{\hbar} E_{\text{eff}} \cdot \nabla_k f_n = Q_n(f_n) + I_n(f_n,f_p), \quad (4.38)$$

$$\partial_t f_p + v_p(k) \cdot \nabla_x f_p + \frac{q}{\hbar} E_{\text{eff}} \cdot \nabla_k f_p = Q_p(f_p) + I_p(f_n,f_p), \quad (4.39)$$

with the velocities

$$v_n(k) = \frac{1}{\hbar}\nabla_k E_n(k), \quad v_p(k) = \frac{1}{\hbar}\nabla_k E_p(k),$$

and E_n and E_p are the conduction and valence band energies, respectively. Denoting by

$$n(x,t) = \int_B f_n(x,k,t) \frac{dk}{4\pi^3} \quad \text{and} \quad p(x,t) = \int_B f_p(x,k,t) \frac{dk}{4\pi^3}$$

the electron and hole densities, respectively, the effective-field equation (4.3) becomes

$$E_{\text{eff}}(x,t) = E_{\text{ext}}(x,t) + \int_{\mathbb{R}^3} (n(\xi,t) - p(\xi,t))E_0(x,\xi)\,d\xi, \quad (4.40)$$

since electrons and holes have charges with opposite sign. Equations (4.38), (4.39), and (4.40), together with the collision operators (4.2), and (4.35), (4.36) are called the *semi-classical bipolar Boltzmann model*.

The bipolar model has an additional nonlinearity due to the coupling between the electron and hole distribution functions through (4.35) and (4.36). In particular, the total number of each type of particles is not conserved anymore since recombination–generation effects can take place. However, the total space charge $n - p - C$ is conserved if the doping atoms are immobile, i.e., C is a function of the space variable x only. Indeed, taking the difference of the Boltzmann equations (4.38) and (4.39) and integrating over $(x, k) \in \mathbb{R}^3 \times B$ yields

$$\frac{d}{dt} \int_{\mathbb{R}^3} (n - p - C(x)) \, dx = \int_{\mathbb{R}^3} \int_B (I_n(f_n, f_p) - I_p(f_n, f_p)) \, dk \, dx = 0,$$

by arguing as in the proof of the first part of Proposition 4.1.

Finally, we discuss two special cases. In the low-density approximation $f_n, f_p \ll 1$ we can write the recombination–generation operators as

$$(I_n(f_n, f_p))(x, k, t) = \int_B g(x, k', k) \left(1 - e^{(E_n(k) - E_p(k'))/k_B T} f_n f_p' \right) dk', \qquad (4.41)$$

$$(I_p(f_n, f_p))(x, k, t) = \int_B g(x, k, k') \left(1 - e^{(E_n(k') - E_p(k))/k_B T} f_n' f_p \right) dk', \qquad (4.42)$$

using (4.37). In the case of Coulomb forces in $\mathbb{R}^3$, the effective field is given by

$$E_{\text{eff}}(x, t) = \frac{1}{4\pi \varepsilon_s} \int_{\mathbb{R}^3} \rho(y, t) \frac{x - y}{|x - y|^3} \, dy,$$

where the total space charge ρ is the sum of the electron density n, the hole density p, and the densities N_d, N_a of the implanted positively charged donor ions and the negatively charged acceptor ions, respectively, with which the semiconductor material is doped (see Sect. 1.6), weighted by the corresponding charges q or $-q$,

$$\rho = q(-n + p - N_a + N_d).$$

Defining the electrostatic potential V by $E_{\text{eff}} = -\nabla V$ and the doping profile $C = N_d - N_a$, we can replace the effective-field equation (4.40) by the Poisson equation

$$\varepsilon_s \Delta V = q(n - p - C) \quad \text{in } \mathbb{R}^3. \qquad (4.43)$$

The Boltzmann equations (4.38) and (4.39) and the Poisson equation (4.43) constitute the so-called *bipolar Boltzmann–Poisson system*. For a mathematical analysis of the bipolar model including recombination–generation terms (*H*-theorems, existence and uniqueness of smooth solutions), we refer to [11].

References

1. H. Babovsky. *Die Boltzmann-Gleichung.* Teubner, Stuttgart, 1998.
2. C. Cercignani, R. Illner, and M. Pulvirenti. *The Mathematical Theory of Dilute Gases.* Springer, New York, 1994.
3. L. Boltzmann. Weitere Studien über das Wärmegleichgewicht unter Gasmolekülen. *Sitzungsberichte Akad. Wiss. Wien* 66 (1872), 275–370. Translation: Further studies on the thermal equilibrium of gas molecules. In: S. Brush (ed.), *Kinetic Theory*, Vol. 2, 88–174. Pergamon Press, Oxford, 1966.
4. T. Carleman. Sur la théorie de l'équation intégro-différentielle de Boltzmann. *Acta Mathematica* 60 (1933), 91–146.
5. L. Arkeryd. On the Boltzmann equation. *Arch. Rat. Mech. Anal.* 45 (1971), 1–34.
6. R. DiPerna and P.-L. Lions. On the Cauchy problem for Boltzmann equations: global existence and weak stability. *Ann. Math.* 130 (1989), 321–366.
7. F. Golse, B. Perthame, P.-L. Lions, and R. Sentis. Regularity of the moments of the solution of a transport equation. *J. Funct. Anal.* 76 (1988), 110–125.
8. P. Gérard. Solutions globales du problème de Cauchy pour l'équation de Boltzmann (d'après R. Di Perna et P.-L. Lions). Séminaire Bourbaki, Vol. 1988–89, *Astérisque* 161–162, Exp. No. 699 (1989), 257–281.
9. P.-L. Lions. Global solutions of kinetic models and related problems. In: C. Cercignani and M. Pulvirenti (eds.), *Nonequilibrium Problems in Many-Particle Systems*, Lecture Notes in Math. 1551, 58–86. Springer, Berlin, 1992.
10. C. Villani. A review of mathematical topics in collisional kinetic theory. In: S. Friedlander and D. Serre (eds.), *Handbook of Mathematical Fluid Dynamics*, Vol. 1, 71–305. Elsevier, Amsterdam, 2002.
11. F. Poupaud. On a system of nonlinear Boltzmann equations of semiconductors physics. *SIAM J. Appl. Math.* 50 (1990), 1593–1606.
12. F. Mustieles. Global existence of solutions for the nonlinear Boltzmann equation of semiconductor physics. *Rev. Mat. Iberoamer.* 6 (1990), 43–59.
13. F. Mustieles. Global existence of weak solutions for a system of nonlinear Boltzmann equations in semiconductor physics. *Math. Meth. Appl. Sci.* 14 (1991), 139–153.
14. H. Andréasson. Global existence of smooth solutions in three dimensions for the semiconductor Vlasov-Poisson-Boltzmann equation. *Nonlin. Anal.: Theory Meth. Appl.* 28 (1990), 1193–1211.
15. A. Majorana and S. Marano. Space homogeneous solutions to the Cauchy problem for semiconductor Boltzmann equations. *SIAM J. Math. Anal.* 28 (1997), 1294–1308.
16. A. Majorana and S. Marano. On the Cauchy problem for spatially homogeneous semiconductor Boltzmann equations: existence and uniqueness. *Annali Math.* 184 (2005), 275–296.
17. A. Majorana and C. Milazzo. Space homogeneous solutions of the linear semiconductor Boltzmann equation. *J. Math. Anal. Appl.* 259 (2001), 609–629.
18. G. Bird. *Molecular Gas Dynamics and Direct Simulation of Gas Flows.* Clarendon Press, Oxford, 1994.
19. M. Fischetti and S. Laux. Monte Carlo analysis of electron transport in small semiconductor devices including band-structure and space-charge effects. *Phys. Rev. B* 38 (1988), 9721–9745.
20. K. Nanbu. Direct simulation scheme derived from the Boltzmann equation. I. Monocomponent gases. *J. Phys. Soc. Japan* 52 (1983), 2042–2049.
21. A. Gnudi, D. Ventura, and G. Baccarani. Modeling impact ionization in a BJT by means of a spherical harmonics expansion of the Boltzmann equation. *IEEE Trans. Computer-Aided Design* 12 (1993), 1706–1713.
22. N. Goldsman, L. Henrickson, and J. Frey. A physics-based analytical/numerical solution to the Boltzmann transport equation for use in device simulation. *Solid State Electr.* 34 (1991), 389–396.

23. C. Gray and H. Ralph. Solution of Boltzmann's equation for semiconductors using a spherical harmonic expansion. *J. Phys. C: Solid State Phys.* 5 (1972), 55–62.
24. C. Buet. A discrete velocity scheme for the Boltzmann operator of rarefied gas dynamics. *Transp. Theory Stat. Phys.* 25 (1996), 33–60.
25. L. Pareschi and B. Perthame. A Fourier spectral method for homogeneous Boltzmann equations. *Transp. Theory Stat. Phys.* 25 (1996), 369–383.
26. E. Gabetta, L. Pareschi, and G. Toscani. Relaxation schemes for nonlinear kinetic equations. *SIAM J. Numer. Anal.* 34 (1997), 2168–2194.
27. C. Auer, A. Majorana, and F. Schürrer. Numerical schemes for solving the non-stationary Boltzmann–Poisson system for two-dimensional semiconductor devices. In: T. Goudon, E. Sonnendrücker, and D. Talay (eds.), *ESAIM: Proceedings* 15 (2005), 75–86.
28. M. Cáceres, J. A. Carrillo, and A. Majorana. Deterministic simulation of the Boltzmann–Poisson system in GaAs-based semiconductors. *SIAM J. Sci. Comput.* 27 (2006), 1981–2009.
29. M. Galler and F. Schürrer. A direct multigroup-WENO solver for the 2D non-stationary Boltzmann–Poisson system for GaAs devices: GaAs-MESFET. *J. Comput. Phys.* 212 (2006), 778–797.
30. G. Ossig and F. Schürrer. Simulation of non-equilibrium electron transport in silicon quantum wires. *J. Comput. Electr.* 7 (2008), 367–370.
31. V. Aristov. *Direct Methods for Solving the Boltzmann Equation and Study of Nonequilibrium Flows.* Kluwer, Dordrecht, 2001.
32. L. Pareschi. Computational methods and fast algorithms for Boltzmann equations. In: N. Bellomo (ed.), *Lecture Notes on the Discretization of the Boltzmann Equation*, Series Adv. Math. Appl. Sci. 63, Chapter 7. World Scientific, Singapore, 2003.
33. K. Brennan. *The Physics of Semiconductors.* Cambridge University Press, Cambridge, 1999.
34. H. Grahn. *Introduction to Semiconductor Physics.* World Scientific, Singapore, 1999.
35. M. Lundstrom. *Fundamentals of Carrier Transport.* 2nd edition, Cambridge University Press, Cambridge, 2000.
36. K. Seeger. *Semiconductor Physics. An Introduction.* Springer, Berlin, 2004.
37. V. Gantmakher and Y. Levinson. *Carrier Scattering in Metals and Semiconductors.* North Holland, New York, 1987.
38. B. Ridley. *Quantum Processes in Semiconductors.* Clarendon Press, Oxford, 1982.
39. W. Zawadzki. Mechanics of electron scattering in semiconductors. In: T. Moss (ed.), *Handbook of Semiconductors*, Vol. 1, Chapter 12. North-Holland, New York, 1982.
40. W. Wenckebach. *Essentials of Semiconductor Physics.* John Wiley & Sons, Chichester, 1999.
41. N. Ben Abdallah and P. Degond. On a hierarchy of macroscopic models for semiconductors. *J. Math. Phys.* 37 (1996), 3308–3333.
42. L. Reggiani (ed.). *Hot Electron Transport in Semiconductors.* Springer, Berlin, 1985.
43. C. Kittel. *Introduction to Solid State Physcis.* 7th edition, John Wiley & Sons, New York, 1996.
44. N. Ben Abdallah, P. Degond, and S. Génieys. An energy-transport model for semiconductors derived from the Boltzmann equation. *J. Stat. Phys.* 84 (1996), 205–231.
45. N. Ashcroft and N. Mermin. *Solid State Physics.* Sanners College, Philadelphia, 1976.
46. A. Majorana. Equilibrium solutions of the non-linear Boltzmann equations for an electron gas in a semiconductor. *Il Nuovo Cimento B* 108 (1993), 871–877.
47. F. Poupaud. Mathematical theory of kinetic equations for transport modelling in semiconductors. In: B. Perthame (ed.), *Advances in Kinetic Theory and Computing: Selected Papers*, Ser. Adv. Math. Appl. Sci. 22, 141–168. World Scientific, Singapore, 1994.
48. Y. Sone. *Kinetic Theory and Fluid Dynamics.* Birkhäuser, Boston, 2002.
49. P. Markowich, C. Ringhofer, and C. Schmeiser. *Semiconductor Equations.* Springer, Vienna, 1990.

Part III
Macroscopic Semi-Classical Models

The following chapters are concerned with the formal derivation of semi-classical macroscopic transport models for semiconductors. As detailed in Chap. 2, there are two classes of macroscopic equations: diffusive and hydrodynamic models whose complexity is distinguished by the number of moments involved in the derivation. We begin with the derivation of diffusive equations, starting from the most simple ones, the drift-diffusion equations which involve a single moment, the particle density. Then the energy-transport equations for two moments, the particle density and the energy density, are studied. Employing more than two moments leads to so-called higher-order diffusive moment models. Furthermore, the hydrodynamic semiconductor equations involving three moments, the particle density, momentum, and energy density, and their extensions are derived.

Chapter 5
Drift-Diffusion Equations

This and the following chapters are concerned with the formal derivation of semi-classical macroscopic transport models from the semiconductor Boltzmann equation. We start in this chapter with the derivation of drift-diffusion equations, which are the simplest semiconductor model in the hierarchy. It was first derived by van Roosbroeck in 1950 [1]. We derive the model using the moment method introduced in Chap. 2. A derivation using a simple collision operator was presented in Sect. 2.4. In this chapter, we will employ the low-density operator (4.22). The derivation was made rigorous by Poupaud [2].

5.1 Derivation from the Boltzmann Equation

The starting point of the derivation is the semiconductor Boltzmann equation (4.1) for the distribution function $f = f(x,k,t)$,

$$\partial_t f + v(k) \cdot \nabla_x f + \frac{q}{\hbar}\nabla_x V \cdot \nabla_k f = Q(f), \quad x \in \mathbb{R}^3,\ k \in B,\ t > 0, \qquad (5.1)$$

with the initial datum $f(\cdot,\cdot,0) = f_I$ in $\mathbb{R}^3 \times B$, together with the Poisson equation for the electric potential V,

$$\varepsilon_s \Delta V = q(n - C(x)),$$

where $v(k) = \nabla_k E/\hbar$ denotes the group velocity with the energy band $E(k)$ depending on the pseudo-wave vector k, $Q(f)$ the collision operator, $n = \int_B f\,dk/4\pi^3$ the electron density, $C = C(x)$ the doping profile, and B the Brillouin zone.

Scaling of the Boltzmann–Poisson system. Before we explain the assumptions needed to derive the drift-diffusion model, we scale the Boltzmann equation in order to identify small parameters. We proceed similarly as in Sect. 2.4. We introduce the domain diameter L, the *mean free path* λ, which is the distance a particle travels

Jüngel, A.: *Drift-Diffusion Equations.* Lect. Notes Phys. **773**, 99–127 (2009)
DOI 10.1007/978-3-540-89526-8_5 © Springer-Verlag Berlin Heidelberg 2009

between two consecutive scattering events, and the reference length $\lambda_0 = \sqrt{L\lambda}$. Furthermore, we define the reference velocity $v_0 = \sqrt{k_B T_L / m^*}$ and the reference potential U_T, where T_L is the lattice temperature and $U_T = k_B T_L / q$ the thermal voltage. This means that the kinetic energy $m^* v_0^2 / 2$, the electric energy $q U_T$, and the thermal energy $k_B T_L$ are of the same order. Thus, we consider the case of small electric fields only. The time, which a particle with the typical velocity v_0 needs to cross the domain, equals $\tau_0 = L / v_0$, and the typical time between two consecutive collisions is $\tau = \lambda / v_0$. We use the reference wave vector $k_0 = m^* v_0 / \hbar$, corresponding to the momentum $\hbar k_0 = m^* v_0$. Thus, the variables are scaled as follows:

$$t = \tau_0 t_s, \quad x = \lambda_0 x_s, \quad k = k_0 k_s, \quad v(k) = v_0 v_s(k_s), \quad V = U_T V_s,$$

where t_s, x_s, k_s, etc., are the scaled quantities. Finally, the collision operator is assumed to be of order $1/\tau$:

$$Q(f) = \frac{1}{\tau} Q_s(f).$$

With this scaling, the Boltzmann equation (5.1) becomes, omitting the index "s",

$$\alpha^2 \partial_t f + \alpha \left(v(k) \cdot \nabla_x f + \nabla_x V \cdot \nabla_k f \right) = Q(f), \tag{5.2}$$

where $\alpha = \lambda / \lambda_0 = \sqrt{\lambda / L}$ is the ratio between the mean free path and the reference length. Scaling the particle and doping concentrations by k_0, the Poisson equation becomes

$$\lambda_D^2 \Delta V = n - C(x), \tag{5.3}$$

where

$$\lambda_D^2 = \frac{\varepsilon_s U_T}{q \lambda_0^2 k_0} \tag{5.4}$$

is the (squared) scaled *Debye length*.

Now, we are able to specify our assumptions, following [2] (also see [3]):

1. The energy band is given by the parabolic band approximation $E(k) = E_c + \hbar^2 |k|^2 / 2m^*$, where E_c is the conduction band minimum.
2. The collision operator is given by the low-density approximation (see (4.22))

$$(Q(f))(x,k,t) = \int_B \sigma(x,k,k')(M f' - M' f)\, dk', \qquad .$$

where the collision cross-section $\sigma(x,k,k')$ is positive and symmetric in k' and k, $M(k) = e^{-E(k)/k_B T_L}$ is the Maxwellian, and $f' = f(k'), M' = M(k')$.
3. The mean free path is very small compared to the device diameter, i.e., $\alpha \ll 1$.

The first assumption implies that the mean velocity can be written as $v(k) = \hbar k / m^*$. Furthermore, as explained in Sect. 1.6, we can extend the Brillouin zone to the whole space and write $B = \mathbb{R}^3$ in the continuum limit. The second condition prescribes a linear collision operator which excludes nondegenerate materials. The diffusion approximation of the Boltzmann equation with degenerate Fermi–Dirac statistics was performed by Golse and Poupaud [4]. The third hypothesis means that collisions occur frequently in the material.

In the parabolic band approximation, the scaled velocity becomes $v(k) = k$ and the scaled collision operator can be written as

$$(Q(f))(x,k,t) = \int_{\mathbb{R}^3} s(x,k,k')(Mf' - M'f)\,dk', \tag{5.5}$$

where we have set $s(x,k,k') = Ne^{-E_c/k_B T_L} \sigma(x,k,k')$, the scaled Maxwellian equals $M(k)$ equals $N^{-1}e^{-|k|^2/2}$, and $N = 2(2\pi)^{-3/2}$ is such that the Maxwellian is normalized, i.e., $\int_{\mathbb{R}^3} M(k)\,dk/4\pi^3 = 1$.

Properties of the collision operator. In Chap. 2 we have mentioned that the derivation of diffusive models is based on some properties of the collision operator Q. More precisely, we need to analyze its kernel $N(Q)$ and its range $R(Q)$, where

$$N(Q) = \{f : Q(f) = 0\}, \quad R(Q) = \{g : \text{ there exists } f \text{ such that } Q(f) = g\}.$$

To this end, we introduce, for fixed $x \in \mathbb{R}^3$, the total cross-section,

$$S(k) = \int_{\mathbb{R}^3} s(x,k,k')M(k')\,dk', \quad k \in \mathbb{R}^3,$$

and the Banach spaces X and Y, which consist of all measurable functions $f : \mathbb{R}^3 \to \mathbb{R}$ such that $\|f\|_X$ and $\|f\|_Y$ are finite, where

$$\|f\|_X^2 = \int_{\mathbb{R}^3} f(k)^2 S(k)M(k)^{-1}\,dk,$$

$$\|f\|_Y^2 = \int_{\mathbb{R}^3} f(k)^2 S(k)^{-1}M(k)^{-1}\,dk.$$

By Proposition 4.2 (3), the kernel of Q is spanned by the Maxwellian $M(k)$. For the analysis of the range of Q, we employ the following version of the *Fredholm alternative*.

Lemma 5.1 (Fredholm alternative). *Let X be a Hilbert space and $Q : X \to X$ be a linear, continuous, and closed operator (i.e., $R(Q)$ is closed). Then*

$$Q(f) = g \text{ has a solution if and only if } g \in N(Q^*)^{\perp}.$$

Here, Q^* denotes the adjoint operator of Q and $N(Q^*)^{\perp}$ is the orthogonal complement of $N(Q^*)$. We refer to, e.g., [5] for the functional analytical details and a proof of the Fredholm alternative.

The following lemma is a consequence of the Fredholm alternative.

Lemma 5.2. *Let the collision operator Q be given by (5.5).*
(1) The equation $Q(f) = g$ has a solution $f \in X$ if and only if

$$\int_{\mathbb{R}^3} g(k)\,dk = 0.$$

In this situation, any solution of $Q(f) = g$ can be written as $f + nM$, where $n = n(x)$ is a parameter.

(2) The solution $f \in X$ of $Q(f) = g$ is unique if the orthogonality relation

$$\int_{\mathbb{R}^3} f(k)S(k)\,dk = 0 \tag{5.6}$$

is satisfied.

Proof. (1) First we symmetrize the collision operator (5.5) by setting $f_s = \sqrt{S/M}\,f$ and $Q_s(f_s) = (SM)^{-1/2}Q(f)$. Then

$$Q_s(f_s) = \frac{1}{\sqrt{SM}}\left(M\int_{\mathbb{R}^3} s(x,k,k')f'\,dk' - Sf\right)$$

$$= \int_{\mathbb{R}^3} s(x,k,k')\left(\frac{MM'}{SS'}\right)^{1/2} f_s'\,dk' - f_s.$$

Since s is symmetric in k and k' by assumption, the operator $Q_s : L^2(\mathbb{R}^3) \to L^2(\mathbb{R}^3)$ is symmetric and self-adjoint. It is possible to prove that Q_s is closed. By the Fredholm alternative, $Q_s(f_s) = g_s$ has a solution in $L^2(\mathbb{R}^3)$ if and only if $g_s \in N(Q_s^*)^\perp = N(Q_s)^\perp$ which means that

$$\int_{\mathbb{R}^3} g_s h\,dk = 0 \quad \text{for all } h \in N(Q_s).$$

The kernel of Q_s is spanned by $\sqrt{S/M}M = \sqrt{SM}$, such that the above relation is equivalent to

$$0 = \int_{\mathbb{R}^3} g_s \sqrt{SM}\,dk = \int_{\mathbb{R}^3} g\,dk,$$

since in the original variables we have $g = Q(f) = \sqrt{SM}Q_s(f_s) = \sqrt{SM}g_s$.

Let f_1 and f_2 be two solutions of $Q(f) = g$. Then, since Q is linear, $Q(f_1 - f_2) = 0$ and thus, $f_1 - f_2 \in N(Q)$. This shows that $f_2 = f_1 + nM$ for some $n = n(x)$.

(2) We only give a sketch of the proof and refer to [2] for details. It is possible to show that $-Q_s$ is coercive on $N(Q_s)^\perp$:

$$-\int_{\mathbb{R}^3} Q_s(f_s)f_s\,dk \geq c\|f_s\|_{L^2(\mathbb{R}^3)}^2 \quad \text{for all } f_s \in N(Q_s)^\perp,$$

where $c > 0$ is some constant. The proof is based on the fact that $\mathrm{Id} + Q_s$ is a Hilbert–Schmidt operator. The coercivity property implies that Q_s (and also Q) is one to one on $N(Q_s)^\perp$, which proves the uniqueness of solutions on this set. $\quad\square$

Finally, we prove some properties of the solution of the operator equations:

$$Q(h_i) = k_i M(k), \quad i = 1, 2, 3. \tag{5.7}$$

These equations have solutions, due to Lemma 5.2, since the functions $k_i \mapsto k_i M(k)$ are odd and hence their integrals vanish. The solutions are unique in the space of functions h_i satisfying $\int_{\mathbb{R}^3} h_i(k) S(k)\, dk = 0$. We set $h = (h_1, h_2, h_3)^\top$.

Lemma 5.3. *Assume that the collision cross-section is invariant with respect to isometric operations, i.e., for all isometric matrices $A \in \mathbb{R}^{3 \times 3}$ it holds:*

$$\sigma(x, Ak, Ak') = \sigma(x, k, k') \quad \text{for all } x, k, k' \in \mathbb{R}^3. \tag{5.8}$$

Then there exists a scalar function $\mu_0(x) \geq 0$ such that the solutions h_i of (5.7) satisfy

$$\int_{\mathbb{R}^3} k \otimes h \frac{dk}{4\pi^3} = -\mu_0(x) \mathrm{Id}, \tag{5.9}$$

where $k \otimes h$ is the matrix with components $k_i h_j$ and Id is the identity matrix in $\mathbb{R}^{3 \times 3}$.

In the statement of the lemma, we have omitted some technical (regularity) assumptions on the collision cross-section σ. Details can be found in [2].

Proof. Let A be the matrix of a rotation about the axis k_1. Then $(Ak)_1 = k_1$ for all $k \in \mathbb{R}^3$. Since A is isometric, i.e., $|Ak| = |k|$, we obtain $M(Ak) = N^{-1} e^{-|Ak|^2/2} = N^{-1} e^{-|k|^2/2} = M(k)$ and $(Ak)_1 M(Ak) = k_1 M(k)$. This implies, together with assumption (5.8) and the transformation $w = Ak'$ with $dw = |\det A|\, dk' = dk'$,

$$
\begin{aligned}
(Q(h_1 \circ A))(k) &= \int_{\mathbb{R}^3} s(x, k, k') \left(M(k) h_1(Ak') - M(k') h_1(Ak) \right) dk' \\
&= \int_{\mathbb{R}^3} s(x, Ak, Ak') \left(M(Ak) h_1(Ak') - M(Ak') h_1(Ak) \right) dk' \\
&= \int_{\mathbb{R}^3} s(x, Ak, w) \left(M(Ak) h_1(w) - M(w) h_1(Ak) \right) dw \\
&= (Q(h_1))(Ak) = (Ak)_1 M(Ak) = k_1 M(k) = (Q(h_1))(k),
\end{aligned}
$$

and thus $Q(h_1 \circ A - h_1) = 0$. Another computation yields

$$
\begin{aligned}
\int_{\mathbb{R}^3} h_1(Ak) S(k)\, dk &= \int_{\mathbb{R}^3} \int_{\mathbb{R}^3} s(x, k, k') h_1(Ak) M(k')\, dk'\, dk \\
&= \int_{\mathbb{R}^3} \int_{\mathbb{R}^3} s(x, Ak, Ak') h_1(Ak) M(Ak')\, dk'\, dk \\
&= \int_{\mathbb{R}^3} \int_{\mathbb{R}^3} s(x, v, w) h_1(v) M(w)\, dw\, dv = \int_{\mathbb{R}^3} h_1(v) S(v)\, dv
\end{aligned}
$$

and hence

$$\int_{\mathbb{R}^3} (h_1 \circ A - h_1) S\, dk = 0.$$

This is the orthogonality condition (5.6) which ensures the uniqueness of the solution of $Q(h_1 \circ A - h_1) = 0$. Therefore, $h_1 \circ A - h_1 = 0$. We conclude that h_1 remains invariant under a rotation about the axis k_1. In particular, we can write $h_1(k) = H_1(k_1, |k|^2 - k_1^2)$ for some function H_1.

Now, let A be the isometric matrix corresponding to the linear mapping $k \mapsto (-k_1, k_2, k_3)$. Since $k \mapsto k_1 M(k)$ is odd, a similar computation as above gives $Q(h_1 \circ A) = -Q(h_1)$ and

$$\int_{\mathbb{R}^3} (h_1 \circ A + h_1) S \, dk = 0.$$

This implies that $h_1 \circ A + h_1 = 0$. Thus, h_1 is an odd function with respect to k_1.

In a similar way, we can show that $h_i(k) = H_i(k_i, |k|^2 - k_i^2)$ for $i = 2, 3$ and H_i are odd functions with respect to the first argument. In fact, all the functions H_i are equal since, for instance, exchanging k_1 and k_2 in

$$Q(H_1(k_1, k_2^2 + k_3^2)) = k_1 M(k_1^2 + k_2^2 + k_3^2)$$

(with a slight abuse of notation) leads to

$$Q(H_1(k_2, k_1^2 + k_3^2)) = k_2 M(k_1^2 + k_2^2 + k_3^2) = Q(H_2(k_2, k_1^2 + k_3^2))$$

or $Q(H_1 - H_2) = 0$ and thus, $H_1 = H_2$. We set $H = H_1$.

Since H is odd with respect to its first argument and $|k|^2 - k_j^2$ does not depend on k_j, we obtain

$$\int_{\mathbb{R}^3} k_i h_j(k) \, dk = \int_{\mathbb{R}^3} k_i H(k_j, |k|^2 - k_j^2) \, dk = 0 \quad \text{for all } i \neq j. \tag{5.10}$$

Furthermore,

$$\int_{\mathbb{R}^3} k_i h_i(k) \, dk = \int_{\mathbb{R}^3} k_i H(k_i, |k|^2 - k_i^2) \, dk = \int_{\mathbb{R}^3} k_j H(k_j, |k|^2 - k_j^2) \, dk$$
$$= \int_{\mathbb{R}^3} k_j h_j(k) \, dk \quad \text{for all } i, j.$$

This means that the integral is independent of i, and we can set

$$\mu_0 = -\int_{\mathbb{R}^3} k_1 h_1(k) \frac{dk}{4\pi^3} = -\int_{\mathbb{R}^3} Q(h_1) h_1 M^{-1} \frac{dk}{4\pi^3}.$$

The parameter μ_0 depends on x since h_1 depends on x through Q. We have proved in (4.24) that

$$\int_{\mathbb{R}^3} Q(f) \chi(f/M) \, dk \leq 0 \quad \text{for all } f$$

and all nondecreasing functions χ. Choosing $\chi(x) = x$ shows that

$$\mu_0(x) = -\int_{\mathbb{R}^3} Q(h_1) h_1 M^{-1} \frac{dk}{4\pi^3} \geq 0,$$

and, in view of (5.10), we have shown (5.9).   □

Derivation of the drift-diffusion equations. Now, we are in the position to derive the drift-diffusion model. The general strategy is explained in Chap. 2. The derivation consists of three steps. Let (f_α, V_α) be a solution of the scaled Boltzmann–Poisson system (5.2) and (5.3). We assume that f_α, V_α, and their derivatives converge, as α to 0, to f, V, and their corresponding derivatives.

First, we perform the (formal) limit $\alpha \to 0$ in (5.2). This yields

$$Q(f) = 0. \tag{5.11}$$

Since the kernel is spanned by the Maxwellian, $f = n(x,t)M$, where $n(x,t) = \int_{\mathbb{R}^3} f(x,k,t)\, dk/4\pi^3$ (notice that the Maxwellian is normalized).

For the second step, we introduce the *Chapman–Enskog expansion* $f_\alpha = nM + \alpha g_\alpha$. (This equation has to be considered as a definition of g_α.) Inserting the expansion in the Boltzmann equation (5.2) gives, after division by α,

$$\alpha \partial_t (nM + \alpha g_\alpha) + (k \cdot \nabla_x (nM) + \nabla_x V_\alpha \cdot \nabla_k (nM))$$
$$+ \alpha (k \cdot \nabla_x g_\alpha + \nabla_x V_\alpha \cdot \nabla_k g_\alpha) = Q(g_\alpha),$$

since $Q(nM) = nQ(M) = 0$. We perform the limit $\alpha \to 0$ to obtain

$$Q(g) = k \cdot \nabla_x (nM) + \nabla_x V \cdot \nabla_k (nM) = (\nabla_x n - n\nabla_x V) \cdot kM, \tag{5.12}$$

where $g = \lim_{\alpha \to 0} g_\alpha$. This operator equation is of the form (5.7), since $\nabla_x n - n\nabla_x V$ is a function of (x,t) only and Q is linear.

In the third step, we derive the evolution equations. Integrating the Boltzmann equation (5.2) and inserting the Chapman–Enskog expansion give, employing the notation $\langle f \rangle = \int_{\mathbb{R}^3} f\, dk/4\pi^3$,

$$\partial_t \langle nM \rangle + \alpha \partial_t \langle g_\alpha \rangle + \alpha^{-1} \text{div}_x \langle k(nM) \rangle + \text{div}_x \langle k g_\alpha \rangle = 0.$$

Here, we have used that $\langle \nabla_x V \cdot \nabla_k f_\alpha \rangle = 0$ and $\langle Q(f_\alpha) \rangle = 0$. Since $k \mapsto kM(k)$ is an odd function, also the integral $\langle k(nM) \rangle$ vanishes. Thus, performing the limit $\alpha \to 0$,

$$\partial_t n + \text{div}_x \langle kg \rangle = 0. \tag{5.13}$$

The flux $J_n = -\langle kg \rangle$ can be computed in terms of n and $\nabla_x V$. More precisely, we have the following result.

Theorem 5.4 (Drift-diffusion equations). *Let the assumptions at the beginning of this section hold, assume (5.8), and let (f_α, V_α) be a solution of the Boltzmann–Poisson system (5.2) and (5.3). Then the limit $f = \lim_{\alpha \to 0} f_\alpha$, $V = \lim_{\alpha \to 0} V_\alpha$ satisfies the* drift-diffusion equations

$$\partial_t n - \text{div} J_n = 0, \quad J_n = \mu_0 (\nabla n - n\nabla V), \tag{5.14}$$
$$\lambda_D^2 \Delta V = n - C(x), \quad x \in \mathbb{R}^3, \, t > 0,$$

where $\mu_0(x) \geq 0$ comes from Lemma 5.3 and $n = \int_{\mathbb{R}^3} f\, dk/4\pi^3$ is the electron density, satisfying $n(\cdot, 0) = \int_{\mathbb{R}^3} f_I\, dk/4\pi^3$.

The current density is the sum of the *drift current* $-\mu_0 n \nabla V$ and the *diffusion current* $\mu_0 \nabla n$, which explains the name of the model. The quantity μ_0 is called the (scaled) *electron mobility*.

Proof. In view of (5.13), we have already proved the first equation in (5.14). In order to prove the second one, we employ Lemma 5.3. Let h_i be a solution of $Q(h_i) = k_i M$ ($i = 1, 2, 3$) and set $h = (h_1, h_2, h_3)^\top$. Then $g = (\nabla_x n - n \nabla_x V) \cdot h$ solves (5.12) and, by (5.9),

$$J_n = -\langle kg \rangle = -\langle k \otimes h \rangle (\nabla_x n - n \nabla_x V) = \mu_0 (\nabla_x n - n \nabla_x V).$$

The limiting Poisson equation follows directly from (5.3) for $V = V_\alpha$ in the formal limit $\alpha \to 0$. □

The unscaled equations are obtained by scaling back to the physical variables. Employing the reference electron mobility $\mu_{n,\text{ref}} = q\tau/m^*$ and the reference current density $J_{n,\text{ref}} = qU_T k_0 \mu_{n,\text{ref}}/\lambda_0$, a calculation yields

$$\partial_t n - q^{-1} \text{div} J_n = 0, \quad J_n = q\mu_n (U_T \nabla n - n \nabla V),$$
$$\varepsilon_s \Delta V = q(n - C(x)), \quad x \in \mathbb{R}^3, \, t > 0.$$

These equations have to be complemented by initial conditions for the particle density:

$$n(x, 0) = n_I(x), \quad x \in \mathbb{R}^3.$$

Remark 5.5 (Hilbert expansion). The derivation of the drift-diffusion equations in [2] is slightly different from ours. In fact, the model can also be derived from the *Hilbert expansion method*. The idea of this technique is to develop the solution f_α of the Boltzmann equation formally in terms of powers of α:

$$f_\alpha = f_0 + \alpha f_1 + \alpha^2 f_2 + \cdots.$$

Inserting this expansion into the Boltzmann equation (5.2), using the linearity of the collision operator, and identifying coefficients of equal powers of α yield the following equations:

$$\text{terms in } \alpha^0: \quad Q(f_0) = 0, \tag{5.15}$$
$$\text{terms in } \alpha^1: \quad Q(f_1) = k \cdot \nabla_x f_0 + \nabla_x V \cdot \nabla_k f_0, \tag{5.16}$$
$$\text{terms in } \alpha^2: \quad Q(f_2) = \partial_t f_0 + k \cdot \nabla_x f_1 + \nabla_x V \cdot \nabla_k f_1. \tag{5.17}$$

The first equation (5.15) corresponds to (5.11) and shows that f_0 is a multiple of the Maxwellian, $f_0 = n(x)M$. By Lemma 5.3, the second equation (5.16), which corresponds to (5.12), can be inverted, and the proof of Theorem 5.4 shows that the first moment of f_1 is given by $\langle kf_1 \rangle = -\mu_0 (\nabla_x n - n \nabla_x V)$. Finally, the conservation law (5.13) is obtained from the third equation (5.17) after integration over $k \in \mathbb{R}^3$ since this equation is solvable, by Lemma 5.2, if and only if its integral with respect to k vanishes. □

The derivation of the drift-diffusion equations from the semiconductor Boltzmann equation is extensively studied in the mathematical literature. Poupaud has proved the convergence toward the linear drift-diffusion model [2]. In the one-dimensional case, this convergence result was extended by Ben Abdallah and Tayeb to the Boltzmann equation with Poisson coupling [6]. A generalization of [6] to the multi-dimensional Boltzmann–Poisson system was performed by Masmoudi and Tayeb [7].

In the above mentioned papers, the low-density collision operator with smooth collision cross-section was considered. The diffusion approximation from the Boltzmann equation for degenerate Fermi–Dirac statistics was studied by Poupaud and Schmeiser [8] and for the degenerate Boltzmann–Poisson system by Golse and Poupaud [4]. Singular cross-sections, modeling electron–phonon scattering, are admissible in the papers of Markowich and Schmeiser [9] for linear collision operators and of Markowich, Poupaud, and Schmeiser [10] for nonlinear operators. Furthermore, Ben Abdallah and Tayeb [11] have proved the diffusion limit from the Boltzmann equation with a spatially oscillating electric potential with an oscillation period being of the same order as the mean free path. Finally, we mention that the drift-diffusion model can also be derived from the Vlasov–Fokker–Planck system [12].

The drift-diffusion equations were first proposed by Van Roosbroeck in 1950 [1]. The first computational solution was presented in 1964 by Gummel [13] and improved some years later by Scharfetter and Gummel [14]. The developed Scharfetter–Gummel scheme was interpreted as a mixed finite-element method [15, 16]. Later, finite-volume discretizations were developed [17, 18]. The first mathematical papers devoted to the drift-diffusion model appeared at the beginning of the 1970s. Mock analyzed the stationary equations in [19] and the transient problem with Neumann boundary conditions in [20]. The global existence and uniqueness of solutions under realistic physical and geometrical conditions was proved by Gajewski and Gröger [21]. A drift-diffusion model involving Fermi–Dirac statistics was analyzed in [22] (existence of global solutions) and [23] (uniqueness of stationary solutions). We refer to [24, 25] for more analytical results for this model and [26–28] for numerical reviews. The modeling aspects are summarized in, for instance, [29, 30].

The above drift-diffusion model is formulated in the whole space. Since a semiconductor occupies a bounded domain, one may ask what happens with the kinetic boundary conditions in the diffusion limit. Let $\Omega \subset \mathbb{R}^3$ be a bounded domain and define as in Sect. 4.3 the inflow boundary

$$\Gamma^- = \{(x,k) \in \partial\Omega \times \mathbb{R}^3 : k \cdot \eta(x) < 0\},$$

where $\eta(x)$ is the exterior unit normal vector at $x \in \partial\Omega$. The kinetic boundary condition then reads as

$$f(x,k,t) = f_D(x,k,t), \quad (x,k) \in \Gamma^-, \, t > 0.$$

Assuming that the boundary function can be written as $f_D(x,k,t) = g(x,t)M(k)$ for some function $g(x,t)$, Poupaud [2] showed that the solution f_α of (5.2) converges to $n(x,t)M$ in the sense

$$\|(f_\alpha - nM)(t)\|_{L^1(\Omega \times \mathbb{R}^3)} \le c\alpha \quad \text{for } 0 < t \le T,$$

where $c > 0$ depends on T, and the boundary conditions become $n = g$ on $\partial\Omega$. A similar result was proved in [7]. The proof is based on the solution of a half-space (Milne) problem, which provides a boundary layer correction. With this correction, the electron density solves Dirichlet boundary conditions. This result can be improved by correcting the Dirichlet condition by a term of the order of the mean free path proportional to the current density (first-order boundary correction). It was shown by Yamnahakki [31] that the electron density satisfies the Robin boundary conditions

$$n - \gamma J_n \cdot \eta = g \quad \text{on } \partial\Omega, \, t > 0,$$

where the (scaled) extrapolation length $\gamma > 0$ is determined from the solution of the Milne problem. A second-order boundary correction was derived and numerically compared to lower-order corrections in [32]. However, we are not aware of rigorous derivations of mixed Dirichlet–Neumann boundary conditions from the kinetic boundary.

The derivation of the drift-diffusion model is mainly based on the following hypotheses:

- The mean free path λ between two consecutive scattering events is much smaller than the reference length λ_0. Typically, the mean free path is of the order 10^{-7} m.
- The electric potential is of the order of $U_T = 0.026$ V (at $T = 300$ K).

Thus, the drift-diffusion model is appropriate for semiconductor devices with characteristic lengths not much smaller than about $1\,\mu$m$= 10^{-6}$ m and applied voltages much smaller than 1 V. However, in applications this model is also used for higher applied voltages (including high-field corrections; see Sect. 5.4). It gives reasonable results as long as the characteristic length is not much smaller than about $1\,\mu$m.

5.2 The Bipolar Model

The bipolar drift-diffusion equations are derived from the bipolar Boltzmann–Poisson system (4.38), (4.39), and (4.43) in the vanishing scaled mean free path limit similarly as in Sect. 5.1. In this section, we sketch the derivation of the bipolar model by emphasizing the differences from the previous section.

One main difference concerns the definition of the reference time τ_R, which is now given by the typical time between two consecutive recombination–generation events. Defining as in the previous section the reference velocity $v_0 = \sqrt{k_B T_L / m^*}$, the mean free path between two recombination–generation processes is $\lambda_R = \tau_R v_0$.

With the mean free path λ between two collision events, we define the reference length by $\lambda_0 = \sqrt{\lambda_R \lambda}$ (instead of the geometric average of the device diameter and the mean free path λ, as was done in Sect. 5.1). The reference time τ_R is of the order 10^{-9} s, whereas the collision time $\tau = \lambda/v_0$ is of the order 10^{-12} s [3, p. 86]. Thus, the ratio $\alpha^2 = (\lambda/\lambda_0)^2 = \tau/\tau_R$ is much smaller than one.

We assume that the energy band is approximated by the parabolic band diagram, that the collision operator is given by the low-density approximation, and that $\alpha \ll 1$. The collision operator is assumed to be of order $1/\tau$, and the recombination–generation terms are supposed to be of order $1/\tau_R$:

$$Q_j(f_j) = \frac{1}{\tau}Q_{j,s}(f_j), \quad I_j(f_n, f_p) = \frac{1}{\tau_R}I_{j,s}(f_n, f_p), \quad j = n, p,$$

where the scaled quantities are denoted by the index "s". Then the scaled bipolar Boltzmann–Poisson system for the electron and hole distribution functions f_n and f_p, respectively, and the electric potential V reads as follows (omitting the index "s"):

$$\alpha^2 \partial_t f_n + \alpha\,(k \cdot \nabla_x f_n + \nabla_x V \cdot \nabla_k f_n) = Q_n(f_n) + \alpha^2 I_n(f_n, f_p), \tag{5.18}$$

$$\alpha^2 \partial_t f_p + \alpha\,(\beta k \cdot \nabla_x f_p - \nabla_x V \cdot \nabla_k f_p) = Q_p(f_p) + \alpha^2 I_p(f_n, f_p), \tag{5.19}$$

$$\lambda_D^2 \Delta V_\alpha = n - p - C(x), \quad x, k \in \mathbb{R}^3, \, t > 0, \tag{5.20}$$

where $\beta = m_e^*/m_h^*$ is the ratio of the effective masses of electrons and holes, $n = \int_{\mathbb{R}^3} f_n\,dk/4\pi^3$ the electron density, and $p = \int_{\mathbb{R}^3} f_p\,dk/4\pi^3$ the hole density. The linear collision operators are given by

$$Q_j(f_j) = \int_{\mathbb{R}^3} s_j(x, k, k')\,(M_j f_j' - M_j' f_j)\,dk', \quad j = n, p,$$

where

$$s_n(x, k, k') = N_n e^{-E_c/k_B T_L}\sigma_n(x, k, k'), \quad s_p(x, k, k') = N_p e^{-E_v/k_B T_L}\sigma_p(x, k, k'),$$

the scaled and normalized Maxwellians read as $M_n(k) = N_n^{-1} e^{-|k|^2/2}$, $M_p(k) = N_p^{-1} e^{-\beta|k|^2/2}$, and $N_n = 2(2\pi)^{-3/2}$, $N_p = 2(2\pi/\beta)^{-3/2}$ are the normalization constants. We recall that E_c and E_v are the conduction band minimum and the valence band maximum, respectively. Finally, the recombination–generation terms are

$$(I_n(f_n, f_p))(x, k, t) = \int_{\mathbb{R}^3} g(x, k', k)\left(1 - e^{(E_c - E_v)/k_B T_L}\frac{f_n f_p'}{N_n N_p M_n M_p'}\right)dk',$$

$$(I_p(f_n, f_p))(x, k, t) = \int_{\mathbb{R}^3} g(x, k', k')\left(1 - e^{(E_c - E_v)/k_B T_L}\frac{f_n' f_p}{N_n N_p M_n' M_p}\right)dk',$$

where $g(x, k, k')$ is the generation rate. Then, Lemmas 5.2 and 5.3 still hold for the collision operators Q_n and Q_p, i.e., the kernel of Q_j is spanned by M_j, the equation

$Q_j(f_j) = g_j$ has a solution if and only if $\int_{\mathbb{R}^3} g_j \, dk = 0$, and the solutions of the equations $Q_n(h_n) = kM_n(k)$ and $Q_p(h_p) = \beta kM_p(k)$ satisfy

$$\int_{\mathbb{R}^3} k \otimes h_n \frac{dk}{4\pi^3} = -\mu_{0,n} \, \mathrm{Id}, \quad \int_{\mathbb{R}^3} \beta k \otimes h_p \frac{dk}{4\pi^3} = -\mu_{0,p} \, \mathrm{Id}. \qquad (5.21)$$

Let $(f_{n,\alpha}, f_{p,\alpha}, V_\alpha)$ be a solution of (5.18), (5.19), and (5.20). We perform the limit $\alpha \to 0$ again in three steps. The first step is the limit $\alpha \to 0$ in (5.18) and (5.19), leading to

$$Q_n(f_n) = 0, \quad Q_p(f_p) = 0,$$

where $f_j = \lim_{\alpha \to 0} f_{j,\alpha}$ $(j = n, p)$. Thus, $f_n = nM_n$ and $f_p = pM_p$, where $n = \int_{\mathbb{R}^3} f_n \, dk/4\pi^3$ and $p = \int_{\mathbb{R}^3} f_p \, dk/4\pi^3$. In the second step we insert the Chapman–Enskog expansion $f_{n,\alpha} = nM_n + \alpha g_{n,\alpha}$ in the Boltzmann equation:

$$\alpha \partial_t (nM_n + \alpha g_{n,\alpha}) + (k \cdot \nabla_x (nM_n) + \nabla_x V \cdot \nabla_k (nM_n))$$
$$+ \alpha (k \cdot \nabla_x g_{n,\alpha} + \nabla_x V \cdot \nabla_k g_{n,\alpha}) = Q_n(g_{n,\alpha}) + \alpha I_n(f_{n,\alpha}, f_{p,\alpha}).$$

The limit $\alpha \to 0$ yields

$$Q_n(g_n) = (\nabla_x n - n\nabla_x V) \cdot kM_n,$$

where $g_n = \lim_{\alpha \to 0} g_{n,\alpha}$. In a similar way, we obtain

$$Q_p(g_p) = \beta (\nabla_x p + p\nabla_x V) \cdot kM_p,$$

where $g_p = \lim_{\alpha \to 0} g_{p,\alpha}$.

Defining the particle current densities by $J_n = \mu_{0,n}(\nabla_x n - n\nabla_x V)$ and $J_p = -\beta \mu_{0,p}(\nabla_x p + p\nabla_x V)$, we have to solve the operator equations

$$Q_n(g_n) = \frac{J_n}{\mu_{0,n}} \cdot kM_n(k) \quad \text{and} \quad Q_p(g_p) = -\frac{J_p}{\mu_{0,p}} \cdot kM_p(k).$$

Since Q_j is linear, we conclude that $g_n = \mu_{0,n}^{-1} J_n \cdot h_n + c_n M_n$ and $g_p = -\mu_{0,p}^{-1} J_p \cdot h_p + c_p M_p$ for some real constants c_n and c_p. In view of (5.21), we infer that

$$\langle kg_n \rangle = \mu_{0,n}^{-1} \langle k \otimes h_n \rangle \cdot J_n = -J_n, \quad \beta \langle kg_p \rangle = -\beta \mu_{0,p}^{-1} \langle k \otimes h_p \rangle \cdot J_p = J_p.$$

In the third step, the balance equations are derived. Here, the recombination–generation integrals appear. We integrate the Boltzmann equations (5.18) and (5.19) over the wave-vector space and insert the Chapman–Enskog expansion:

$$\partial_t \langle nM_n \rangle + \alpha \partial_t \langle g_{n,\alpha} \rangle + \mathrm{div}_x \langle kg_{n,\alpha} \rangle = \langle I_n(f_{n,\alpha}, f_{p,\alpha}) \rangle,$$
$$\partial_t \langle pM_p \rangle + \alpha \partial_t \langle g_{p,\alpha} \rangle + \beta \mathrm{div}_x \langle kg_{p,\alpha} \rangle = \langle I_p(f_{n,\alpha}, f_{p,\alpha}) \rangle.$$

The limit $\alpha \to 0$ then leads to

$$\partial_t n + \mathrm{div}_x \langle k g_n \rangle = \langle I_n(nM_n, pM_p) \rangle,$$
$$\partial_t p + \beta \mathrm{div}_x \langle k g_p \rangle = \langle I_p(nM_n, pM_p) \rangle.$$

The limiting recombination–generation terms can be expressed in terms of the particle densities:

$$\langle I_n(nM_n, pM_p) \rangle = \int_{\mathbb{R}^3} \int_{\mathbb{R}^3} g(x, k', k) \left(1 - e^{(E_c - E_v)/k_B T_L} \frac{np}{N_n N_p} \right) dk' \, dk$$
$$= -A(x)(np - n_i^2),$$

where

$$A(x) = \frac{1}{n_i^2} \int_{\mathbb{R}^3} \int_{\mathbb{R}^3} g(x, k', k) \, dk \, dk', \quad n_i = 2(2\pi)^{-3/2} \beta^{3/4} e^{-(E_c - E_v)/2k_B T_L}. \quad (5.22)$$

In a similar way, we obtain

$$\langle I_p(nM_n, pM_p) \rangle = \int_{\mathbb{R}^3} \int_{\mathbb{R}^3} g(x, k, k') \left(1 - e^{(E_c - E_v)/k_B T_L} \frac{np}{N_n N_p} \right) dk' \, dk$$
$$= -A(x)(np - n_i^2).$$

Thus, we have proved the following result.

Theorem 5.6 (Bipolar drift-diffusion equations). *Let the assumptions at the beginning of this section hold, assume (5.8), and let $(f_{n,\alpha}, f_{p,\alpha}, V_\alpha)$ be a solution of the bipolar Boltzmann–Poisson system (5.18), (5.19), and (5.20). Then the limit functions $f_n = \lim_{\alpha \to 0} f_{n,\alpha}$, $f_p = \lim_{\alpha \to 0} f_{p,\alpha}$, and $V = \lim_{\alpha \to 0} V_\alpha$ are a solution to the bipolar drift-diffusion equations:*

$$\partial_t n - J_n = -R(n, p), \quad J_n = \mu_{0,n}(\nabla n - n\nabla V),$$
$$\partial_t p + J_p = -R(n, p), \quad J_p = -\mu_{0,p}(\nabla p + p\nabla V),$$
$$\lambda_D^2 \Delta V = n - p - C(x), \quad x \in \mathbb{R}^3, \, t > 0,$$

where

$$R(n, p) = A(x)(np - n_i^2),$$

and $A(x)$ and n_i are defined in (5.22).

Scaling back to the physical variables, we obtain the equations

$$\partial_t n - \frac{1}{q} J_n = -R(n, p), \quad J_n = q\mu_n(U_T \nabla n - n\nabla V), \quad (5.23)$$

$$\partial_t p + \frac{1}{q} J_p = -R(n, p), \quad J_p = -q\mu_p(U_T \nabla p + p\nabla V), \quad (5.24)$$

$$\varepsilon_s \Delta V = q(n - p - C(x)), \quad x \in \mathbb{R}^3, \, t > 0. \quad (5.25)$$

The unscaled recombination–generation rate remains unchanged, but now,

$$A(x) = \left(\frac{\hbar^2}{m_e^* k_B T_L}\right)^3 \int_{\mathbb{R}^3} \int_{\mathbb{R}^3} g(x, k, k') \, dk' \, dk,$$

and the *intrinsic density* n_i reads as

$$n_i = 2 \left(\frac{\sqrt{m_e^* m_h^*} k_B T_L}{2\pi\hbar^2}\right)^{3/2} \exp\left(-\frac{E_c - E_v}{2k_B T_L}\right).$$

This is exactly the value derived in Sect. 1.6 (see (1.61)).

Remark 5.7 (Shockley–Read–Hall recombination). The above expression for the re-combination–generation term is a simplified version of the so-called *Shockley–Read–Hall recombination term*

$$R(n, p) = \frac{np - n_i^2}{\tau_p(n + n_d) + \tau_n(p + p_d)},$$

where τ_n and τ_p are the carrier lifetimes, and the densities n_d and p_d are defined by

$$n_d = N_c \exp\left(\frac{E_t - E_c}{k_B T_L}\right), \quad p_d = N_v \exp\left(\frac{E_v - E_t}{k_B T_L}\right).$$

Here, E_t denotes the trap energy level, and N_c and N_v are the carrier effective densities of states defined in (1.58). Notice that $n_d p_d = n_i^2$. By "trap level", we mean energy levels in the forbidden band region, caused by crystal impurities. They facilitate the generation of electron–hole pairs, since the jump from the valence to the conduction band can be split into two parts, each of which requires less energy than the gap energy. The Shockley–Read–Hall model is usually derived by assuming one trap level and quasi-stationarity of the dynamics of the trapped electrons (see [33, Chap. 10] or [34, Chap. 10]). A generalization to a distribution of trapped states across the forbidden region was given in [35]. $\square$

5.3 Thermal Equilibrium State and Boundary Conditions

When we consider the drift-diffusion equations in a bounded domain $\Omega \subset \mathbb{R}^3$, we need to impose some boundary conditions. Their definition is based on the notion of thermal equilibrium, which we explain first.

Let (n, p, V) be a solution of the unscaled drift-diffusion equations (5.23), (5.24), and (5.25). The thermal equilibrium state is a steady state with no current flow, i.e.,

$$\partial_t n = \partial_t p = 0 \quad \text{and} \quad J_n = J_p = 0 \quad \text{in } \Omega.$$

This implies $R(n, p) = 0$ or $np = n_i^2$ in Ω and

$$0 = U_T \nabla n - n \nabla V = n \nabla (U_T \ln n - V),$$
$$0 = U_T \nabla p + p \nabla V = p \nabla (U_T \ln p + V).$$

It is physically reasonable to assume that the particle densities n and p are positive in Ω. (In fact, this can be proved by employing maximum principle arguments, assuming that the boundary data are positive; see, e.g., [25].) This yields

$$\alpha = U_T \ln n - V = \text{const.}, \quad \beta = U_T \ln p + V = \text{const.} \tag{5.26}$$

or

$$n = e^{\alpha/U_T} e^{V/U_T}, \quad p = e^{\beta/U_T} e^{-V/U_T} \quad \text{in } \Omega.$$

We wish to determine the constants α and β. For this, we use the equation $np = n_i^2$ and the fact that V is determined only up to an additive constant. Then the sum of the two equations in (5.26) yields $\alpha + \beta = U_T \ln(np) = 2U_T \ln n_i$ and replacing V by $V + \gamma$, where $\gamma = U_T \ln n_i - \alpha$, gives

$$n = e^{(\alpha+\gamma)/U_T} e^{V/U_T} = n_i e^{V/U_T} \tag{5.27}$$

and

$$p = e^{(\beta-\gamma)/U_T} e^{-V/U_T} = e^{(\beta+\alpha-U_T \ln n_i)/U_T} e^{-V/U_T} = n_i e^{-V/U_T}. \tag{5.28}$$

The equilibrium potential satisfies the semilinear elliptic equation

$$\varepsilon_s \Delta V = q(n - p - C) = q(n_i e^{V/U_T} - n_i e^{-V/U_T} - C)$$
$$= q\left(2n_i \sinh \frac{V}{U_T} - C\right) \quad \text{in } \Omega. \tag{5.29}$$

Now, we can define the boundary conditions. We assume that the boundary of the semiconductor domain consists of two parts: one part, called the Dirichlet boundary, on which the particle densities and the potential are prescribed,

$$n = n_D, \quad p = p_D, \quad V = V_D \quad \text{on } \Gamma_D, \tag{5.30}$$

and the other part, the Neumann boundary, which models the insulating boundary segments on which the normal components of the current densities and the electric field vanish,

$$J_n \cdot \eta = J_p \cdot \eta = \nabla V \cdot \eta = 0 \quad \text{on } \Gamma_N.$$

In view of expressions (5.23) and (5.24) for J_n and J_p, this is equivalent to

$$\nabla n \cdot \eta = \nabla p \cdot \eta = \nabla V \cdot \eta = 0 \quad \text{on } \Gamma_N. \tag{5.31}$$

It remains to determine the boundary functions n_D, p_D, and V_D. We make the following assumptions:

- The total space charge vanishes on Γ_D: $n_D - p_D - C(x) = 0$.
- The densities are in equilibrium on Γ_D: $n_D p_D = n_i^2$.
- The boundary potential is the superposition of the *built-in potential* V_{bi} and the applied voltage U: $V_D = V_{\text{bi}} + U$.

Clearly, in thermal equilibrium we have $U = 0$. The built-in potential is the potential corresponding to the equilibrium densities given by (5.27) and (5.28):

$$n_D = n_i e^{V_{bi}/U_T}, \quad p_D = n_i e^{-V_{bi}/U_T}. \tag{5.32}$$

Substituting the equation $n_D - p_D - C = 0$ into $n_D p_D = n_i^2$ leads to the quadratic equation $n_D^2 - C n_D = n_i^2$ whose solution is given by

$$n_D = \frac{1}{2}\left(C + \sqrt{C^2 + 4n_i^2}\right), \quad p_D = \frac{1}{2}\left(-C + \sqrt{C^2 + 4n_i^2}\right) \quad \text{on } \Gamma_D.$$

Thus we infer from (5.32) that

$$V_{bi} = U_T \ln\frac{n_D}{n_i} = U_T \ln\left(\frac{C}{2n_i} + \sqrt{\frac{C^2}{4n_i^2} + 1}\right) = U_T \operatorname{arsinh}\frac{C}{2n_i}. \tag{5.33}$$

Therefore, the thermal equilibrium state (n_{eq}, p_{eq}, V_{eq}) is the (unique) solution of (5.29) with the boundary conditions

$$V_{eq} = V_{bi} \quad \text{on } \Gamma_D, \quad \nabla V_{eq} \cdot \eta = 0 \quad \text{on } \Gamma_N,$$

where V_{bi} is given by (5.33), and

$$n_{eq} = n_i e^{V_{eq}/U_T}, \quad p_{eq} = n_i e^{-V_{eq}/U_T} \quad \text{in } \Omega.$$

Furthermore, the drift-diffusion equations (5.23), (5.24), and (5.25) are solved with the boundary conditions (5.30) and (5.31), where n_D and p_D are defined in (5.32) and $V_D = V_{bi} + U$, with V_{bi} given in (5.33).

5.4 High-Field Models

The drift-diffusion model in Sect. 5.1 is derived under the assumption that the applied voltage is of the order of the thermal voltage whose value, at room temperature, is about 26 mV. High electric fields are thus excluded. In this section, we derive a drift-diffusion model from the semi-classical Boltzmann equation

$$\partial_t f + v(k) \cdot \nabla_x f + \frac{q}{\hbar}\nabla_x V \cdot \nabla_k f = Q(f), \quad x \in \mathbb{R}^3, \ k \in B, \ t > 0,$$

with the initial datum $f(\cdot, \cdot, 0) = f_I$ in $\mathbb{R}^3 \times B$ for high electric fields. The collision operator is given by the low-density approximation (4.22),

$$(Q(f))(x, k, t) = \int_B \sigma(x, k, k')(Mf' - M'f)\, dk',$$

where M is the Maxwellian and the collision cross-section $\sigma(x, k, k')$ is assumed to be symmetric in k and k'. The following derivation is formally also valid for more general scattering terms, but the rigorous results of Poupaud [36] are only available for the low-density operator. We assume a parabolic band structure,

$$v(k) = \hbar k/m^*, \quad k \in \mathbb{R}^3.$$

Then the Maxwellian reads as $M = \frac{1}{2}(2\pi)^{3/2}e^{-|k|^2/2}$. Up to the solution of certain operator equations (see (5.36) below), the derivation formally holds for general band diagrams.

First, we scale the Boltzmann equation. We introduce the average relaxation time τ,

$$\frac{1}{\tau} = \int_{\mathbb{R}^3} \sigma(x,k,k')MM' \, dk',$$

the thermal velocity $v_0 = \sqrt{k_B T_L/m^*}$, and the mean free path $\lambda = v_0 \tau$. The reference length (for instance, the device diameter) is denoted by λ_0, and we define the reference time $\tau_0 = \lambda_0/v_0$ and the reference wave vector $k_0 = m^* v_0/\hbar$. Then the scaled mean free path is given by $\alpha = \lambda/\lambda_0 = \tau/\tau_0$. The electric field $-\nabla_x V$ is scaled by U_T/λ, where $U_T = k_B T_L/q$ is the thermal voltage. Notice that this corresponds to a high-field scaling, since in Sect. 5.1 we have employed the scaling $E_0 = U_T/\lambda_0$ for the electric field. For small scaled mean free paths $\alpha \ll 1$, it holds $U_T/\lambda = E_0/\alpha \gg E_0$. With this scaling the Boltzmann equation with given electric field $E = -\nabla_x V$ becomes

$$\alpha \partial_t f + \alpha k \cdot \nabla_x f - E \cdot \nabla_k f = Q(f). \tag{5.34}$$

The derivation of the high-field model consists of the formal limit $\alpha \to 0$ and a Chapman–Enskog expansion. Our presentation is based on the work [36] by Poupaud. Let f_α be a solution of the Boltzmann equation (5.34). In the first step of the derivation we perform the limit $\alpha \to 0$ in (5.34):

$$E \cdot \nabla_k f + Q(f) = 0, \tag{5.35}$$

where $f = \lim_{\alpha \to 0} f_\alpha$. Thus, we have to deal with the solution of operator equations which are of the form

$$E \cdot \nabla_k f + Q(f) = h, \tag{5.36}$$

where h is some function. In contrast to the solution of the operator equation $Q(f) = h$, which was needed in Sect. 5.1 for the derivation of the (low-field) drift-diffusion model, the above equation may not have a solution. Indeed, the following result holds.

Lemma 5.8. *Let $E \in \mathbb{R}^3$ and let*

$$\omega(k) = \int_{\mathbb{R}^3} \sigma(k,k')M' \, dk' \tag{5.37}$$

be the collision frequency. If there exist constants $K > 0$ and $0 < \gamma < 1$ such that

$$\omega(k) \geq K(1 + |k|^\gamma)^{-1} \quad \text{for all } k \in \mathbb{R}^3, \tag{5.38}$$

or if $\lim_{|k|\to\infty} |k|\omega(k) > |E|$, *the operator equation* (5.35) *has a unique solution* f *satisfying* $\int_{\mathbb{R}^3} f \, dk/4\pi^3 = 1$. *However, if for some constants* $K > 0$ *and* $\gamma > 1$,

$$\omega(k) \leq K(1+|k|^\gamma)^{-1} \quad \text{for all } k \in \mathbb{R}^3,$$

or if $\lim_{|k|\to\infty} |k|\omega(k) < |E|$, *then* (5.35) *has no solution. Moreover, under condition* (5.38), *the nonhomogeneous operator equation* (5.36) *has a solution* f *satisfying* $\int_{\mathbb{R}^3} f \, dk = 0$ *if and only if*

$$\int_{\mathbb{R}^3} h \, dk = 0.$$

For a proof of this lemma, we refer to [36, Thm. 2, Thm. 3, Prop. 2]. It is shown in [37, 38] that if the operator equation (5.35) has no solution, there exist traveling waves. This is called the *runaway phenomenon*. It means that the electrons acquire larger and larger velocities with no upper bound, and the probability density of any fixed velocity tends to zero.

In the following, let $\langle g \rangle = \int_{\mathbb{R}^3} g(k) \, dk/4\pi^3$. By Lemma 5.8, there exists a unique solution f of (5.35) if (5.38) holds, and we can write

$$f(x,k,t) = \rho_0(x,t) F_{E(x,t)},$$

where F_E satisfies $F_E \geq 0$, $\langle F_E \rangle = 1$, and ρ_0 is given by $\rho_0 = \langle f \rangle$. Integrating the Boltzmann equation (5.34) over $k \in \mathbb{R}^3$ gives the moment equation

$$\partial_t \langle f_\alpha \rangle + \text{div}_x \langle k f_\alpha \rangle = 0,$$

since the integrals over the force and collision terms vanish. The limit $\alpha \to 0$ in this equation leads to

$$\partial_t \langle f \rangle + \text{div}_x \langle k f \rangle = 0.$$

Inserting the expression $f = \rho_0 F_E$, we obtain

$$\partial_t \rho_0 + \text{div}_x(\bar{v}(E)\rho_0) = 0, \tag{5.39}$$

where $\bar{v}(E) = \langle k F_E \rangle$ is the averaged velocity. In fact, for general energy bands, we have $\bar{v}(E) = \langle v(k) F_E \rangle$. This is the so-called *high-field drift equation*.

In order to derive a diffusive term, we perform in the second step the Chapman–Enskog expansion $f_\alpha = f + \alpha g_\alpha$. We insert this expansion into the Boltzmann equation (5.34) and employ (5.35):

$$\partial_t f + k \cdot \nabla_x f + \alpha(\partial_t g_\alpha + k \cdot \nabla_x g_\alpha) - (E \cdot \nabla_k g_\alpha + Q(g_\alpha)) = 0. \tag{5.40}$$

In the limit $\alpha \to 0$, we have the nonhomogeneous operator equation

$$E \cdot \nabla_k g + Q(g) = \partial_t f + k \cdot \nabla_x f, \tag{5.41}$$

where $g = \lim_{\alpha\to 0} g_\alpha$. This equation is solvable, according to Lemma 5.8, since the integral of its right-hand side over $k \in \mathbb{R}^3$ vanishes. We write the right-hand side in a more convenient way. Using the drift equation (5.39), we obtain after an elementary computation

$$\partial_t f + k \cdot \nabla_x f = \partial_t(\rho_0 F_E) + k \cdot \nabla_x(\rho_0 F_E) = \rho_0 G_E + \nabla_x \rho_0 \cdot H_E,$$

where

$$H_E = k F_E - \bar{v}(E) F_E, \quad G_E = \partial_t F_E + \mathrm{div}_x H_E.$$

Then (5.41) becomes

$$E \cdot \nabla_k g + Q(g) = \rho_0 G_E + \nabla_x \rho_0 \cdot H_E.$$

This formulation allows us to write the solution g semi-explicitly as

$$g = \rho_0 \phi + \nabla_x \rho_0 \cdot \chi + \rho_1 F_E, \tag{5.42}$$

where ϕ and χ are solutions of the operator equations

$$E \cdot \nabla_k \phi + Q(\phi) = G_E, \quad E \cdot \nabla_k \chi + Q(\chi) = H_E, \tag{5.43}$$

and ρ_1 is not specified. Notice that the necessary solvability condition is satisfied since $\langle H_E \rangle = \bar{v}(E) - \bar{v}(E)\langle F_E \rangle = 0$ and $\langle G_E \rangle = \partial_t \langle F_E \rangle + \mathrm{div}_x \langle H_E \rangle = 0$. By Lemma 5.8, we can choose ϕ and χ such that

$$\langle \phi \rangle = 0, \quad \langle \chi \rangle = 0. \tag{5.44}$$

In the third step, we determine an evolution equation for g. An integration of (5.40) over $k \in \mathbb{R}^3$ yields, in view of the drift equation (5.39),

$$\partial_t \langle g_\alpha \rangle + \mathrm{div}_x \langle k g_\alpha \rangle = 0.$$

In the limit $\alpha \to 0$ this becomes

$$\partial_t \langle g \rangle + \mathrm{div}_x \langle k g \rangle = 0.$$

Taking into account (5.42) and (5.44), we obtain $\langle g \rangle = \rho_1$ and

$$\langle k g \rangle = \rho_0 \langle k \phi \rangle + \langle k \otimes \chi \rangle \nabla_x \rho_0 + \rho_1 \bar{v}(E).$$

Thus, setting $\overline{w}(E) = -\langle k \phi \rangle$ and $D(E) = -\langle k \otimes \chi \rangle$, the function ρ_1 solves the evolution equation

$$\partial_t \rho_1 - \mathrm{div}_x \left(D(E)\nabla_x \rho_0 + \overline{w}(E)\rho_0 - \bar{v}(E)\rho_1 \right) = 0.$$

Adding (5.39) and the above equation, the particle density $n = \rho_0 + \alpha \rho_1$ is a formal solution of

$$\partial_t n - \mathrm{div}_x \left(\alpha D(E)\nabla_x n - (\bar{v}(E) - \alpha \overline{w}(E))n \right) = \mathcal{O}(\alpha^2).$$

We have shown the following result.

Theorem 5.9 (High-field drift-diffusion equations). *Let f_α be a solution of the parabolic-band Boltzmann equation (5.34) and let the collision frequency (5.37) be such that the operator equation (5.35) is solvable (see Lemma 5.8 for a sufficient solvability condition). Then, in the formal limit $\alpha \to 0$, f_α converges to $f = \langle f \rangle F_E$ and $g_\alpha = (f_\alpha - f)/\alpha$ converges to g, satisfying up to order $\mathcal{O}(\alpha^2)$ the* high-field drift-diffusion equations *for the electron density $n = \langle f + \alpha g \rangle$ ($\alpha > 0$)*

$$\partial_t n - \operatorname{div} J_n = 0, \quad J_n = \alpha D(E) \nabla n - \bar{v}_\alpha(E) n, \quad x \in \mathbb{R}^3, \, t > 0,$$

with initial datum $n(\cdot, 0) = \langle f_I \rangle$. The diffusivity tensor $D(E)$ and the averaged velocity $\bar{v}_\alpha(E)$ are defined by

$$D(E) = -\langle k \otimes \chi \rangle, \quad \bar{v}_\alpha(E) = \langle k F_E \rangle + \alpha \langle k \phi \rangle,$$

and ϕ and χ are the unique solutions of the operator equations (5.43).

It is shown in [36] that the diffusivity tensor $D(E)$ is positive definite. The averaged velocity $\bar{v} = \langle k F_E \rangle$ has the following properties.

Lemma 5.10. *Let the collision cross-section $\sigma(x, k, k')$ be isometrically invariant (see Lemma 5.3). Then there exists $\mu_0(x) \geq 0$ such that*

$$\bar{v}(0) = 0, \quad \nabla_E \otimes \bar{v}(0) = -\mu_0(x) \operatorname{Id}.$$

Moreover, if the collision cross-section is constant, $\sigma(x, k, k') = 1/\tau$, the collision operator becomes $Q(f) = (M - f)/\tau$, and hence, $\bar{v}(E) = -\tau E$.

The function μ_0 is the low-field mobility of the drift-diffusion model derived in Sect. 5.1. The above lemma shows that the velocity is, at least close to $E = 0$, decreasing in the electric field, which is physically reasonable.

Proof. We prove first that $\bar{v}(0) = 0$. The operator equation (5.35) with $E = 0$ has the solution $F_0 = M$. This shows that $\bar{v}(0) = \langle k F_0 \rangle = \langle k M \rangle = 0$. Differentiating (5.35) with respect to E gives

$$\nabla_k F_E + Q(\nabla_E F_E) = 0,$$

and at $E = 0$, $Q(\nabla_E F_0) = kM$. It is proved in Lemma 5.3 that there exists $\mu_0(x) \geq 0$ such that $\langle k \otimes \nabla_E F_0 \rangle = -\mu_0 \operatorname{Id}$. Therefore, $\nabla_E \otimes \bar{v}(0) = \langle k \otimes \nabla_E F_0 \rangle = -\mu_0 \operatorname{Id}$.

If the collision operator is of relaxation-time form, we multiply (5.35), written as

$$0 = E \cdot \nabla_k F_E + Q(F_E) = E \cdot \nabla_k F_E + \frac{1}{\tau}(M - F_E),$$

by k and integrate over the wave-vector space. This gives, by integrating by parts,

$$\bar{v}(E) = \langle k F_E \rangle = \langle k M + \tau k(E \cdot \nabla_k F_E) \rangle = -\tau \langle F_E \rangle E = -\tau E.$$

The lemma is proved. $\quad \square$

We can even show that $\overline{v}_\alpha(0) = 0$. Indeed, the operator equation for ϕ at $E = 0$ becomes

$$Q(\phi) = G_0 = \partial_t F_0 + \mathrm{div}_x(kF_0 - \overline{v}(0)F_0) = \partial_t M + \mathrm{div}_x(kM) = 0.$$

Its solution is (a multiple of) the Maxwellian implying that $\overline{w}(0) = -\langle k\phi \rangle = -\langle kM \rangle = 0$ and hence $\overline{v}_\alpha(0) = \overline{v}(0) - \alpha\overline{w}(0) = 0$. On the other hand, the computation of $\nabla_E \otimes \overline{v}_\alpha(0)$ seems to be more involved.

For constant collision cross-sections and $\alpha = 0$, we recover the drift term of the drift-diffusion model of Sect. 5.1, $\overline{v}_0(E)n = -\tau En = \tau n\nabla V$. Defining the electron mobility $\mu(E)$ by $\overline{v}_\alpha(E) = -\mu(E)E$, this result shows that $\mu(E)$ is constant if $\alpha = 0$.

Unfortunately, it seems to be difficult to derive an explicit dependence of $\overline{v}_\alpha$ on the field $E \neq 0$ for general scattering operators. Assuming that the distribution function can be decomposed as $f = f_0 + (E \cdot k)f_1$, for some functions f_0 and f_1, and closing the hierarchy of moment equations by making an ansatz for the heat flux, Hänsch et al. [39] have derived the following mobility model:

$$\mu(E) = \frac{2\mu_0}{1 + \sqrt{1 + (2\mu_0|E|/v_{\mathrm{sat}})^2}}, \qquad (5.45)$$

where μ_0 is the low-field mobility and v_{sat} the saturation velocity, satisfying

$$\lim_{|E|\to\infty} |\overline{v}_\alpha(E)| = \lim_{|E|\to\infty} |\mu(E)E| = v_{\mathrm{sat}}.$$

The diffusivity is often computed from the Einstein relation $D = \mu k_B T_L$. In [40], the high-field mobility and diffusivity were derived from an energy-transport model (see Sect. 6.1), yielding the following scaled high-field model:

$$\partial_t n - \mathrm{div}\left(\frac{\mu_0}{2 - \mu(E)/\mu_0}\nabla n + \mu(E)En\right) = 0,$$

where the mobility $\mu(E)$ is given by (5.45). For small fields, the diffusion coefficient converges to the low-field electron mobility μ_0,

$$\frac{\mu_0}{2 - \mu(E)/\mu_0} \to \mu_0 \quad \text{as } |E| \to 0,$$

whereas in the high-field limit, the mobility is reduced:

$$\frac{\mu_0}{2 - \mu(E)/\mu_0} \to \frac{\mu_0}{2} \quad \text{as } |E| \to \infty.$$

For related high-field drift-diffusion or mobility models, we refer, for instance, to [41–43].

The above approach for a linear collision operator was revised by Cercignani et al. in [44, 45], including the coupling with the Poisson equation. In Goudon, Nieto, and co-workers [46, 47], the analysis of the high-field limit coupled with the

Poisson equation was carried out for the Fokker–Planck equation. High-field asymptotics for degenerate semiconductors were derived by Ben Abdallah and Chaker [48]. Finally, a nonlinear collision operator similar to (4.9) was considered by Ben Abdallah, Chaker, and Schmeiser [49].

5.5 Drift-Diffusion Models Using Fermi–Dirac Statistics

In the previous sections, we have assumed that the equilibrium distribution is given by the Maxwellian. In this section, we derive drift-diffusion models employing Fermi–Dirac statistics. More precisely, we suppose that the collision operator is linear and that its kernel consists of the Fermi–Dirac distribution functions

$$F(x,k,t) = \frac{1}{\eta + \exp(-\lambda_0(x,t) + |k|^2/2)}, \quad k \in \mathbb{R}^3, \tag{5.46}$$

where $\eta \geq 0$ is the degeneracy parameter. The Fermi–Dirac distribution is obtained for the choice $\eta = 1$, whereas we recover for $\eta = 0$ the Maxwellian. The function λ_0 is chosen such that $\langle F \rangle = \int_{\mathbb{R}^3} F \, dk/4\pi^3 = n$ is fulfilled for a given particle density n. A simple scattering operator satisfying the above hypothesis is the relaxation-time operator

$$Q(f) = \frac{1}{\tau}(F[f] - f), \tag{5.47}$$

where $F = F[f]$ is such that $\langle F[f] \rangle = \langle f \rangle$ and $\tau > 0$. The kernel of the scattering operator (4.2) also consists only of Fermi–Dirac distributions, according to Proposition 4.1.

As in Sect. 5.1, we assume that the energy band is given by the parabolic band approximation. The following arguments are also valid for more general energy bands but the formulas would be less explicit.

In order to derive the drift-diffusion equations, let (f_α, V_α) be a solution of the scaled Boltzmann–Poisson system

$$\alpha^2 \partial_t f_\alpha + \alpha (k \cdot \nabla_x f_\alpha + \nabla_x V_\alpha \cdot \nabla_k f_\alpha) = Q(f_\alpha), \quad \lambda_D^2 \Delta V_\alpha = \langle f_\alpha \rangle - C(x), \tag{5.48}$$

for $x, k \in \mathbb{R}^3$, $t > 0$. The initial condition is $f_\alpha(\cdot,\cdot,0) = f_I$ in $\mathbb{R}^3 \times \mathbb{R}^3$. The derivation is based on three steps. First, we let formally $\alpha \to 0$ in the above equations, leading to

$$Q(f) = 0, \quad \lambda_D^2 \Delta V = \langle f \rangle - C(x),$$

where $f = \lim_{\alpha \to 0} f_\alpha$ and $V = \lim_{\alpha \to 0} V_\alpha$. By assumption, f equals the Fermi–Dirac function $F = F[f]$. In the second step, we insert the Chapman–Enskog expansion $f_\alpha = F[f_\alpha] + \alpha g_\alpha$ in the Boltzmann equation and employ $Q(F[f_\alpha]) = 0$:

$$\alpha \partial_t (F[f_\alpha] + \alpha g_\alpha) + (k \cdot \nabla_x F[f_\alpha] + \nabla_x V \cdot \nabla_k F[f_\alpha])$$
$$+ \alpha (k \cdot \nabla_x g_\alpha + \nabla_x V \cdot \nabla_k g_\alpha) = Q(g_\alpha).$$

In the limit $\alpha \to 0$ we obtain

$$Q(g) = k \cdot \nabla_x F + \nabla_x V \cdot \nabla_k F = \nabla_x(\lambda_0 - V) \cdot kF(1 - \eta F).$$

We assume that there exists a solution $h = (h_1, h_2, h_3)^\top$ of the operator equations

$$Q(h_i) = k_i F(1 - \eta F), \quad i = 1, 2, 3, \tag{5.49}$$

which is unique in the orthogonal complement of the kernel of Q. Then we can write

$$g = \nabla_x(\lambda_0 - V) \cdot h + cF, \quad c \in \mathbb{R}. \tag{5.50}$$

In the third step, we insert the Chapman–Enskog expansion in the moment equation

$$\partial_t \langle f_\alpha \rangle + \alpha^{-1} \mathrm{div}_x \langle k f_\alpha \rangle = 0,$$

observe that the integral $\langle kF[f_\alpha] \rangle$ vanishes, and pass to the limit $\alpha \to 0$. The result reads as

$$\partial_t \langle F \rangle + \mathrm{div}_x \langle kg \rangle = 0.$$

The first integral $\langle F \rangle$ is the electron density, and the second integral $J_n = -\langle kg \rangle$ can be interpreted as the particle current density.

Theorem 5.11 (Drift-diffusion equations using Fermi–Dirac statistics I). *We assume that the kernel $N(Q)$ of the collision operator consists of Fermi–Dirac distributions and that the operator equation (5.49) is uniquely solvable in the orthogonal complement $N(Q)^\perp$. Furthermore, let (f_α, V_α) be a solution of the Boltzmann–Poisson system (5.48). Then the formal limit functions $f = \lim_{\alpha \to 0} f_\alpha$ and $V = \lim_{\alpha \to 0} V_\alpha$ satisfy the drift-diffusion equations*

$$\partial_t n - \mathrm{div}\, J_n = 0, \quad J_n = D\nabla(\lambda_0 - V), \quad x \in \mathbb{R}^3, \ t > 0, \tag{5.51}$$

$$\lambda_D^2 \Delta V = n - C(x), \quad n = \eta^{-1} N F_{1/2}(\lambda_0 + \log \eta), \tag{5.52}$$

where $N = 2(2\pi)^{-3/2}$ is a normalization constant, $F_{1/2}$ is the Fermi integral introduced in (1.59), and the diffusion matrix D is given by $D = -\int_{\mathbb{R}^3} k \otimes h \, dk/4\pi^3$, where h is a solution of (5.49). Finally, the initial datum of n is given by $n(\cdot, 0) = \int_{\mathbb{R}^3} f_I \, dk/4\pi^3$.

Proof. It remains to compute the expressions for the densities n and J_n. Since g is given by (5.50), the formula for J_n follows immediately from the definition of g, since $\langle kF \rangle = 0$:

$$J_n = -\langle kg \rangle = -\langle k \otimes h \rangle \nabla_x(\lambda_0 - V).$$

Next, we compute the expression for n, employing spherical coordinates and the substitution $x = r^2/2$:

$$n = \frac{1}{4\pi^3} \int_{\mathbb{R}^3} \frac{dk}{\eta + e^{-\lambda_0 + |k|^2/2}} = \frac{1}{\pi^2} \int_0^\infty \frac{r^2 dr}{\eta + e^{-\lambda_0 + r^2/2}}$$

$$= \frac{1}{\eta \pi^2} \int_0^\infty \frac{\sqrt{2x} \, dx}{1 + e^{-\log\eta - \lambda_0 + x}} = \frac{2}{\eta(2\pi)^{3/2}} F_{1/2}(\lambda_0 + \log\eta).$$

The conclusion follows. □

The diffusion matrix can be computed explicitly if the collision operator is given by the relaxation-time approximation (5.47).

Proposition 5.12 (Drift-diffusion equations using Fermi–Dirac statistics II). *Let the collision operator be given by* (5.47). *Then the drift-diffusion equations* (5.51) *and* (5.52) *can be formulated as*

$$\partial_t n - \operatorname{div} J_n = 0, \quad J_n = \tau n \nabla(\lambda_0 - V), \quad x \in \mathbb{R}^3, \, t > 0, \tag{5.53}$$

$$\lambda_D^2 \Delta V = n - C(x), \quad n = \eta^{-1} N F_{1/2}(\lambda_0 + \log\eta). \tag{5.54}$$

Proof. We have to show that $D = \tau n \operatorname{Id}$. Since Q is of the special form (5.47), the solution of (5.49) is given by

$$h_i = F - \tau k_i F(1 - \eta F), \quad i = 1, 2, 3.$$

Thus, the diffusion coefficients become

$$D_{ij} = \tau \int_{\mathbb{R}^3} k_i k_j F(1 - \eta F) \frac{dk}{4\pi^3}.$$

Passing to spherical coordinates $k = r\omega$ with $|\omega| = 1$, we have

$$(D_{ij}) = \frac{\tau}{4\pi^3} \int_0^\infty \int_{|\omega|=1} \frac{e^{-\lambda_0 + r^2/2}}{(\eta + e^{-\lambda_0 + r^2/2})^2} r^4 \omega \otimes \omega \, d\omega \, dr.$$

A computation shows that the integral of $\omega \otimes \omega$ over $\{|\omega| = 1\}$ has the value $(4\pi/3) \operatorname{Id}$, and hence

$$(D_{ij}) = \frac{\tau}{3\pi^2} \int_0^\infty \frac{e^{-\lambda_0 + r^2/2}}{(\eta + e^{-\lambda_0 + r^2/2})^2} r^4 \, dr \operatorname{Id}.$$

Thus, the diffusion matrix is diagonal with identical entries on the diagonal. We write $D_0 = D_{ii}$ for $i = 1, 2, 3$. The substitution $x = r^2/2$ yields

$$D_0 = \frac{2^{3/2} \tau}{3\pi^2 \eta} \int_0^\infty \frac{e^{-z+x}}{(1 + e^{-z+x})^2} x^{3/2} \, dx, \quad z = \log\eta + \lambda_0.$$

Observing that

$$\frac{d}{dx} \frac{1}{1 + e^{-z+x}} = -\frac{e^{-z+x}}{(1 + e^{-z+x})^2},$$

an integration by parts shows that

$$D_0 = \frac{2^{3/2}\tau}{3\pi^2\eta} \int_0^\infty \frac{1}{1+e^{-z+x}} \frac{3}{2} x^{1/2}\, dx = \frac{2\tau}{\eta(2\pi)^{3/2}} F_{1/2}(\lambda_0 + \log\eta) = \tau n,$$

which proves the proposition. □

The drift-diffusion model (5.53) and (5.54) was formulated by Bonch-Bruevich and Kalashnikov [50] and mathematically analyzed by Gajewski and Gröger [22]. A numerical treatment can be found in [51]. Golse and Poupaud [4] derived a drift-diffusion model incorporating Fermi–Dirac statistics from the Boltzmann equation which is similar to the above equations.

The relaxation-time operator (5.47) is a strong simplification. From a physical point of view, it is preferable to assume that the collision cross-section in the general collision operator (4.2) is constant such that we obtain

$$Q(f) = \frac{1}{\tau} \int_{\mathbb{R}^3} \left(Mf'(1-\eta f) - M'f(1-\eta f') \right)\, dk'.$$

It is essentially shown in Proposition 4.1 that the kernel of this operator consists of the Fermi–Dirac distributions. The diffusion approximation for a given potential V leads to the following drift-diffusion model [8]:

$$\partial_t n - \operatorname{div} J_n = 0, \quad J_n = D\nabla(\lambda_0 - V),$$

where the relation between n and λ_0 is as above, the diffusion coefficient is given by

$$D = Ne^{\lambda_0}\left(1 + \frac{\rho_0}{n}F_{-1/2}(\lambda_0)\right),$$

ρ_0 is a constant, and $F_{-1/2}$ is the Fermi integral with index $-1/2$.

There are two interesting limiting situations for the model (5.53) and (5.54). When $\eta \to 0$, the Fermi–Dirac distribution (5.46) reduces to the Maxwellian $M = e^{\lambda_0 - |k|^2/2}$, and we expect that the corresponding drift-diffusion model is the same as in Sect. 5.1. In some sense, this corresponds to a *low-density limit* (see Remark 1.12). On the other hand, when η becomes very large, $\eta \gg 1$, we claim that the limiting model corresponds to a *high-density limit*. These two cases are analyzed in the following examples.

Example 5.13 (Low-density limit). We can set $\eta = 0$ in the following integral:

$$\eta^{-1} F_{1/2}(\lambda_0 + \log\eta) = \frac{2}{\sqrt{\pi}} \int_0^\infty \frac{\sqrt{x}\,dx}{\eta + e^{-\lambda_0 + x}},$$

yielding

$$n = \frac{2N}{\sqrt{\pi}} \int_0^\infty \sqrt{x}e^{\lambda_0 - x}\, dx = \frac{2N}{\sqrt{\pi}} e^{\lambda_0}\Gamma\left(\frac{3}{2}\right) = Ne^{\lambda_0},$$

where Γ is the Gamma function introduced in (1.60). Thus,

$$J_n = \tau n \nabla(\lambda_0 - V) = \tau(\nabla n - n \nabla V).$$

The balance equation $\partial_t n - \mathrm{div}\, J_n = 0$ together with the above formula for the current density is exactly the drift-diffusion model derived in Sect. 5.1 (see (5.14)). □

Example 5.14 (High-density limit). For large values of η, $\eta \gg 1$, we employ the asymptotics $F_{1/2}(z) \sim (4/3\sqrt{\pi})z^{3/2}$ $(z \to \infty)$. Then we can write the electron density approximately as

$$n = \frac{1}{\eta}\frac{2\sqrt{2}}{3\pi^2}(\lambda_0 + \log \eta)^{3/2}.$$

Since

$$\nabla(n^{5/3}) = \frac{1}{\eta^{5/3}}\left(\frac{2\sqrt{2}}{3\pi^2}\right)^{5/3}\frac{5}{2}(\lambda_0 + \log \eta)^{3/2}\nabla\lambda_0 = \frac{1}{\eta^{2/3}}\left(\frac{2\sqrt{2}}{3\pi^2}\right)^{2/3}\frac{5}{2}n\nabla\lambda_0,$$

the current density can be formulated approximately as

$$J_n = \tau n \nabla(\lambda_0 - V) = \tau(\nabla(N_\eta n^{5/3}) - n\nabla V),$$

where $N_\eta = (9\pi^4\eta^2/250)^{1/3}$. Thus, the *high-density drift-diffusion equations* read as follows:

$$\partial_t n - \tau\mathrm{div}\left(\nabla(N_\eta n^{5/3}) - n\nabla V\right) = 0, \quad \lambda_D^2\Delta V = n - C(x), \quad x \in \mathbb{R}^3,\, t > 0,$$

with initial conditions for n. This model was rigorously derived from the drift-diffusion model (5.53) and (5.54) in [52] and mathematically analyzed in [53–55]. In these papers, the above model is referred to as the *degenerate drift-diffusion equations* since the diffusion term $\Delta(n^{5/3})$ is mathematically of degenerate type and the model is valid for degenerate semiconductor materials. A numerical discretization in one and two space dimensions was presented in [56, 57] employing mixed finite elements and in [58, 59] using a finite-volume approximation. The high-density equations coupled to a heat equation with power dissipation were studied by Guan and Wu [60]. We mention that Poupaud and Schmeiser [8] rigorously derived a high-density model from the Boltzmann equation, but the model differs from the above equations.

Interestingly, the diffusion term $\nabla(n^{5/3})$ can be interpreted thermodynamically. Treating the electrons as particles of an ideal gas, its pressure is given by $P = nT$, where T is the particle temperature. In the adiabatic and hence isentropic case, for particles without spin, the temperature depends on the particle density, $T = T_0 n^{2/3}$ [61], which implies that $\nabla P = T_0 \nabla(n^{5/3})$. □

Figure 5.1 summarizes the drift-diffusion models derived in this and the previous sections.

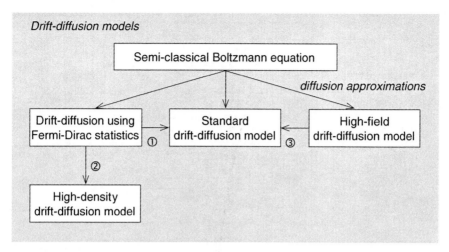

Fig. 5.1 Relations between the drift-diffusion models of Sects. 5.1, 5.4, and 5.5. The arrow (1) denotes the low-density limit $\eta \to 0$, where η is the degeneracy parameter. The arrow (2) signifies the high-density limit $\eta \gg 1$. The arrow (3) denotes the low-field limit $|E| \to 0$, where $\mu(E) \to \mu_0$

References

1. W. van Roosbroeck. Theory of flow of electron and holes in germanium and other semiconductors. *Bell Syst. Techn. J.* 29 (1950), 560–607.
2. F. Poupaud. Diffusion approximation of the linear semiconductor Boltzmann equation: analysis of boundary layers. *Asympt. Anal.* 4 (1991), 293–317.
3. P. Markowich, C. Ringhofer, and C. Schmeiser. *Semiconductor Equations.* Springer, Vienna, 1990.
4. F. Golse and F. Poupaud. Limite fluide des équations de Boltzmann des semiconducteurs pour une statistique de Fermi–Dirac. *Asympt. Anal.* 6 (1992), 135–160.
5. E. Zeidler. *Nonlinear Functional Analysis and Its Applications*, Vol. II. Springer, New York, 1990.
6. N. Ben Abdallah and M. Tayeb. Diffusion approximation for the one dimensional Boltzmann–Poisson system. *Discrete Contin. Dyn. Sys. B* 4 (2004), 1129–1142.
7. N. Masmoudi and M. Tayeb. Diffusion limit of a semiconductor Boltzmann–Poisson system. *SIAM J. Math. Anal.* 38 (2007), 1788–1807.
8. F. Poupaud and C. Schmeiser. Charge transport in semiconductors with degeneracy effects. *Math. Meth. Appl. Sci.* 14 (1991), 301–318.
9. P. Markowich and C. Schmeiser. Relaxation time approximation for electron–phonon interaction in semiconductors. *Math. Models Meth. Appl. Sci.* 5 (1995), 519–527.
10. P. Markowich, F. Poupaud, and C. Schmeiser. Diffusion approximation of nonlinear electron phonon collision mechanisms. *RAIRO Modél. Math. Anal. Numér.* 29 (1995), 857–869.
11. N. Ben Abdallah and M. Tayeb. Diffusion approximation and homogenization of the semiconductor Boltzmann equation. *SIAM Multiscale Model. Simul.* 4 (2005), 896–914.
12. F. Poupaud and J. Soler. Parabolic limit and stability of the Vlasov-Fokker–Planck system. *Math. Models Meth. Appl. Sci.* 10 (2000), 1027–1045.
13. H. Gummel. A self-consistent iterative scheme for one-dimensional steady state transistor calculations. *IEEE Trans. Electr. Devices* ED-11 (1964), 455–465.
14. D. Scharfetter and H. Gummel. Large signal analysis of a silicon Read diode oscillator. *IEEE Trans. Electr. Devices* ED-16 (1969), 64–77.

15. F. Brezzi, L. Marini, and P. Pietra. Méthodes d'éléments finis mixtes et schéma de Scharfetter–Gummel. *C. R. Acad. Sci. Paris, Sér. I* 305 (1987), 599–604.
16. F. Brezzi, L. Marini, and P. Pietra. Two-dimensional exponential fitting and applications to drift-diffusion models. *SIAM J. Numer. Anal.* 26 (1989), 1342–1355.
17. C. Chainais-Hillairet, J.-G. Liu, and Y.-J. Peng. Finite volume scheme for multi-dimensional drift-diffusion equations and convergence analysis. *Math. Model. Numer. Anal.* 37 (2003), 319–338.
18. R. Sacco and F. Saleri. Mixed finite volume methods for semiconductor device simulation. *Numer. Meth. Part. Diff. Eqs.* 13 (1997), 215–236.
19. M. Mock. On equations describing steady-state carrier distributions in a semiconductor device. *Commun. Pure Appl. Math.* 25 (1972), 781–792.
20. M. Mock. An initial value problem from semiconductor device theory. *SIAM J. Math. Anal.* 5 (1974), 597–612.
21. H. Gajewski and K. Gröger. On the basic equations for carrier transport in semiconductors. *J. Math. Anal. Appl.* 113 (1986), 12–35.
22. H. Gajewski and K. Gröger. Semiconductor equations for variable mobilities based on Boltzmann statistics or Fermi–Dirac statistics. *Math. Nachr.* 140 (1989), 7–36.
23. K. Gröger and J. Rehberg. Uniqueness for the two-dimensional semiconductor equations in case of high carrier densities. *Math. Z.* 213 (1993), 523–530.
24. J. Jerome. *Analysis of Charge Transport. A Mathematical Study of Semiconductor Devices.* Springer, Berlin, 1996.
25. P. Markowich. *The Stationary Semiconductor Device Equations.* Springer, Vienna, 1986.
26. F. Brezzi, L. Marini, P. Markowich, and P. Pietra. On some numerical problems in semiconductor device simulation. In: G. Toscani, V. Boffi, and S. Rionero (eds.), *Mathematical Aspects of Fluid and Plasma Dynamics (Salice Terme, 1988)*, Lecture Notes in Math. 1460, 31–42. Springer, Berlin, 1991.
27. F. Brezzi, L. Marini, S. Micheletti, P. Pietra, R. Sacco, and S. Wang. Discretization of semiconductor device problems. In: W. Schilders and E. ter Maten, *Handbook of Numerical Analysis, Vol. 13: Numerical Methods in Electromagnetics*, 317–441. North-Holland, Amsterdam, 2005.
28. J. Jerome. The approximation problem for drift-diffusion systems. *SIAM Rev.* 37 (1995), 552–572.
29. W. Hänsch. *The Drift-Diffusion Equation and Its Applications in MOSFET Modeling.* Springer, Vienna, 1991.
30. S. Selberherr. *Analysis and Simulation of Semiconductor Devices.* Springer, Vienna, 1984.
31. A. Yamnahakki. Second order boundary conditions for the drift-diffusion equations of semiconductors. *Math. Models Meth. Appl. Sci.* 5 (1995), 429–455.
32. S. Taguchi and A. Jüngel. Kinetic theory of a two-surface problem of electron flow in a semiconductor. *J. Stat. Phys.* 130 (2007), 313–342.
33. K. Brennan. *The Physics of Semiconductors.* Cambridge University Press, Cambridge, 1999.
34. M. Grundmann. *The Physics of Semiconductors.* Springer, Berlin, 2006.
35. T. Goudon, V. Miljanovic, and C. Schmeiser. On the Shockley–Read–Hall model: generation–recombination in semiconductors. *SIAM J. Appl. Math.* 67 (2007), 1183–1201.
36. F. Poupaud. Runaway phenomena and fluid approximation under high fields in semiconductor kinetic theory. *Z. Angew. Math. Mech.* 72 (1992), 359–372.
37. L. Arlotti and G. Frosali. Runaway particles for a Boltzmann-like transport equation. *Math. Models Meth. Appl. Sci.* 2 (1992), 203–221.
38. G. Frosali and C. Van der Mee. Scattering theory in the linear transport theory of particle swarms. *J. Stat. Phys.* 56 (1989), 139–148.
39. W. Hänsch and M. Miura-Mattausch. The hot-electron problem in small semiconductor devices. *J. Appl. Phys.* 60 (1986), 650–656.
40. P. Degond and A. Jüngel. High-field approximations of the energy-transport model for semiconductors with non-parabolic band structure. *Z. Angew. Math. Phys.* 52 (2001), 1053–1070.
41. E. Kan, U. Ravaioli, and T. Kerkhoven. Calculation of velocity overshoot in submicron devices using an augmented drift-diffusion model. *Solid State Electr.* 34 (1991), 995–999.

42. K. Thornber. Current equations for velocity overshoot. *IEEE Electr. Device Letters* 3 (1983), 69–71.
43. N. Zakhleniuk. Nonequilibrium drift-diffusion transport in semiconductors in presence of strong inhomogeneous electric fields. *Appl. Phys. Letters* 89 (2006), 252112.
44. C. Cercignani, I. Gamba, and C. Levermore. High-field approximations to a Boltzmann–Poisson system and boundary conditions in a semiconductor. *Appl. Math. Letters* 10 (1997), no. 4, 111–117.
45. C. Cercignani, I. Gamba, and C. Levermore. A drift-collision balance for a Boltzmann–Poisson system in bounded domains. *SIAM J. Appl. Math.* 61 (2001), 1932–1958.
46. T. Goudon, J. Nieto, F. Poupaud, and J. Soler. Multidimensional high-field limit of the electrostatic Vlasov-Poisson-Fokker–Planck system. *J. Diff. Eqs.* 213 (2005), 418–442.
47. J. Nieto, F. Poupaud, and J. Soler. High-field limit for the Vlasov-Poisson-Fokker–Planck system. *Arch. Rat. Mech. Anal.* 158 (2001), 29–59.
48. N. Ben Abdallah and H. Chaker. The high field asymptotics for degenerate semiconductors. *Math. Models Meth. Appl. Sci.* 11 (2001), 1253–1272.
49. N. Ben Abdallah, H. Chaker, and C. Schmeiser. The high field asymptotics for a fermionic Boltzmann equation: entropy solutions and kinetic shock profiles. *J. Hyperb. Diff. Eqs.* 4 (2007), 679–704.
50. V. Bonch-Bruevich and S. Kalashnikov. *Halbleiterphysik.* VEB Deutscher Verlag der Wissenschaften, Berlin, 1982.
51. J. Viallet and S. Mottet. Transient simulation of heterostructure. In: J. Miller et al. (eds.), *NASECODE IV Conference Proceedings.* Dublin, Boole Press, 1985.
52. A. Jüngel. Asymptotic analysis of a semiconductor model based on Fermi–Dirac statistics. *Math. Meth. Appl. Sci.* 19 (1996), 401–424.
53. J. I. Díaz, G. Galiano, and A. Jüngel. On a quasilinear degenerate system arising in semiconductor theory. Part I: existence and uniqueness of solutions. *Nonlin. Anal.: Real-World Appl.* 2 (2001), 305–336.
54. A. Jüngel. On the existence and uniqueness of transient solutions of a degenerate nonlinear drift-diffusion model for semiconductors. *Math. Models Meth. Appl. Sci.* 4 (1994), 677–703.
55. A. Jüngel. A nonlinear drift-diffusion system with electric convection arising in semiconductor and electrophoretic modeling. *Math. Nachr.* 185 (1997), 85–110.
56. A. Jüngel. Numerical approximation of a drift-diffusion model for semiconductors with nonlinear diffusion. *Z. Angew. Math. Mech.* 75 (1995), 783–799.
57. A. Jüngel and P. Pietra. A discretization scheme for a quasi-hydrodynamic semiconductor model. *Math. Models Meth. Appl. Sci.* 7 (1997), 935–955.
58. C. Chainais-Hillairet and F. Filbet. Asymptotic behavior of a finite volume scheme for the transient drift-diffusion model. *IMA J. Numer. Anal.* 27 (2007), 689–716.
59. C. Chainais-Hillairet and Y.-J. Peng. Finite volume approximation for degenerate drift-diffusion system in several space dimensions. *Math. Models Meth. Appl. Sci.* 14 (2004), 461–481.
60. P. Guan and B. Wu. Existence of weak solutions to a degenerate time-dependent semiconductor equations with temperature effects. *J. Math. Anal. Appl.* 332 (2007), 367–380.
61. R. Courant and K. Friedrichs. *Supersonic Flow and Shock Waves.* Interscience, New York, 1967.

Chapter 6
Energy-Transport Equations

The drift-diffusion equations are derived by the moment method by employing only the zeroth-order moment $\langle M \rangle = \int_B M \, dk / 4\pi^3$, where the Maxwellian M describes the equilibrium state. As explained in Sect. 2.4, we obtain more general diffusion equations by taking into account higher-order moments. The energy-transport equations are derived by choosing the moments $n = \langle M \rangle$ (particle density) and $ne = \langle E(k)M \rangle$ (energy density), where $E(k)$ is the energy band. The results of Sect. 2.4 are valid only for a simple BGK collision operator. In this chapter, we will assume more realistic scattering terms including elastic, carrier–carrier, and inelastic collision processes. In the following we proceed as in [1] and [2].

6.1 Derivation from the Boltzmann Equation

We consider the semi-classical Boltzmann equation

$$\partial_t f + v(k) \cdot \nabla_x f + \frac{q}{\hbar} \nabla_x V \cdot \nabla_k f = Q(f), \quad x \in \mathbb{R}^3, \ k \in B, \ t > 0,$$

with the initial datum $f(\cdot, \cdot, 0) = f_I$ in $\mathbb{R}^3 \times B$. We assume that the collision operator models elastic collisions of phonons and ionized impurites, electron–electron scattering, and inelastic collisions,

$$Q(f) = Q_{el}(f) + Q_{ee}(f) + Q_{in}(f),$$

where Q_{el} and Q_{ee} are given by

$$(Q_{el}(f))(x,k,t) = \int_B \sigma_{el}(x,k,k')\delta(E(k') - E(k))(f' - f)\, dk', \tag{6.1}$$

$$(Q_{ee}(f))(x,k,t) = \int_{B^3} \sigma_{ee}(x,k,k',k_1,k_1')\delta(E' + E_1' - E - E_1) \tag{6.2}$$
$$\times \left(f'f_1'(1-f)(1-f_1) - ff_1(1-f')(1-f_1') \right) dk' \, dk_1 \, dk_1',$$

Jüngel, A.: *Energy-Transport Equations.* Lect. Notes Phys. **773**, 129–155 (2009)
DOI 10.1007/978-3-540-89526-8_6

(see (4.11) and (4.12)), where $f' = f(x, k', t)$, $f_1 = f(x, k_1, t)$, and $f'_1 = f(x, k'_1, t)$. We suppose that the scattering rates σ_{el} and σ_{ee} are symmetric in the sense of (4.29) and (4.31), respectively. The inelastic collision operator models, for instance, inelastic phonon scattering with transition rate (4.7). For the derivation of the energy-transport model, we do not need a specific form for Q_{in}, but we assume that inelastic collisions conserve mass, i.e., $\int_B Q_{in}(f) \, dk / 4\pi^3 = 0$ for all functions f. Later, we will specify Q_{in} by modeling inelastic phonon scattering (see (6.16) below).

Scaling of the Boltzmann equation. The Boltzmann equation is scaled similarly as in the previous chapter. The reference wave vector k_0, velocity v_0, and voltage U_T are chosen as in Sect. 5.1. The mean free paths of elastic, electron–electron, and inelastic collisions are denoted by λ_{el}, λ_{ee}, and λ_{in}, respectively. We assume that the typical time is given by the time between two inelastic collisions, $\tau_0 = \lambda_{in}/v_0$, and that the typical length scale is the geometric average of the elastic and inelastic mean free paths, $\lambda_0 = \sqrt{\lambda_{el}\lambda_{in}}$. Furthermore, we suppose that the length scale of the carrier–carrier interactions is of the order of the reference length, from which $\lambda_{ee} = \lambda_0$ follows. Then, introducing the parameter $\alpha = \sqrt{\lambda_{el}/\lambda_{in}}$, we obtain the dimensionless Boltzmann equation

$$\alpha^2 \partial_t f + \alpha \left(v(k) \cdot \nabla_x f + \nabla_x V \cdot \nabla_k f \right) = Q_{el}(f) + \alpha Q_{ee}(f) + \alpha^2 Q_{in}(f), \qquad (6.3)$$

where $x \in \mathbb{R}^3$, $k \in B$, and $t > 0$. Elastic collisions are assumed to be dominant compared to electron–electron and inelastic scattering such that $\alpha \ll 1$.

In the literature, energy-transport models were derived employing different scalings. Ben Abdallah, Degond, and Génieys have performed a Hilbert expansion for the scaled Boltzmann equation (6.3) in which the carrier–carrier scattering term is assumed to be of order one [3]. However, elastic scattering in semiconductors occurs usually more frequently than carrier–carrier collisions and thus, the order of the electron–electron collision term should be less than that of the elastic scattering integral. Moreover, this scaling leads in the diffusion approximation to a diffusion matrix which involves the solution of the operator equation $(Q_{el} + L)(f) = g$, where $L = DQ_{ee}(F)$ denotes the linearization of Q_{ee} at the equilibrium state F. It can be shown that this equation has a solution, but generally, it cannot be computed explicitly. Therefore, Ben Abadallah and Degond have proposed a scaling in which the electron–electron collisions are of the same order α^2 as the inelastic scattering [1]. The diffusion approximation then yields the so-called spherical harmonics expansion (SHE) model for a distribution function which depends on the energy, but not on the wave vector anymore (see Chap. 7). A further diffusion limit gives the energy-transport model. Here, the computation of the diffusion matrix requires the solution of an equation for the elastic collision operator only, $Q_{el}(f) = g$. Under certain assumptions on the energy band and the scattering rate, this equation can be explicitly solved providing explicit expressions for the diffusion coefficients. The advantages of the above scaling, leading to (6.3) and suggested by Degond, Levermore, and Schmeiser [2], are that the energy-transport model can be derived without passing through the SHE model and that the computation of the diffusion matrix is based on the solution of the equation involving Q_{el} only.

Properties of the collision operators. We have already shown some properties of the elastic and electron–electron collision terms in Sect. 4.2 (see Propositions 4.5 and 4.6). We recall that Q_{el} and Q_{ee} conserve mass and energy, the kernel of Q_{el} consists of functions depending on the energy only, and the kernel of Q_{ee} consists of the Fermi–Dirac distributions. Moreover, $-Q_{el}$ is self-adjoint on $L^2(B)$. For the diffusion approximation of the Boltzmann equation, we need two additional properties.

Lemma 6.1. *The equation $Q_{el}(f) = h$ has a solution if and only if*

$$\int_B h(k)\delta(E(k) - \varepsilon)\,dk = 0 \quad for\ all\ \varepsilon \in R(E), \tag{6.4}$$

where $R(E)$ is the range of the energy band $E(k)$.

We recall that integrals involving delta distributions can be reformulated by means of the coarea formula to integrals over isoenergetic surfaces (see Sect. 4.2).

Proof. We show that the orthogonal complement $N(Q_{el})^\perp$ in $L^2(B)$ only consists of functions satisfying (6.4). Then the self-adjointness of $-Q_{el}$ and the Fredholm alternative (Lemma 5.1) give the conclusion. Let $h \in N(Q_{el})^\perp$ and $f \in N(Q_{el})$. Then, by Proposition 4.5 in Sect. 4.2, we can write $f(k) = F(E(k))$ for some function F. By definition of the delta distribution, we compute

$$0 = \int_B h(k)f(k)\,dk = \int_B h(k)F(E(k))\,dk$$
$$= \int_B h(k) \int_{\mathbb{R}} F(\varepsilon)\delta(E(k) - \varepsilon)\,d\varepsilon\,dk$$
$$= \int_{R(E)} \left(\int_B h(k)\delta(E(k) - \varepsilon)\,dk \right) F(\varepsilon)\,d\varepsilon.$$

This equation holds for any function F. Hence, the expression in the brackets vanishes, proving (6.4). On the other hand, if h is a function satisfying (6.4), the same arguments as above show that $h \in N(Q_{el})^\perp$. $\square$

Lemma 6.2. *Let*

$$S(\varepsilon) = \int_B (Q_{ee}(F))(k)\delta(E(k) - \varepsilon)\,dk, \quad \varepsilon \in R(E).$$

Then it holds

$$\int_{\mathbb{R}} S(\varepsilon)\,d\varepsilon = \int_{\mathbb{R}} S(\varepsilon)\varepsilon\,d\varepsilon = 0.$$

Furthermore, if $S(\varepsilon) = 0$ for all $\varepsilon \in R(E)$, then F is a Fermi–Dirac distribution, i.e., there exist parameters μ and T such that $F(k) = F_{\mu,T}(k) = 1/(1 + e^{(E(k)-\mu)/T})$.

The parameter μ is called the *chemical potential*, T the *temperature*.

Proof. The proof is very similar to the proof of Proposition 4.6 (2) in Sect. 4.2. We have for all functions G,

$$\int_{\mathbb{R}} S(\varepsilon)G(\varepsilon)\,d\varepsilon = \int_{B^4} (Q_{ee}(F))(k)G(E(k))\,d^4k\Big|_{E(k)=\varepsilon}$$

$$= -\frac{1}{4}\int_{B^4} \sigma(k,k',k_1,k_1')\delta(E+E_1-E'-E_1')(G'+G_1'-G-G_1)$$

$$\times \left(F'F_1'(1-F)(1-F') - FF_1(1-F')(1-F_1')\right)\,d^4k,$$

where $d^4k = dk\,dk'\,dk_1\,dk_1'$. The first assertion follows after taking $G(\varepsilon)=1$ and $G(\varepsilon)=\varepsilon$. The above integral is nonpositive if we choose $G = \log F - \log(1-F)$. Hence, if $F \in N(Q_{ee})$, the integrand vanishes and one proves as for Proposition 4.6 that $\log F - \log(1-F)$ is an affine function of the energy. Thus, F is a Fermi–Dirac distribution. $\square$

Derivation of the energy-transport equations. We proceed now to the diffusion limit $\alpha \to 0$. Let (f_α, V_α) be a solution of the Boltzmann–Poisson system (6.3) and (5.3). As explained in Sect. 2.4, the limit is divided into three steps. The first step is the formal limit $\alpha \to 0$ in (6.3). Setting $f = \lim_{\alpha\to 0} f_\alpha$, we infer that

$$Q_{el}(f) = 0.$$

By Proposition 4.5, the kernel of Q_{el} is spanned by functions which depend only on the energy, $f(x,k,t) = F(x,E(k),t)$.

For the second step, we insert the Chapman–Enskog expansion $f_\alpha = F + \alpha g_\alpha$ into the Boltzmann equation (6.3):

$$\alpha\partial_t(F+\alpha g_\alpha) + (v(k)\cdot\nabla_x F + \nabla_x V\cdot\nabla_k F)$$
$$+ \alpha\left(v(k)\cdot\nabla_x g_\alpha + \nabla_x V\cdot\nabla_x g_\alpha\right) = Q_{el}(g_\alpha) + Q_{ee}(f_\alpha) + \alpha Q_{in}(f_\alpha).$$

Here we have used that $Q_{el}(F) = 0$, since $F \in N(Q_{el})$. The limit $\alpha \to 0$ then shows that

$$Q_{el}(g) = v(k)\cdot\nabla_x F + \nabla_x V\cdot\nabla_k F - Q_{ee}(F), \qquad (6.5)$$

where $g = \lim_{\alpha\to 0} g_\alpha$. By Lemma 6.1, this equation has a solution if and only if

$$\int_B (v(k)\cdot\nabla_x F + \nabla_x V\cdot\nabla_k F - Q_{ee}(F))\,\delta(E(k)-\varepsilon)\,dk = 0$$

for all $\varepsilon \in R(E)$. Since $v(k) = \nabla_k E(k)$, we obtain

$$\int_B (v(k)\cdot\nabla_x F + \nabla_x V\cdot\nabla_k F)\,\delta(E(k)-\varepsilon)\,dk$$

$$= (\nabla_x F + \partial_E F\nabla_x V)(\varepsilon)\cdot\int_B \nabla_k E(k)\delta(E(k)-\varepsilon)\,dk.$$

We claim that the last integral vanishes. A heuristic argument is as follows. Let H be the Heaviside function, defined by $H(x) = 0$ for $x < 0$ and $H(x) = 1$ for $x > 0$. Then $H' = \delta$ and

$$\int_B \nabla_k E(k)\delta(E(k)-\varepsilon)\,dk = \int_B \nabla_k H(E(k)-\varepsilon)\,dk = 0, \qquad (6.6)$$

since $E(k)$ is periodic on B. Thus, the solvability condition (6.4) becomes

$$\int_B Q_{ee}(F)\delta(E(k) - \varepsilon)\,dk = 0 \quad \text{for all } \varepsilon \in R(E).$$

By Lemma 6.2, it follows that F is a Fermi–Dirac distribution, $F = F_{\mu,T}$ for some functions $\mu(x,t)$ and $T(x,t)$. As $F_{\mu,T}$ lies in the kernel of Q_{ee}, by Proposition 4.6, the operator equation (6.5) can be written as

$$
\begin{aligned}
Q_{el}(g) &= v(k) \cdot \nabla_x F + \nabla_x V \cdot \nabla_k F \\
&= F(1 - F)v(k) \cdot \left(\nabla_x \left(\frac{\mu}{T} \right) - \frac{\nabla_x V}{T} - E\nabla_x \left(\frac{1}{T} \right) \right),
\end{aligned}
\tag{6.7}
$$

and we know that it is solvable.

The third step is concerned with the limit $\alpha \to 0$ in the moment equations. We introduce the notation $\langle g \rangle = \int_B g\,dk/4\pi^3$. As moments, we employ the particle density $n = \langle F \rangle$ and the energy density $ne = \langle E(k)F \rangle$. An integration of the Boltzmann equation (6.3) and the Chapman–Enskog expansion leads to the two moment equations for $j = 0, 1$,

$$
\begin{aligned}
\partial_t \langle E^j(F + \alpha g_\alpha) \rangle &+ \alpha^{-1} \langle E^j(v(k) \cdot \nabla_x F + \nabla_x V \cdot \nabla_k F) \rangle \\
&+ \langle E^j(v(k) \cdot \nabla_x g_\alpha + \nabla_x V \cdot \nabla_k g_\alpha) \rangle \\
&= \alpha^{-2} \langle E^j Q_{el}(f_\alpha) \rangle + \alpha^{-1} \langle E^j Q_{ee}(f_\alpha) \rangle + \langle E^j Q_{in}(f_\alpha) \rangle.
\end{aligned}
$$

Taking into account the operator equation (6.7), the second term on the left-hand side equals $\alpha^{-1} \langle E^j Q_{el}(g) \rangle$ and this vanishes since elastic collisions conserve mass and energy (see Proposition 4.5). By the same reason, the first term on the right-hand side vanishes. This holds also true for the second term on the right-hand side (see Proposition 4.6). Inelastic scattering conserves mass, but generally not energy. Therefore, $\langle Q_{in}(f_\alpha) \rangle = 0$, but $\langle E(k)Q_{in}(f_\alpha) \rangle$ may not vanish. Thus, the limit $\alpha \to 0$ in the above moment equations gives

$$\partial_t \langle E^j F \rangle + \langle E^j(v(k) \cdot \nabla_x g + \nabla_x V \cdot \nabla_k g) \rangle = \langle E^j Q_{in}(F) \rangle, \quad j = 0, 1. \tag{6.8}$$

The second term on the left-hand side can be reformulated:

$$\langle E^j(v(k) \cdot \nabla_x g + \nabla_x V \cdot \nabla_k g) \rangle = \text{div}_x \langle E^j v(k)g \rangle - \nabla_x V \cdot \langle \nabla_k(E^j)g \rangle.$$

We introduce the particle and energy current densities $J_0 = -\langle v(k)g \rangle$ and $J_1 = -\langle v(k)E(k)g \rangle$, respectively. Then, since $\nabla_k E = v$,

$$
\begin{aligned}
\langle v(k) \cdot \nabla_x g + \nabla_x V \cdot \nabla_k g \rangle &= -\text{div}_x J_0, \\
\langle E(v(k) \cdot \nabla_x g + \nabla_x V \cdot \nabla_k g) \rangle &= -\text{div}_x J_1 + \nabla_x V \cdot J_0,
\end{aligned}
$$

and the moment equations become

$$\partial_t n - \mathrm{div}_x J_0 = 0, \quad \partial_t(ne) - \mathrm{div}_x J_1 + \nabla_x V \cdot J_0 = W,$$

where $W = \langle E(k) Q_{\mathrm{in}}(F) \rangle$.

It remains to compute the current densities. Let d_0 be a solution of

$$Q_{\mathrm{el}}(d_0) = -vF(1 - F). \tag{6.9}$$

Notice that d_0 is vector valued since the right-hand side of the above equation lies in $\mathbb{R}^3$. We write $d_0 = (d_{0,1}, d_{0,2}, d_{0,3})^{\top}$. The existence of a solution of this equation follows as for (6.7) since there, the right-hand side is a multiple of $v(k)F(1 - F)$. Due to the linearity of Q_{el}, the general solution of (6.7) is then given by $g = -d_0 \cdot (\nabla_x(\mu/T) - \nabla_x V/T - E\nabla_x(1/T)) + F_1$, where $F_1 \in N(Q_{\mathrm{el}})$. We compute

$$J_j = \langle E^j v \otimes d_0 \rangle \left(\nabla_x \frac{\mu}{T} - \frac{\nabla_x V}{T} \right) - \langle E^{j+1} v \otimes d_0 \rangle \nabla_x \left(\frac{1}{T} \right), \quad j = 0, 1.$$

We have shown the following result.

Theorem 6.3 (Energy-transport equations). *Let the scattering rates σ_{el} and σ_{ee} of the collision operators (6.1) and (6.2) be symmetric in the sense of (4.29) and (4.31), respectively. Furthermore, let (f_α, V_α) be a solution of the Boltzmann–Poisson system (6.3) and (5.3). Then the (formal) limit functions $F = \lim_{\alpha \to 0} f_\alpha$ and $V = \lim_{\alpha \to 0} V_\alpha$ satisfy the* energy-transport equations

$$\partial_t n - \mathrm{div} J_0 = 0, \quad \partial_t(ne) - \mathrm{div} J_1 + J_0 \cdot \nabla V = W(\mu, T), \tag{6.10}$$

$$J_0 = D_{00} \left(\nabla \left(\frac{\mu}{T} \right) - \frac{\nabla V}{T} \right) - D_{01} \nabla \left(\frac{1}{T} \right), \tag{6.11}$$

$$J_1 = D_{10} \left(\nabla \left(\frac{\mu}{T} \right) - \frac{\nabla V}{T} \right) - D_{11} \nabla \left(\frac{1}{T} \right), \tag{6.12}$$

$$\lambda_D^2 \Delta V = n - C(x), \quad x \in \mathbb{R}^3, \, t > 0, \tag{6.13}$$

where the electron and energy densities are given by, respectively,

$$n = n(\mu, T) = \int_B F_{\mu,T} \frac{dk}{4\pi^3}, \quad ne = (ne)(\mu, T) = \int_B F_{\mu,T} E(k) \frac{dk}{4\pi^3},$$

$F = F_{\mu,T} = 1/(1 + e^{(E(k)-\mu)/T})$ is the Fermi–Dirac distribution, the diffusion coefficients are defined by

$$D_{ij} = D_{ij}(\mu, T) = \int_B E^{i+j} v \otimes d_0 \frac{dk}{4\pi^3}, \quad i, j = 0, 1, \tag{6.14}$$

the function d_0 is a solution of (6.9), depending on (μ, T) through F, and the averaged inelastic scattering term equals

$$W(\mu, T) = \int_B E(k) Q_{\mathrm{in}}(F_{\mu,T}) \frac{dk}{4\pi^3}. \tag{6.15}$$

The initial data are given by

$$n(\cdot,0) = \int_B f_I \frac{dk}{4\pi^3}, \quad (ne)(\cdot,0) = \int_B f_I E(k) \frac{dk}{4\pi^3}.$$

The diffusion coefficients D_{ij} are matrices of $\mathbb{R}^{3\times3}$ since $v \otimes d_0$ is a (3×3)-matrix with coefficients $v_\ell d_{0,m}$.

The first energy-transport model was presented by Stratton in 1962 [4]. His approach is based on the relaxation-time approximation of the collision integral to obtain an approximate explicit solution for the distribution function. Often, the energy-transport model is considered as an approximation of the hydrodynamic equations [5, 6] (see Chap. 9). In the physical literature, there exist many versions of the energy-transport equations, usually called the *energy-balance model*, derived under various hypotheses on the relaxation-time model, the band structure, and the degeneracy [7–10]. Some of these versions will be derived in Sect. 6.2.

There are only a few results about the existence and uniqueness of solutions of (6.10), (6.11), (6.12), and (6.13), considered in a bounded domain with appropriate initial and boundary conditions. The first existence result for a heuristic model with particular diffusion coefficients (not being of the form (6.10), (6.11), and (6.12)) is due to Allegretto and Xie [11]. Under the assumption of a uniformly positive definite diffusion matrix (D_{ij}), the existence of solutions of the stationary and transient boundary-value problem was shown in [12–14]. Under physically realistic assumptions, however, the diffusion matrix is a priori only positive semi-definite. If lower positive bounds for the particle density and temperature were available, the positive definiteness of the matrix would follow. Existence results for physically more realistic diffusion coefficients can be found in [15, 16] for the stationary model and in [17, 18] for the transient equations, but only for data close to thermal equilibrium.

The numerical discretization of energy-transport models was investigated in the physical literature since the 1980s (see, for instance, [7–9, 19]). Mathematicians started to pay attention to these models in the 1990s, using finite-difference methods [20, 21], mixed finite-volume schemes [22], mixed finite-element methods [23–27], and essentially non-oscillatory (ENO) numerical schemes [28].

Equations (6.10) can be interpreted as conservation laws of mass and energy, since in the absence of external forces and source terms, an integration of the equations over $\mathbb{R}^3$ shows that the particle mass $\int_{\mathbb{R}^3} n(x,t)\,dx$ and the thermal energy $\int_{\mathbb{R}^3} ne(x,t)\,dx$ are constant in time. In nonequilibrium thermodynamics, the formulation (6.10), (6.11), and (6.12) is well known. Indeed, the so-called *thermodynamic fluxes* depend linearly on the *thermodynamic forces* $X_0 = \nabla(\mu/T) - \nabla V/T$ and $X_1 = -\nabla(1/T)$ [29, 30],

$$J_i = \sum_{j=0}^{1} D_{ij} X_j.$$

The variables μ/T and $-1/T$ are known as the (primal) *entropy variables*.

Properties of the diffusion matrix and inelastic collision term. By Onsager's principle, the diffusion matrix (D_{ij}) should be symmetric and positive definite. This can be proven under an assumption on the energy-band structure.

Proposition 6.4 (Properties of the diffusion matrix). *The diffusion matrix satisfies the following properties:*

(1) The matrix $\mathscr{D} = (D_{ij})$ is symmetric and $D_{01} = D_{10}$.

(2) Let the functions $d_{0,1}$, $d_{0,2}$, $d_{0,3}$, $Ed_{0,1}$, $Ed_{0,2}$, and $Ed_{0,3}$ be linearly independent. Then $\mathscr{D}$ is positive definite for any $\mu \in \mathbb{R}$ and $T > 0$.

The assumption in part (2) of the proposition is a geometric condition on the band structure since d_0 is determined as a solution to (6.9) through $E(k)$. The positive definiteness of the diffusion matrix indicates that the energy-transport model is expected to be of parabolic type (in the sense of the theory of partial differential equations). More precisely, for an existence theory, we also need that the functional $\rho = (n, ne)$ depending on the entropy variables $u = (\mu/T, -1/T)$ is monotone in the sense of $(\rho(u) - \rho(v)) \cdot (u - v) \geq 0$. An elementary computation shows that this is the case since the Jacobian of ρ is positive semi-definite. For details of parabolic systems of the type $\partial_t \rho(u) - \text{div}(\mathscr{D}\nabla u) = f(u, \nabla u)$, we refer to [31] and, involving also the electric potential, to [12]. The structure of the energy-transport model will be discussed in more detail in Sects. 6.3 and 6.4.

Proof. (1) The property $D_{01} = D_{10}$ follows directly from the definition (6.14). The symmetry of $\mathscr{D}$ is a consequence of the symmetry of the elastic collision operator Q_{el} (see Proposition 4.5). Indeed, we compute one component of the matrix D_{ij}, employing the definition (6.9) of d_0:

$$D_{ij,\ell m} = \int_B E^{i+j} v_\ell d_{0,m} \frac{dk}{4\pi^3} = -\int_B E^{i+j} Q_{\text{el}}(d_{0,\ell}) d_{0,m} \frac{dk}{4\pi^3 F(1-F)}$$

$$= -\int_B E^{i+j} d_{0,\ell} Q_{\text{el}}(d_{0,m}) \frac{dk}{4\pi^3 F(1-F)} = D_{ij,m\ell}.$$

To be precise, we have proved in Sect. 4.2 the symmetry of Q_{el} in $L^2(B) = L^2(B; dk)$, but this property also holds in the weighted space $L^2(B; dk/F(1-F))$. This shows that $D_{ij}^\top = D_{ij}$.

(2) We choose a vector $z = (\xi, \eta)^\top \neq 0$ with $\xi, \eta \in \mathbb{R}^3$. Then

$$z^\top \mathscr{D} z = \int_B (\xi + E\eta)^\top (v \otimes d_0)(\xi + E\eta) \frac{dk}{4\pi^3}$$

$$= -\int_B (\xi + E\eta)^\top (Q_{\text{el}}(d_0) \otimes d_0)(\xi + E\eta) \frac{dk}{4\pi^3 F(1-F)}$$

$$= -\int_{B^2} \sigma_{\text{el}}(k, k') \delta(E' - E) \sum_{i,j=0}^{1} (\xi_i + E\eta_i)(\xi_j + E\eta_j)$$

$$\times (d'_{0,i} - d_{0,i}) d_{0,j} \frac{dk' \, dk}{4\pi^3 F(1-F)}$$

$$= \frac{1}{2} \int_{B^2} \sigma_{el}(k,k')\delta(E'-E) \sum_{i,j=0}^{1} (\xi_i + E\eta_i)(\xi_j + E\eta_j)$$

$$\times (d'_{0,i} - d_{0,i})(d'_{0,j} - d_{0,j}) \frac{dk'\,dk}{4\pi^3 F(1-F)}.$$

The last equality follows similarly as in the proof of the second part of Proposition 4.5. Thus,

$$z^\top \mathcal{D} z = \frac{1}{2} \int_{B^2} \sigma_{el}(k,k')\delta(E'-E) \left| \sum_{i=0}^{1} (\xi_i + E\eta_i)(d'_{0,i} - d_{0,i}) \right|^2 \frac{dk'\,dk}{4\pi^3 F(1-F)}$$

$$= \int_{B^2} \sigma_{el}(k,k')\delta(E'-E) \left| z \cdot \begin{pmatrix} d_0 \\ E d_0 \end{pmatrix} \right|^2 \frac{dk'\,dk}{4\pi^3 F(1-F)}.$$

This integral is positive since otherwise, the equation

$$z \cdot \begin{pmatrix} d_0 \\ E d_0 \end{pmatrix} = 0 \quad \text{for some } z \neq 0$$

would imply that d_0, $E d_0$ are linearly dependent which is excluded. □

Now, we specify the inelastic collision integral. Let Q_{in} be given by (4.2) with transition rate (4.7),

$$(Q_{in}(f))(x,k,t) = \int_B \left(s_{ph}(x,k',k)f'(1-f) - s_{ph}(x,k,k')f(1-f') \right) dk', \quad (6.16)$$

where

$$s_{ph}(x,k,k') = \sigma_{ph}(x,k,k') \tag{6.17}$$
$$\times \left((1+N_{ph})\delta(E(k')-E(k)+E_{ph}) + N_{ph}\delta(E(k')-E(k)-E_{ph}) \right),$$

$\sigma_{ph}(x,k,k')$ is symmetric in k, k', $N_{ph} = 1/(e^{E_{ph}} - 1)$, and $E_{ph} \geq 0$ is the phonon energy. We claim that the corresponding averaged inelastic collision integral W is of relaxation-time type, i.e., the temperature of the particles tends to the constant (scaled) lattice temperature $T_L = 1$ if no external forces are present.

Proposition 6.5 (Property of the averaged inelastic collision term). *Let the inelastic collision integral Q_{in} be given by (6.16) with transition rate (6.17). Then the integral W, defined in (6.15), is monotone with respect to T, i.e.,*

$$W(\mu,T)(T-1) \leq 0 \quad \text{for all } \mu \in \mathbb{R}, \, t > 0.$$

Proof. Employing the identity $F = (1-F)M$, where F is the Fermi–Dirac distribution and $M = e^{-(E-\mu)/T}$, we rewrite the collision integral as

$$W(\mu,T) = \int_{B^2} \left(s_{\text{ph}}(k',k)F'(1-F) - s_{\text{ph}}(k,k')F(1-F') \right) E \frac{dk'\,dk}{4\pi^3}$$

$$= \int_{B^2} (1-F)(1-F') \left(s_{\text{ph}}(k',k)M' - s_{\text{ph}}(k,k')M \right) E \frac{dk'\,dk}{4\pi^3}.$$

Definition (6.17) of s_{ph} and the property $\delta(E'-E\pm E_{\text{ph}}) = \delta(E-E'\mp E_{\text{ph}})$ yield

$$s_{\text{ph}}(k',k)M' - s_{\text{ph}}(k,k')M$$

$$= \sigma_{\text{ph}} \left((1+N_{\text{ph}})\delta(E-E'+E_{\text{ph}})M' + N_{\text{ph}}\delta(E-E'-E_{\text{ph}})M' \right.$$

$$\left. -(1+N_{\text{ph}})\delta(E-E'-E_{\text{ph}})M - N_{\text{ph}}\delta(E-E'+E_{\text{ph}})M \right)$$

$$= \sigma_{\text{ph}}\delta(E-E'+E_{\text{ph}}) \left((1+N_{\text{ph}})M' - N_{\text{ph}}M \right)$$

$$+ \sigma_{\text{ph}}\delta(E-E'-E_{\text{ph}}) \left(N_{\text{ph}}M' - (1+N_{\text{ph}})M \right).$$

Employing this expression in the integral of W and exchanging k and k' in the second summand, we obtain

$$W = \int_{B^2} (1-F)(1-F')\sigma_{\text{ph}} \left(\delta(E-E'+E_{\text{ph}})E((1+N_{\text{ph}})M' - N_{\text{ph}}M) \right.$$

$$\left. + \delta(E'-E-E_{\text{ph}})E'(N_{\text{ph}}M - (1+N_{\text{ph}})M') \right) \frac{dk'\,dk}{4\pi^3}.$$

Since $\delta(E'-E-E_{\text{ph}})E' = \delta(E-E'+E_{\text{ph}})(E+E_{\text{ph}})$, some terms cancel and we end up with

$$W = \int_{B^2} (1-F)(1-F')\delta(E-E'+E_{\text{ph}})E_{\text{ph}} \left(N_{\text{ph}}M - (1+N_{\text{ph}})M' \right) \frac{dk'\,dk}{4\pi^3}.$$

We can write $1+N_{\text{ph}} = e^{E_{\text{ph}}}N_{\text{ph}}$ such that

$$W = \int_{B^2} (1-F)(1-F')\delta(E-E'+E_{\text{ph}})E_{\text{ph}}N_{\text{ph}}(M - e^{E_{\text{ph}}}M') \frac{dk'\,dk}{4\pi^3}.$$

The delta distribution allows us to substitute the energy E in the Maxwellian M by $E'-E_{\text{ph}}$ such that, for all E and E' satisfying $E = E'-E_{\text{ph}}$,

$$M - e^{E_{\text{ph}}}M' = e^{-(E-\mu)/T} - e^{E_{\text{ph}}}e^{-(E'-\mu)/T} = e^{-(E'-\mu)/T} \left(e^{E_{\text{ph}}/T} - e^{E_{\text{ph}}} \right).$$

We conclude that

$$W(\mu,T)(T-1) = \int_{B^2} (1-F)(1-F')\delta(E-E'+E_{\text{ph}})E_{\text{ph}}N_{\text{ph}}M'$$

$$\times \left(e^{E_{\text{ph}}/T} - e^{E_{\text{ph}}} \right)(T-1)\frac{dk'\,dk}{4\pi^3} \le 0,$$

since $(e^{E_{\text{ph}}/T} - e^{E_{\text{ph}}})(T-1) \le 0$ for all $T > 0$. $\square$

The unscaled energy-transport model reads as follows:

$$\partial_t n - \frac{1}{q}\mathrm{div}\, J_n = 0, \quad \partial_t(ne) - \mathrm{div}\, J_e + J_n \cdot \nabla V = W(\mu, T),$$

$$J_n = qD_{00}\left(\nabla\left(\frac{q\mu}{k_B T}\right) - \frac{q\nabla V}{k_B T}\right) - qD_{01}\nabla\left(\frac{1}{k_B T}\right),$$

$$J_e = D_{10}\left(\nabla\left(\frac{q\mu}{k_B T}\right) - \frac{q\nabla V}{k_B T}\right) - D_{11}\nabla\left(\frac{1}{k_B T}\right),$$

$$\varepsilon_s \Delta V = q(n - C(x)),$$

complemented with initial conditions for the electron density and temperature,

$$n(\cdot, 0) = n_I, \quad T(\cdot, 0) = T_I \quad \text{in } \mathbb{R}^3.$$

Given the initial particle density and temperature, the initial chemical potential $\mu(\cdot, 0) = \mu_I$ can be computed from $n_I = \int_B (1 + e^{(E(k) - \mu_I)/T_I})^{-1} \, dk/4\pi^3$. As the derivative $\partial n_I / \partial \mu_I$ is strictly positive, the expression $n_I = n_I(\mu_I)$ can be inverted. Hence, the initial datum for the thermal energy $ne = (ne)(\mu, T)$ can be calculated.

When the model is considered in a bounded domain, we impose mixed Dirichlet–Neumann boundary conditions for the particle density, temperature, and electric potential, similar as for the drift-diffusion model. For this, we assume that $\partial \Omega = \Gamma_D \cup \Gamma_N$ with $\Gamma_D \cap \Gamma_N = \emptyset$. Then

$$n = n_D, \quad T = T_D, \quad V = V_D \quad \text{on } \Gamma_D, \, t > 0,$$
$$J_n \cdot \eta = J_e \cdot \eta = \nabla V \cdot \eta = 0 \quad \text{on } \Gamma_N, \, t > 0,$$

where η is the exterior unit normal to Γ_N.

The energy-transport model (6.10), (6.11), and (6.12) is still not explicit since the diffusion coefficients and the moments depend nonlocally and nonlinearly on the entropy variables μ/T and $-1/T$. In the following section we specify the elastic scattering rate and the energy band in order to derive explicit transport models.

6.2 Some Explicit Models

In this section we derive explicit expressions for the particle density $n(\mu, T)$, the energy density $ne(\mu, T)$, the diffusion coefficients $D_{ij}(\mu, T)$, and the relaxation-time term $W(\mu, T)$ depending on the entropy variables μ/T and $-1/T$ as solutions of (6.10), (6.11), and (6.12). For this, we need some simplifying assumptions on the elastic collision operator and the energy band structure:

1. The Fermi–Dirac distribution $F = F_{\mu, T}$ is approximated by the Maxwellian $M = e^{-(E-\mu)/T}$ and $F(1 - F)$ is approximated by $e^{-(E-\mu)/T}$.
2. The scattering rate of the elastic collision integral only depends on the energy,

$$\sigma_{\mathrm{el}}(x,k,k') = s(x,E(k)) \quad \text{for all } k,\ k' \text{ with } E(k) = E(k').$$

3. The energy band $E(k)$ is spherically symmetric and strictly monotone, i.e., there exists a strictly monotone function $\gamma : \mathbb{R} \to \mathbb{R}$ such that

$$|k|^2 = \gamma(E(|k|)), \quad k \in B = \mathbb{R}^3.$$

We choose the reference point for the energy such that $E(k)$ is always nonnegative. According to Remark 1.12, the Fermi–Dirac distribution can be approximated by the Maxwell–Boltzmann distribution if the energy $E - \mu$ is much larger than the thermal energy T. This is the case for nondegenerate semiconductor materials.

The second condition makes sense since, due to the term $\delta(E(k') - E(k))$ in the definition (6.1) of Q_{el}, the scattering rate needs to be defined only on the surface $\{k' \in B : E(k) = E(k')\}$ of energy $E(k)$. This assumption is reasonable, at least approximately, for many scattering mechanisms. For instance, when the energy band is parabolic and the scattering goes as the density of states, one may write $\sigma_{\mathrm{el}}(x,k,k') = s_1(x)E(k)^\beta$, with $\beta = 1/2$ for nonpolar phonon scattering, $\beta = -3/2$ for moderate ionized impurity scattering in the elastic limit [32, Sect. 2.3.4] (see also [33, Sect. 9.2]), and $\beta = 0$ for acoustic phonon scattering [1, formula (3.37)]. For polar optical phonon scattering, which is a strong collision mechanism in compound semiconductors like GaAs, the scattering rate is more complicated but σ_{el} is a function of $E(k)$ only [32, Sect. 2.8.1]. The assumption of a spherically symmetric energy band seems to be reasonable for the Γ valley of GaAs. In silicon and germanium, the energy band is described by an ellipsoid rather than by a sphere.

Proposition 6.6. *Under Maxwell–Boltzmann statistics and the above assumptions on the scattering rate and the energy band, the particle density, energy density, and diffusion coefficients can be written, respectively, as*

$$n = \frac{1}{2\pi^2} e^{\mu/T} \int_0^\infty e^{-\varepsilon/T} \sqrt{\gamma(\varepsilon)}\gamma'(\varepsilon)\,d\varepsilon, \tag{6.18}$$

$$ne = \frac{1}{2\pi^2} e^{\mu/T} \int_0^\infty e^{-\varepsilon/T} \sqrt{\gamma(\varepsilon)}\gamma'(\varepsilon)\varepsilon\,d\varepsilon, \tag{6.19}$$

$$D_{ij} = \frac{1}{3\pi^3} e^{\mu/T} \int_0^\infty e^{-\varepsilon/T} \frac{\gamma(\varepsilon)\varepsilon^{i+j}}{s(x,\varepsilon)\gamma'(\varepsilon)^2}\,d\varepsilon\,\mathrm{Id}, \quad i,j = 0,1. \tag{6.20}$$

Furthermore, the density of states of energy ε equals

$$N(\varepsilon) = \int_{\mathbb{R}^3} \delta(E(k) - \varepsilon)\frac{dk}{4\pi^3} = \frac{1}{2\pi^2}\sqrt{\gamma(\varepsilon)}\gamma'(\varepsilon), \quad \varepsilon \geq 0. \tag{6.21}$$

The expression (6.20) shows that the diffusion coefficients, which are (3×3)-matrices, can be identified with their diagonal element, and we can write

$$D_{ij} = \frac{1}{3\pi^3} e^{\mu/T} \int_0^\infty e^{-\varepsilon/T} \frac{\gamma(\varepsilon)\varepsilon^{i+j}}{s(x,\varepsilon)\gamma'(\varepsilon)^2}\,d\varepsilon. \tag{6.22}$$

Proof. We will transform the integrals over the wave-vector space $\mathbb{R}^3$ to integrals over the energy space $(0, \infty)$ by means of the coarea formula (1.55). Let $\rho = |k|$. Then differentiation of $\rho^2 = \gamma(E(\rho))$ with respect to ρ yields $2\rho = \gamma'(E(\rho))E'(\rho)$ and

$$E'(\rho) = \frac{2\sqrt{\gamma(\varepsilon)}}{\gamma'(\varepsilon)} \quad \text{with } \varepsilon = E(\rho),$$

and we conclude that

$$\nabla_k E(|k|) = E'(\rho)\nabla_k|k| = \frac{2\sqrt{\gamma(\varepsilon)}}{\gamma'(\varepsilon)}\frac{k}{|k|}. \tag{6.23}$$

Now, the coarea formula gives

$$\begin{aligned}
n &= \int_{\mathbb{R}^3} e^{-(E(|k|)-\mu)/T}\frac{dk}{4\pi^3} = \frac{1}{4\pi^3}\int_0^\infty\int_{\{E(\rho)=\varepsilon\}} e^{-(\varepsilon-\mu)/T}\frac{dS_\varepsilon(k)}{|\nabla_k E(|k|)|}\,d\varepsilon \\
&= \frac{1}{4\pi^3}\int_0^\infty e^{-(\varepsilon-\mu)/T}\frac{\gamma'(\varepsilon)}{2\sqrt{\gamma(\varepsilon)}}\int_{\{E(\rho)=\varepsilon\}} dS_\varepsilon(k)\,d\varepsilon.
\end{aligned}$$

The surface of all k such that $E(\rho) = \varepsilon$ or, equivalently, $\rho^2 = \gamma(E(\rho)) = \gamma(\varepsilon)$ equals $4\pi\rho^2 = 4\pi\gamma(\varepsilon)$ if $\varepsilon = E(\rho)$. Hence,

$$\int_{\{E(\rho)=\varepsilon\}} dS_\varepsilon(k) = 4\pi\gamma(\varepsilon),$$

and the electron density becomes

$$n = \frac{1}{2\pi^2}e^{\mu/T}\int_0^\infty e^{-\varepsilon/T}\sqrt{\gamma(\varepsilon)}\gamma'(\varepsilon)\,d\varepsilon,$$

which is (6.18). In a similar way, the energy density can be computed leading to (6.19). The density of states becomes

$$\begin{aligned}
N(\varepsilon) &= \int_0^\infty\int_{\{E(|k|)=\eta\}} \delta(E-\varepsilon)\frac{dS_\eta(k)}{|\nabla_k E|}\frac{d\eta}{4\pi^3} \\
&= \int_0^\infty\frac{\gamma'(\eta)}{2\sqrt{\gamma(\eta)}}\delta(\eta-\varepsilon)4\pi\gamma(\eta)\frac{d\eta}{4\pi^3} = \frac{1}{2\pi^2}\sqrt{\gamma(\varepsilon)}\gamma'(\varepsilon).
\end{aligned}$$

It remains to compute the diffusion coefficients. The second of the above assumptions allows us to simplify the elastic collision operator. Indeed, we obtain

$$\begin{aligned}
Q_{\text{el}}(f) &= \int_{\mathbb{R}^3} s(x,E)\delta(E'-E)(f'-f)\,dk' \\
&= s(x,E)\left(\int_{\mathbb{R}^3}\delta(E'-E)f'\,dk' - f(k)\int_{\mathbb{R}^3}\delta(E'-E)\,dk'\right) = \frac{[f]-f}{\tau(x,E(k))},
\end{aligned}$$

where

$$\tau(x,E) = \frac{1}{4\pi^3 s(x,E)N(E)} \tag{6.24}$$

is the relaxation time, $N(E) = \int_{\mathbb{R}^3} \delta(E' - E)\,dk'/4\pi^3$ is the density of states of energy E, and

$$[f](x,k,t) = \frac{1}{4\pi^3 N(E(k))} \int_{\mathbb{R}^3} \delta(E' - E)f(x,k',t)\,dk'$$

is the average of f over the energy surface $\{k' : E(k') = E(k)\}$.

With the above expression, we are able to solve the operator equation (6.9) which becomes in the Maxwell–Boltzmann approximation $Q_{\text{el}}(d_0) = -vM$. The solution is given by $d_0 = \tau(x,E)\nabla_k E e^{-(E-\mu)/T}$ since

$$
\begin{aligned}
[d_0] &= \frac{1}{4\pi^3 N(E(k))} \int_{\mathbb{R}^3} \delta(E' - E)\tau(x,E')\nabla_k E' e^{-(E'-\mu)/T}\,dk' \\
&= \frac{\tau(x,E)}{4\pi^3 N(E(k))} e^{-(E-\mu)/T} \int_{\mathbb{R}^3} \delta(E' - E)\nabla_k E'\,dk' = 0
\end{aligned}
$$

and hence,

$$Q_{\text{el}}(d_0) = -\frac{d_0}{\tau(x,E(k))} = -\nabla_k E e^{-(E-\mu)/T} = -vM.$$

With the explicit expression for d_0, the diffusion coefficients (6.14) can be written as

$$D_{ij} = \int_{\mathbb{R}^3} E^{i+j}(\nabla_k E \otimes \nabla_k E)\tau(x,E)e^{-(E-\mu)/T}\,\frac{dk}{4\pi^3}.$$

Then (6.23) and the coarea formula lead to

$$
\begin{aligned}
D_{ij} &= \frac{1}{4\pi^3} \int_{\mathbb{R}} e^{-(\varepsilon-\mu)/T}\varepsilon^{i+j}\tau(\varepsilon) \int_{\{E(k)=\varepsilon\}} \nabla_k E \otimes \nabla_k E \frac{dS_\varepsilon(k)}{|\nabla_k E|}\,d\varepsilon \\
&= \frac{1}{4\pi^3} \int_{\mathbb{R}} e^{-(\varepsilon-\mu)/T}\varepsilon^{i+j}\tau(\varepsilon)\frac{2\sqrt{\gamma(\varepsilon)}}{\gamma'(\varepsilon)} \int_{\{E(|k|)=\varepsilon\}} \frac{k}{|k|} \otimes \frac{k}{|k|}\,dS_\varepsilon(k)\,d\varepsilon.
\end{aligned}
$$

In spherical coordinates, it holds

$$\frac{k}{|k|} = \begin{pmatrix} \sin\theta\cos\phi \\ \sin\theta\sin\phi \\ \cos\theta \end{pmatrix},$$

for angles θ and ϕ, and therefore, after some computations,

$$\int_{\{E(|k|)=\varepsilon\}} \frac{k}{|k|} \otimes \frac{k}{|k|}\,dS_\varepsilon(k) = \int_0^{2\pi}\int_0^\pi \frac{k}{|k|} \otimes \frac{k}{|k|}\gamma(\varepsilon)\sin\theta\,d\theta\,d\phi = \frac{4\pi}{3}\gamma(\varepsilon)\mathrm{Id}.$$

Replacing $\tau(\varepsilon)$ by (6.24) and (6.21) then gives (6.20). $\square$

When the averaged inelastic collision integral is given by a Fokker–Planck approximation [34], we can make the relaxation-time term more explicit.

Proposition 6.7. *Assume that*

$$\int_{\{E(k)=\varepsilon\}} Q_{\text{in}}(F) \frac{\mathrm{d}S_\varepsilon(k)}{|\nabla_k E|}$$

is approximated by the Fokker–Planck term

$$s_0 \frac{\partial}{\partial\varepsilon}\left(\frac{N(\varepsilon)}{\tau(x,\varepsilon)}\left(F+\frac{\partial F}{\partial\varepsilon}\right)\right),$$

where $s_0 > 0$ and $\tau(x,\varepsilon)$ is defined in (6.24). Then the relaxation-time term can be written as

$$W(\mu,T) = -\frac{s_0}{4\pi^4}e^{\mu/T}\left(1-\frac{1}{T}\right)\int_0^\infty s(x,\varepsilon)\gamma(\varepsilon)\gamma'(\varepsilon)^2 e^{-\varepsilon/T}\,\mathrm{d}\varepsilon,$$

Proof. We obtain from (6.15) and (6.21), after integrating by parts,

$$\begin{aligned}
W &= \frac{1}{4\pi^3}\int_{\mathbb{R}} s_0 \frac{\partial}{\partial\varepsilon}\left(\frac{N(\varepsilon)}{\tau(\varepsilon)}\left(F+\frac{\partial F}{\partial\varepsilon}\right)\right)\varepsilon\,\mathrm{d}\varepsilon \\
&= -s_0 \int_0^\infty s(x,\varepsilon)N(\varepsilon)^2\left(1-\frac{1}{T}\right)e^{-(\varepsilon-\mu)/T}\,\mathrm{d}\varepsilon \\
&= -\frac{s_0}{4\pi^4}\int_0^\infty s(x,\varepsilon)\gamma(\varepsilon)\gamma'(\varepsilon)^2\left(1-\frac{1}{T}\right)e^{-(\varepsilon-\mu)/T}\,\mathrm{d}\varepsilon,
\end{aligned}$$

which shows the proposition. $\square$

When the energy band is parabolic and the scattering rate is made explicit, we are able to calculate the integrals over the energy space and to derive local expressions for the densities and coefficients as functions of μ and T.

Example 6.8 (Parabolic band approximation). In addition to the hypotheses at the beginning of this section, we assume that

- the energy band is parabolic, $E(k) = |k|^2/2$, $k \in \mathbb{R}^3$, and
- the scattering rate is given by $s(x,\varepsilon) = s_1(x)\varepsilon^\beta$, $\beta \geq 0$.

The first assumption implies that $\gamma(\varepsilon) = 2\varepsilon$. Then, by (6.21), $N(\varepsilon) = \sqrt{2\varepsilon}/\pi^2$. In view of (6.18) and (6.19), the electron and energy densities become

$$\begin{aligned}
n &= \frac{\sqrt{2}}{\pi^2}e^{\mu/T}\int_0^\infty e^{-\varepsilon/T}\sqrt{\varepsilon}\,\mathrm{d}\varepsilon = \frac{\sqrt{2}}{\pi^2}T^{3/2}e^{\mu/T}\int_0^\infty e^{-z}\sqrt{z}\,\mathrm{d}z \\
&= \frac{\sqrt{2}}{\pi^2}T^{3/2}e^{\mu/T}\Gamma\left(\frac{3}{2}\right) = \frac{2}{(2\pi)^{3/2}}T^{3/2}e^{\mu/T}, \quad (6.25)
\end{aligned}$$

$$ne = \frac{\sqrt{2}}{\pi^2}e^{\mu/T}\int_0^\infty e^{-\varepsilon/T}\varepsilon^{3/2}\,d\varepsilon = \frac{\sqrt{2}}{\pi^2}T^{5/2}e^{\mu/T}\int_0^\infty e^{-z}z^{3/2}\,dz$$

$$= \frac{\sqrt{2}}{\pi^2}T^{5/2}e^{\mu/T}\Gamma\left(\frac{5}{2}\right) = \frac{3}{2}nT,$$

where Γ is the Gamma function, defined in (1.60). Notice that the expression for the particle density is exactly the formula (in scaled form) obtained in Lemma 1.18.

The diffusion coefficients (6.22) can be written as

$$D_{ij} = \frac{e^{\mu/T}}{6\pi^3 s_1(x)}\int_0^\infty e^{-\varepsilon/T}\varepsilon^{i+j+1-\beta}\,d\varepsilon$$

$$= \frac{e^{\mu/T}}{6\pi^3 s_1(x)}T^{i+j+2-\beta}\int_0^\infty e^{-z}z^{i+j+1-\beta}\,dz$$

$$= \frac{e^{\mu/T}}{6\pi^3 s_1(x)}T^{i+j+2-\beta}\Gamma(i+j+2-\beta).$$

Employing the expression for the electron density, we can write the diffusion matrix $\mathscr{D} = (D_{ij})$ as

$$\mathscr{D} = \mu_0\Gamma(2-\beta)nT^{1/2-\beta}\begin{pmatrix} 1 & (2-\beta)T \\ (2-\beta)T & (3-\beta)(2-\beta)T^2 \end{pmatrix}, \tag{6.26}$$

where $\mu_0(x) = 1/(\sqrt{18\pi^3}s_1(x))$. Typical choices for β are $\beta = \frac{1}{2}$, employed by Chen et al. [7], and $\beta = 0$, used by Lyumkis et al. [35], leading to the diffusion matrices

$$\mathscr{D}_{\text{Chen}} = \frac{\sqrt{\pi}}{2}\mu_0 n\begin{pmatrix} 1 & \frac{3}{2}T \\ \frac{3}{2}T & \frac{15}{4}T^2 \end{pmatrix}, \quad \mathscr{D}_{\text{Lyumkis}} = \mu_0 nT^{1/2}\begin{pmatrix} 1 & 2T \\ 2T & 6T^2 \end{pmatrix}.$$

It remains to compute the relaxation-time term (see Proposition 6.7):

$$W = -\frac{2s_0 s_1}{\pi^4}e^{\mu/T}\left(1-\frac{1}{T}\right)\int_0^\infty e^{-\varepsilon/T}\varepsilon^{\beta+1}\,d\varepsilon$$

$$= -\frac{2s_0 s_1}{\pi^4}e^{\mu/T}T^{\beta+1}(T-1)\Gamma(\beta+2) = -\frac{3}{2}\frac{n(T-1)}{\tau_\beta(x,T)},$$

where the energy relaxation time is

$$\tau_\beta(x,T) = \frac{\pi^{5/2}T^{1/2-\beta}}{\sqrt{8}s_0 s_1(x)\Gamma(\beta+2)}. \tag{6.27}$$

Notice that for the choice $\beta = 1/2$ of Chen et al. [7], the relaxation time $\tau_{1/2}(x,T)$ is constant with respect to T. An energy-dependent relaxation time was already suggested by Stratton [4, formula (25)].

We summarize the energy-transport equations in the parabolic band approximation:

$$\partial_t n - \operatorname{div} J_0 = 0, \quad \partial_t \left(\frac{3}{2} nT\right) - \operatorname{div} J_1 + J_0 \cdot \nabla V = W(\mu, T),$$

$$J_0 = D_{00}\left(\nabla\left(\frac{\mu}{T}\right) - \frac{\nabla V}{T}\right) - D_{01}\nabla\left(\frac{1}{T}\right),$$

$$J_1 = D_{10}\left(\nabla\left(\frac{\mu}{T}\right) - \frac{\nabla V}{T}\right) - D_{11}\nabla\left(\frac{1}{T}\right),$$

where the diffusion matrix is given by (6.26) and the relaxation-time term by

$$W(\mu, T) = -\frac{3}{2}\frac{n(T-1)}{\tau_\beta(T)},$$

the energy relaxation time $\tau_\beta(T)$ is defined in (6.27), and the electron density $n = 2(2\pi)^{-3/2}T^{3/2}e^{\mu/T}$ is a function of μ and T. □

Example 6.9 (Nonparabolic band approximation). Let the energy band be given by the nonparabolic approximation of Kane (see (1.31)),

$$E(1+\alpha E) = \frac{|k|^2}{2},$$

where $\alpha > 0$ is the (scaled) nonparabolicity parameter. Then $\gamma(\varepsilon) = 2\varepsilon(1+\alpha\varepsilon)$ and the electron and energy density become

$$n = \frac{\sqrt{2}}{\pi^2}e^{\mu/T}\int_0^\infty e^{-\mu/T}\sqrt{\varepsilon(1+\alpha\varepsilon)}(1+2\alpha\varepsilon)\,d\varepsilon = N_\alpha(T)T^{3/2}e^{\mu/T},$$

$$ne = \frac{3}{2}Q_\beta(T)nT,$$

where

$$N_\alpha(T) = \frac{\sqrt{2}}{\pi^2}\int_0^\infty e^{-z}\sqrt{z(1+\alpha Tz)}(1+2\alpha Tz)\,dz,$$

$$Q_\beta(T) = \frac{2}{3}\frac{\int_0^\infty e^{-z}\sqrt{z(1+\alpha Tz)}(1+2\alpha Tz)z\,dz}{\int_0^\infty e^{-z}\sqrt{z(1+\alpha Tz)}(1+2\alpha Tz)\,dz}. \tag{6.28}$$

The diffusion coefficients can be formulated as

$$D_{ij} = \mu_{ij}(\alpha, T)n, \quad i, j = 0, 1,$$

where

$$\mu_{ij}(\alpha, T) = \mu_0 T^{i+j+1/2-\beta}\int_0^\infty \frac{z^{i+j+1-\beta}(1+\alpha Tz)}{(1+2\alpha Tz)^2}e^{-z}\,dz, \tag{6.29}$$

and μ_0 is defined one line below (6.26). Finally, the relaxation-time term equals

$$W(n,T) = \frac{3}{2} \frac{n(1-T)}{\tau_\beta^\alpha(x,T)},$$

where

$$\tau_\beta^\alpha(x,T)^{-1} = \frac{\sqrt{8}s_0 s_1(x)}{\pi^{5/2}} T^{\beta-1/2} \int_0^\infty z^{1+\beta}(1+\alpha Tz)(1+2\alpha Tz)^2 e^{-z} dz.$$

The above integral can be written as a third-order polynomial in αT with coefficients depending on the Gamma function evaluated at $\beta + j$, where $j = 2,\dots,5$ (see [36, Sect. 4.1.3] for details). A numerical comparison of the nonparabolic and parabolic energy-transport models can be found in [23]. □

6.3 Symmetrization and Entropy

The energy-transport equations (6.10), (6.11), and (6.12) describe the transport of particles due to diffusive, thermal, and electric effects. Similar models were used, for instance, in electro-chemistry [37, 38] and alloy solidification [39, 40]. In fact, the model is also well known in nonequilibrium thermodynamics [29, 30], as mentioned in Sect. 6.1. In thermodynamics, it is known that the convective parts due to the electric field can be eliminated by using so-called dual entropy variables. Interestingly, the existence of such a change of unknowns is (in some sense) equivalent to the existence of an entropy functional which gives information on the long-time behavior of the variables. In this section, we explain the relation to nonequilibrium thermodynamics in detail.

We consider the energy-transport equations (6.10), (6.11), and (6.12), here written in the form

$$\partial_t \rho_j(u) - \operatorname{div} J_j + jJ_0 \cdot \nabla V = W(u), \quad J_j = \sum_{i=0}^1 D_{ji}\nabla u_i + D_{j0}\nabla V u_1, \qquad (6.30)$$

where $j = 0, 1$. The particle density $\rho_0 = n$ and the energy density $\rho_1 = ne$ are functions of μ and T through

$$\rho = \begin{pmatrix} n \\ ne \end{pmatrix} = \int_B \frac{1}{1 + e^{-(E(k)-\mu)/T}} \begin{pmatrix} 1 \\ E(k) \end{pmatrix} \frac{dk}{4\pi^3},$$

and $u_0 = \mu/T$ and $u_1 = -1/T$ are the (primal) entropy variables. We define the vector-valued functions $\rho = (\rho_0, \rho_1)^\top$ and $u = (u_0, u_1)^\top$. The diffusion coefficients D_{ij} and the relaxation-time term $W(u)$ are defined in (6.14) and (6.15), respectively. Proposition 6.4 shows that $\mathscr{D} = (D_{ij})$ is symmetric and positive definite. The symmetry expresses the Onsager principle of thermodynamics, whereas the positive definiteness is related to the second law of thermodynamics [30]. Equations (6.30) are solved in the bounded domain $\Omega \subset \mathbb{R}^3$ with boundary $\partial\Omega = \Gamma_D \cup \Gamma_N$ such that

$\Gamma_D \cap \Gamma_N = \emptyset$ and they are complemented by the mixed Dirichlet–Neumann boundary and initial conditions

$$u = u_D, \quad V = V_D \quad \text{on } \Gamma_D, \, t > 0,$$
$$J_0 \cdot \eta = J_1 \cdot \eta = \nabla V \cdot \eta = 0 \quad \text{on } \Gamma_N, \, t > 0,$$
$$u(\cdot, 0) = u_I \quad \text{in } \Omega.$$

The function $\rho : \mathbb{R}^2 \to \mathbb{R}^2$ has some important properties. First, it is not difficult to verify that it is monotone, i.e., $(\rho(u) - \rho(v)) \cdot (u - v) \geq 0$ for all $u, v \in \mathbb{R}^2$. Moreover, ρ is a gradient, i.e., there exists a function $\chi : \mathbb{R}^2 \to \mathbb{R}$ such that $\nabla_u \chi = \rho$. This function has the form

$$\chi(u) = \int_B \left(-\log(1 + e^{u_0 + E(k)u_1}) + (u_0 + E(k)u_1) \right) \frac{dk}{4\pi^3}.$$

These properties are needed for the (local) well posedness of the energy-transport equations; see [31] for the potential-free case and [12] for the case including the electric potential.

The key of the mathematical analysis performed in [12, 13] was the use of the *dual entropy variables*, defined by

$$w_0 = \frac{\mu - V}{T} = u_0 + V u_1, \quad w_1 = -\frac{1}{T} = u_1. \tag{6.31}$$

Then the system of equations (6.30) is formally equivalent to

$$\partial_t b_j(w, V) - \text{div} I_j = Q_j(w), \quad I_j = \sum_{i=0}^{1} L_{ij}(w, V) \nabla w_i, \quad j = 0, 1, \tag{6.32}$$

where $b_0(w, V) = \rho_0 = n$, $b_1(w, V) = \rho_1 - \rho_0 V = ne - nV$, $Q_0 = 0$, $Q_1 = W$, and

$$L_{00} = D_{00}, \quad L_{01} = L_{10} = D_{01} - D_{00}V,$$
$$L_{11} = D_{11} - 2D_{01}V + D_{00}V^2.$$

The new diffusion matrix $\mathcal{L} = (L_{ij})$ is symmetric and positive definite, since $\mathcal{L} = P^\top \mathcal{D} P$ and $P = (P_{ij})$ is the invertible matrix

$$P = \begin{pmatrix} 1 & -V \\ 0 & 1 \end{pmatrix}.$$

The dual entropy variables w_0 and w_1 satisfy mixed Dirichlet–Neumann boundary conditions with $w_0 = w_{0,D} = (u_0 + V u_1)|_{\Gamma_D}$ and $w_1 = w_{1,D} = u_1|_{\Gamma_D}$ on Γ_D. The transformed set of equations (6.32) is "symmetrized" in the sense that the convective terms $D_{j0} \nabla V / T$ and $J_0 \cdot \nabla V$ are eliminated. This simplifies the mathematical analysis and it is useful for numerical approximations [24, 27].

The transformation of variables (6.31) is well known in nonequilibrium thermodynamics [41, Sect. 53]. It can also be employed for the drift-diffusion model (see

Sect. 5.1). Indeed, with the electron density $\rho(u)$ and the entropy variable $u = \log n$ (here called the chemical potential), we can symmetrize the expression for the current density,

$$J_n = \mu_0(\nabla n - n\nabla V) = \mu_0 n\nabla(\log n - V) = \mu_0 n\nabla w,$$

where $w = u - V$ can be interpreted as a dual entropy variable. In the context of the drift-diffusion model, w is referred to as the quasi-Fermi potential. This symmetrization property was already observed by Albinus [42].

The existence of a symmetrizing change of unknowns implies the existence of an entropy functional or, more precisely, of a monotone free energy functional,

$$E(u(t)) = \int_\Omega (\rho(u) \cdot (u - u_D) - (\chi(u) - \chi(u_D)))\, dx + \frac{1}{2T_D} \int_\Omega |\nabla(V - V_D)|^2\, dx,$$

where T_D is the electron temperature on the contacts Γ_D. We recall that $\nabla_u \chi = \rho$. The first integral represents the internal or thermodynamic energy, the second integral the electric energy. The first integrand can be interpreted in some sense as the Legendre transform of χ. The free energy $E(u(t))$ is nonnegative, since ρ is monotone, and monotoneously decreasing in time, since

$$\frac{dE}{dt} + c \int_\Omega \left(|\nabla(w_0 - w_{0,D})|^2 + |\nabla(w_1 - w_{1,D})|^2 \right)\, dx \le 0, \quad t > 0, \tag{6.33}$$

where c is a constant depending on the lower bounds for the particle density and temperature. If these bounds are positive, then $c > 0$, otherwise we have only $c \ge 0$. Inequality (6.33) is also called an *entropy inequality* and the second integral on the left-hand side the *entropy production*. The entropy inequality was proved in [12, 36] under the assumption that the temperature $T_D = -1/w_{1,D}$ is constant on Γ_D. This condition is satisfied if T_D is equal to the (constant) room temperature at the contacts. If T_D is not constant, a similar inequality as above exists but we need to add a constant depending on T_D on the right-hand side. Inequality (6.33) provides uniform bounds for the entropy production term $\sum_j |\nabla(w_j - w_{j,D})|^2$. If the boundary data w_D is given by the thermal equilibrium state, it can be shown that (6.33) implies the convergence of $u(t)$ to the thermal equilibrium state as $t \to \infty$ [12]. We prove an entropy inequality for a more general diffusive moment model in Sect. 8.4.

The fact that the existence of a symmetric formulation is formally equivalent to the existence of an entropy functional is well known in the theory of hyperbolic conservation laws and was first formulated mathematically by Kawashima and Shizuta [43]. In [44], this equivalence was analyzed for parabolic systems. In fact, the above considerations are also valid for models describing the flow of M components of a fluid or gas of charged particles with particle density ρ_j for the j-component, $j = 0, \ldots, M$ [12].

6.4 Drift-Diffusion Formulation

The formulation of the energy-transport model in terms of the dual entropy variables (6.31) has the advantage that all terms involving the electric field are eliminated and that the elliptic differential operator is symmetric. On the other hand, the resulting system is of cross-diffusion type, i.e., the diffusion matrix is not diagonal and the equations are strongly coupled. For numerical purposes it may be advantageous to have a "decoupled" formulation which allows for a fixed-point strategy to solve the system of equations. In other words, we are seeking for a formulation in which the diffusion matrix is diagonal. In this section, we show that such a formulation exists if Maxwell–Boltzmann statistics are assumed, i.e., the Fermi–Dirac distribution $F = F_{\mu,T}$ is approximated by the Maxwellian $M = e^{-(E-\mu)/T}$ and $F(1-F)$ is approximated by $e^{-(E-\mu)/T}$.

We recall that the current densities are defined by

$$J_j = D_{j0}\left(\nabla\left(\frac{\mu}{T}\right) - \frac{\nabla V}{T}\right) - D_{j1}\nabla\left(\frac{1}{T}\right), \quad j = 0, 1, \tag{6.34}$$

with the diffusion coefficients $D_{ij} = \langle E^{i+j}v \otimes d_0\rangle$ (see (6.11), (6.12), and (6.14)), and the function d_0 is a solution of $Q_{\mathrm{el}}(d_0) = -vM$.

Proposition 6.10. *Under the assumption of Maxwell–Boltzmann statistics, the current densities (6.11) and (6.12) can be formulated as*

$$J_j = \nabla D_{j0} - \frac{D_{j0}}{T}\nabla V, \quad j = 0, 1. \tag{6.35}$$

Proof. First, we claim that $\nabla_x d_0 = (\nabla(\mu/T) - E\nabla(1/T))d_0 + F_1$, where $F_1 \in N(Q_{\mathrm{el}})^3$. By Lemma 4.5, F_1 is a function of $E(k)$ (and x, t) only. Since $Q_{\mathrm{el}}(d_0)$ is linear, a formal differentiation gives

$$Q_{\mathrm{el}}(\nabla_x d_0) = \nabla_x Q_{\mathrm{el}}(d_0) = -v\nabla_x M = -\left(\nabla\left(\frac{\mu}{T}\right) - E\nabla\left(\frac{1}{T}\right)\right)vM$$

$$= Q_{\mathrm{el}}\left(\left(\nabla\left(\frac{\mu}{T}\right) - E\nabla\left(\frac{1}{T}\right)\right)d_0\right).$$

This shows the claim. We notice that in this step, we need to suppose Maxwell–Boltzmann statistics. Indeed, in the general case, we have $Q_{\mathrm{el}}(\nabla_x d_0) = -v\nabla_x (F(1-F))$ and the right-hand side is more complicated.

Next, we compute, for $j = 0, 1$,

$$\nabla_x D_{j0} = \langle E^j v \otimes \nabla_x d_0\rangle = \left\langle \left(\nabla\left(\frac{\mu}{T}\right) - E\nabla\left(\frac{1}{T}\right)\right) E^j v \otimes d_0 \right\rangle + \langle E^j v \otimes F_1\rangle$$

$$= D_{j0}\nabla\left(\frac{\mu}{T}\right) - D_{j1}\nabla\left(\frac{1}{T}\right).$$

The last equality follows since all matrix components of $\langle E^j v \otimes F_1 \rangle$ vanish:

$$\langle E^j (v \otimes F_1)_{\ell m} \rangle = \int_B E^j \frac{\partial E}{\partial k_\ell} F_{1,m}(E(k)) \frac{dk}{4\pi^3} = \int_B \frac{\partial}{\partial k_\ell} G_m(E(k)) \frac{dk}{4\pi^3} = 0,$$

where G_m is such that $(dG_m/d\varepsilon)(\varepsilon) = \varepsilon^j F_{1,m}(\varepsilon)$. Therefore, the assertion follows from (6.34). $\square$

Expression (6.35) for the current densities can be interpreted as a drift-diffusion formulation with the diffusion current ∇D_{i0} and the drift current $-D_{i0}\nabla V/T$. The variables which "diagonalize" the diffusion matrix are given by $g_0 = D_{00}$ and $g_1 = D_{10}$. Then, the temperature is a function of the new variables, $T = T(g_0, g_1)$. We have to explain how it can be computed from g_0 and g_1. For this, we show that the following function can be inverted if the diffusion matrix is positive definite:

$$f(T) = \frac{g_1}{g_0} = \frac{\langle v \otimes d_0 \rangle}{\langle E v \otimes d_0 \rangle}, \tag{6.36}$$

and d_0 solves $Q_{\mathrm{el}}(d_0) = -vM$.

Lemma 6.11. *Under Maxwell–Boltzmann statistics, the derivative of the function f, defined in (6.36), is given by*

$$f'(T) = \frac{\det \mathscr{D}}{(TD_{00})^2},$$

where $\mathscr{D} = (D_{ij})$ is the diffusion matrix.

Thus, the invertibility of (6.36) is equivalent to the positive definiteness of the diffusion matrix $\mathscr{D}$.

Proof. Since

$$Q_{\mathrm{el}}\left(\frac{\partial d_0}{\partial T}\right) = \frac{\partial}{\partial T} Q_{\mathrm{el}}(d_0) = -vM \frac{E-\mu}{T^2} = Q_{\mathrm{el}}\left(\frac{E-\mu}{T^2} d_0\right),$$

we conclude that $\partial d_0/\partial T = d_0(E-\mu)/T^2 + F_1(E(k))$, where $F_1 \in N(Q_{\mathrm{el}})^3$. Thus, the derivative of D_{j0}, $j = 0, 1$, with respect to T becomes

$$\frac{\partial D_{j0}}{\partial T} = \left\langle E^j v \otimes \frac{\partial d_0}{\partial T} \right\rangle = T^{-2}(D_{j1} - \mu D_{j0}),$$

since, as in the proof of Proposition 6.10, $\langle E^j v \otimes F_1 \rangle = 0$. Therefore,

$$f'(T) = \frac{1}{D_{00}^2}\left(\frac{\partial D_{10}}{\partial T} D_{00} - D_{10}\frac{\partial D_{00}}{\partial T}\right) = \frac{D_{11}D_{00} - D_{01}D_{10}}{D_{00}^2 T^2},$$

which proves the lemma. $\square$

Summarizing, the energy-transport equations in the drift-diffusion formulation read as follows:

$$\partial_t n(g_0,g_1) - \operatorname{div} J_0 = 0, \quad \partial_t (ne)(g_0,g_1) - \operatorname{div} J_1 + J_0 \cdot \nabla V = W(g_0,g_1),$$

$$J_0 = \nabla g_0 - \frac{g_0}{T(g_0,g_1)} \nabla V, \quad J_1 = \nabla g_1 - \frac{g_1}{T(g_0,g_1)} \nabla V,$$

$$\lambda_D^2 \Delta V = n(g_0,g_1) - C(x),$$

where $T(g_0,g_1)$ is the unique solution of (6.36), the densities $n(g_0,g_1)$ and (ne) (g_0, g_1) are computed from

$$n(g_0,g_1) = e^{\mu/T} \int_B e^{-E(k)/T(g_0,g_1)} \frac{dk}{4\pi^3},$$

$$(ne)(g_0,g_1) = e^{\mu/T} \int_B e^{-E(k)/T(g_0,g_1)} E(k) \frac{dk}{4\pi^3},$$

and $u = \mu/T$ is, for given T, the unique solution of the nonlinear equation $D_{00}(u) = g_0$. This equation is uniquely solvable. Indeed, since $Q_{el}(\partial d_0/\partial u) = -vM/\partial u = -vM = Q_{el}(d_0)$ and hence, $\partial d_0/\partial u = d_0 + F_1$, where $F_1 \in N(Q_{el})^3$, we obtain $(dD_{00}/du)(u) = \langle v \otimes (\partial d_0/\partial u) \rangle = D_{00}$ and this is positive if $\mathscr{D}$ is positive definite.

We consider some examples.

Example 6.12 (Parabolic band approximation). In the parabolic band approximation, according to Example 6.8, the diffusion coefficients are given by

$$D_{00} = \mu_0 \Gamma(2-\beta) n T^{1/2-\beta}, \quad D_{10} = \mu_0 \Gamma(3-\beta) n T^{3/2-\beta}$$

(see (6.26)). Moreover, $ne = \frac{3}{2} nT$ and $W = 3n(1-T)/\tau_\beta(x,T)$ with $\tau_\beta(x,T)$ as in (6.27). Thus, the current densities in the drift-diffusion formulation read as

$$J_0 = \mu_0 \Gamma(2-\beta) \left(\nabla(n T^{1/2-\beta}) - n T^{-1/2-\beta} \nabla V \right),$$

$$J_1 = \mu_0 \Gamma(3-\beta) \left(\nabla(n T^{3/2-\beta}) - n T^{1/2-\beta} \nabla V \right).$$

The cases $\beta = 1/2$, $\beta = 0$, and $\beta = -1/2$ are of particular interest.

For $\beta = 1/2$, the energy-transport model reads as

$$\partial_t n - \operatorname{div} J_0 = 0, \quad \partial_t \left(\frac{3}{2} nT \right) - \operatorname{div} J_1 + J_0 \cdot \nabla V = W(n,T),$$

$$J_0 = \mu^* \left(\nabla n - n T^{-1} \nabla V \right), \quad J_1 = \frac{3}{2} \mu^* \left(\nabla(nT) - n \nabla V \right),$$

where $\mu^* = \sqrt{\pi} \mu_0/2$. This is the model studied by Chen et al. [7], mentioned in Example 6.8. Finite-element approximations for this so-called *Chen model* can be found, for instance, in [23, 25, 45].

In the case $\beta = 0$, we obtain the current densities (the evolution equations remain unchanged)

$$J_0 = \mu_0 \left(\nabla(nT^{1/2}) - nT^{-1/2}\nabla V \right), \quad J_1 = 2\mu_0 \left(\nabla(nT^{3/2}) - nT^{1/2}\nabla V \right).$$

The corresponding energy-transport model, called the *Lyumkis model*, was numerically solved by Lyumkis et al. [35], as mentioned in Example 6.8, and mathematically analyzed in [17]. However, it seems that the Chen model is preferred in numerical simulations.

Another choice is $\beta = -1/2$. Then the current densities can be written as

$$J_0 = \mu^* \left(\nabla(nT) - n\nabla V \right), \quad J_1 = \frac{5}{2}\mu^* \left(\nabla(nT^2) - nT\nabla V \right),$$

where now $\mu^* = 3\sqrt{\pi}\mu_0/4$. These expressions can be derived from the hydrodynamic equations in the diffusion limit if the heat conductivity is given by $\kappa = \frac{5}{2}nT$ (see Sect. 9.3). □

Example 6.13 (Nonparabolic band approximation). The energy-transport model in the nonparabolic band approximation $|k|^2/2 = E(1 + \alpha E)$ and drift-diffusion formulation is given by

$$\partial_t n - \operatorname{div} J_0 = 0, \quad \partial_t \left(\frac{3}{2}Q_\beta(T)nT \right) - \operatorname{div} J_1 + J_0 \cdot \nabla V = W(n,T),$$

$$J_0 = \nabla(\mu_{00}(\alpha,T)n) - \mu_{00}(\alpha,t)nT^{-1}\nabla V,$$

$$J_1 = \nabla(\mu_{10}(\alpha,T)n) - \mu_{10}(\alpha,t)nT^{-1}\nabla V,$$

where $Q_\beta(T)$ is defined in (6.28), μ_{ij} is introduced in (6.29), and the relaxation-time term $W(n,T)$ is made explicit in Example 6.9. □

Remark 6.14 (Drift-diffusion approximation for general energy bands). When inelastic phonon scattering is very strong, we recover the drift-diffusion equations from Sect. 5.1. In order to make this statement precise, we assume that the relaxation-time term $W(\mu,T)$ from (6.15) is strongly monotone in T and of order $\mathscr{O}(\alpha^{-1})$, where α^{-1} expresses the strength of inelastic collisions. Thus, we can rewrite (6.10) as

$$\partial_t n - \operatorname{div} J_0 = 0, \quad \partial_t(ne) - \operatorname{div} J_1 + J_0 \cdot \nabla V = \alpha^{-1}W(\mu,T).$$

Formally, the limit $\alpha \to 0$ in the second equation leads to $W(\mu,T) = 0$. Since, by Proposition 6.5, $W(\mu,1) = 0$ and since W is strongly monotone in T, this implies that $T = 1$. Hence, we can write the electron current density (6.34) as

$$J_0 = D_{00}\nabla(\mu - V),$$

and we have derived the *drift-diffusion equation*

$$0 = \partial_t n(\mu) - \operatorname{div} J_0 = \partial_t n(\mu) - \operatorname{div}(D_{00}\nabla(\mu - V)),$$

where the electron density $n = n(\mu)$ is a function of the chemical potential μ through

$$n(\mu) = \int_B \frac{1}{1 + e^{E(k)-\mu}} \frac{dk}{4\pi^3}$$

and the diffusion coefficient D_{00} is defined by

$$D_{00} = \int_B \nabla_k E(k) \otimes d_0 \frac{dk}{4\pi^3},$$

where d_0 is a solution of

$$Q_{el}(d_0) = -\nabla_k E(k) F(1 - F) = -\frac{\nabla_k E e^{E-\mu}}{(1 + e^{E-\mu})^2}.$$

The drift-diffusion equation is of parabolic type since $\partial n/\partial \mu$ is positive. It is a generalization of the drift-diffusion model (5.51) and (5.52), derived in Sect. 5.5, to general energy bands.

We can make the model more explicit if the energy band is spherically symmetric and if nondegenerate semiconductor materials are considered (such that the Fermi–Dirac distribution F can be approximated by the Maxwellian $e^{-(E(k)-\mu)}$). Then, by (6.18), $n = N_\gamma e^\mu$, and by (6.22), $D_{00} = D_{\gamma,s} n$, where N_γ only depends on the energy band $\gamma(\varepsilon)$ and $D_{\gamma,s}$ depends on the energy band and the scattering rate. We obtain $\nabla n = n\nabla\mu$ and hence,

$$\partial_t n - \text{div} J_0 = 0, \quad J_0 = D_{\gamma,s}(\nabla n - n\nabla V).$$

This is exactly of the form (5.14), derived from a different elastic collision operator. In contrast to the drift-diffusion equations derived in Sect. 5.1, the above model is valid for more general band structures. $\quad\square$

References

1. N. Ben Abdallah and P. Degond. On a hierarchy of macroscopic models for semiconductors. *J. Math. Phys.* 37 (1996), 3308–3333.
2. P. Degond, C. Levermore, and C. Schmeiser. A note on the energy-transport limit of the semi-conductor Boltzmann equation. In: N. Ben Abdallah et al. (eds.), *Proceedings of Transport in Transition Regimes* (Minneapolis, 2000), IMA Math. Appl. 135, 137–153. Springer, New York, 2004.
3. N. Ben Abdallah, P. Degond, and S. Génieys. An energy-transport model for semiconductors derived from the Boltzmann equation. *J. Stat. Phys.* 84 (1996), 205–231.
4. R. Stratton. Diffusion of hot and cold electrons in semiconductor barriers. *Phys. Rev.* 126 (1962), 2002–2014.
5. Y. Apanovich, E. Lyumkis, B. Polski, A. Shur, and P. Blakey. A comparison of energy balance and simplified hydrodynamic models for GaAs simulation. *COMPEL* 12 (1993), 221–230.
6. M. Rudan, A. Gnudi, and W. Quade. A generalized approach to the hydrodynamic model of semiconductor equations. In: G. Baccarani (ed.), *Process and Device Modeling for Microelectronics*, 109–154. Elsevier, Amsterdam, 1993.

7. D. Chen, E. Kan, U. Ravaioli, C. Shu, and R. Dutton. An improved energy transport model including nonparabolicity and non-Maxwellian distribution effects. *IEEE Electr. Device Lett.* 13 (1992), 26–28.

8. A. Forghieri, R. Guerrieri, P. Ciampolini, A. Gnudi, M. Rudan, and G. Baccarani. A new discretization strategy of the semiconductor equations comprising momentum and energy balance. *IEEE Trans. Computer-Aided Design Integr. Circuits Sys.* 7 (1988), 231–242.

9. K. Souissi, F. Odeh, H. Tang, and A. Gnudi. Comparative studies of hydrodynamic and energy transport models. *COMPEL* 13 (1994), 439–453.

10. D. Woolard, H. Tian, R. Trew, M. Littlejohn, and K. Kim. Hydrodynamic electron-transport: Nonparabolic corrections to the streaming terms. *Phys. Rev. B* 44 (1991), 11119–11132.

11. W. Allegretto and H. Xie. Nonisothermal semiconductor systems. In: X. Liu and D. Siegel (eds.), *Comparison Methods and Stability Theory*. Lect. Notes Pure Appl. Math. 162, 17–24. Marcel Dekker, New York, 1994.

12. P. Degond, S. Génieys, and A. Jüngel. A system of parabolic equations in nonequilibrium thermodynamics including thermal and electrical effects. *J. Math. Pures Appl.* 76 (1997), 991–1015.

13. P. Degond, S. Génieys, and A. Jüngel. A steady-state system in nonequilibrium thermodynamics including thermal and electrical effects. *Math. Meth. Appl. Sci.* 21 (1998), 1399–1413.

14. A. Jüngel. Regularity and uniqueness of solutions to a parabolic system in nonequilibrium thermodynamics. *Nonlin. Anal.* 41 (2000), 669–688.

15. W. Fang and K. Ito. Existence of stationary solutions to an energy drift-diffusion model for semiconductor devices. *Math. Models Meth. Appl. Sci.* 11 (2001), 827–840.

16. J. Griepentrog. An application of the implicit function theorem to an energy model of the semiconductor theory. *Z. Angew. Math. Mech.* 79 (1999), 43–51.

17. L. Chen and L. Hsiao. The solution of Lyumkis energy transport model in semiconductor science. *Math. Meth. Appl. Sci.* 26 (2003), 1421–1433.

18. L. Chen, L. Hsiao, and Y. Li. Large time behavior and energy relaxation time limit of the solutions to an energy transport model in semiconductors. *J. Math. Anal. Appl.* 312 (2005), 596–619.

19. Y. Apanovich, P. Blakey, R. Cottle, E. Lyumkis, B. Polsky, A. Shur, and A. Tcherniaev. Numerical simulations of submicrometer devices including coupled nonlocal transport and non-isothermal effects. *IEEE Trans. Electr. Devices* 42 (1995), 890–897.

20. M. Fournié. Numerical discretization of energy-transport model for semiconductors using high-order compact schemes. *Appl. Math. Letters* 15 (2002), 727–734.

21. C. Ringhofer. An entropy-based finite difference method for the energy transport system. *Math. Models Meth. Appl. Sci.* 11 (2001), 769–796.

22. F. Bosisio, R. Sacco, F. Saleri, and E. Gatti. Exponentially fitted mixed finite volumes for energy balance models in semiconductor device simulation. In: H. Bock et al. (eds.), *Proceedings of ENUMATH 97*, 188–197. World Scientific, Singapore, 1998.

23. P. Degond, A. Jüngel, and P. Pietra. Numerical discretization of energy-transport models for semiconductors with nonparabolic band structure. *SIAM J. Sci. Comput.* 22 (2000), 986–1007.

24. S. Gadau and A. Jüngel. A 3D mixed finite-element approximation of the semiconductor energy-transport equations. *SIAM J. Sci. Comput.* 31 (2008), 1120–1140.

25. S. Holst, A. Jüngel, and P. Pietra. An adaptive mixed scheme for energy-transport simulations of field-effect transistors. *SIAM J. Sci. Comput.* 25 (2004), 1698–1716.

26. C. Lab and P. Caussignac. An energy-transport model for semiconductor heterostructure devices: Application to AlGaAs/GaAs MODFETs. *COMPEL* 18 (1999), 61–76.

27. A. Marrocco and P. Montarnal. Simulation de modèles "energy transport" à l'aide des éléments finis mixtes. *C. R. Acad. Sci. Paris, Sér. I* 323 (1996), 535–541.

28. J. Jerome and C.-W. Shu. Energy models for one-carrier transport in semiconductor devices. In: W. Coughran et al. (eds.), *Semiconductors*, Part II, IMA Math. Appl. 59, 185–207. Springer, New York, 1994.

29. S. de Groot and P. Mazur. *Nonequilibrium Thermodynamics*. Dover Publications, New York, 1984.
30. H. Kreuzer. *Nonequilibrium Thermodynamics and Its Statistical Foundation*. Clarondon Press, Oxford, 1981.
31. H. W. Alt and S. Luckhaus. Quasilinear elliptic-parabolic differential equations. *Math. Z.* 183 (1983), 311–341.
32. M. Lundstrom. *Fundamentals of Carrier Transport*. 2nd edition, Cambridge University Press, Cambridge, 2000.
33. K. Brennan. *The Physics of Semiconductors*. Cambridge University Press, Cambridge, 1999.
34. C. Schmeiser and A. Zwirchmayr. Elastic and drift-diffusion limits of electron–phonon interaction in semiconductors. *Math. Models Meth. Appl. Sci.* 8 (1998), 37–53.
35. E. Lyumkis, B. Polsky, A. Shur, and P. Visocky. Transient semiconductor device simulation including energy balance equation. *COMPEL* 11 (1992), 311–325.
36. A. Jüngel. *Quasi-hydrodynamic Semiconductor Equations*. Birkhäuser, Basel, 2001.
37. Y. Choi and R. Lui. Multi-dimensional electrochemistry model. *Arch. Rat. Mech. Anal.* 130 (1995), 315–342.
38. Z. Deyl (ed.). *Electrophoresis: A Survey of Techniques and Applications*. Elsevier, Amsterdam, 1979.
39. A. Bermudez and C. Saguez. Mathematical formulation and numerical solution of an alloy solidification problem. In: A. Fasano (ed.), *Free Boundary Problems: Theory and Applications*, Vol. 1, 237–247. Pitman, Boston, 1983.
40. R. Hills, D. Loper, and P. Roberts. A thermodynamically consistent model of a mushy zone. *Quart. J. Mech. Appl. Math.* 36 (1983), 505–539.
41. S. de Groot. *Thermodynamik irreversibler Prozesse*. Bibliographisches Institut, Mannheim, 1960.
42. G. Albinus. A thermodynamically motivated formulation of the energy model of semiconductor devices. Preprint No. 210, WIAS Berlin, Germany, 1995.
43. S. Kawashima and Y. Shizuta. On the normal form of the symmetric hyperbolic-parabolic systems associated with the conservation laws. *Tohoku Math. J., II. Ser.* 40 (1988), 449–464.
44. P. Degond, S. Génieys, and A. Jüngel. Symmetrization and entropy inequality for general diffusion equations. *C. R. Acad. Sci. Paris, Sér. I* 325 (1997), 963–968.
45. S. Holst, A. Jüngel, and P. Pietra. A mixed finite-element discretization of the energy-transport equations for semiconductors. *SIAM J. Sci. Comput.* 24 (2003), 2058–2075.

Chapter 7
Spherical Harmonics Expansion Equations

The spherical harmonics expansion (SHE) model can be derived from the Boltzmann equation by the three-step procedure introduced in Sect. 2.4. In contrast to the previous chapters, we do not integrate the Boltzmann equation over the whole wave-vector space but only over the isoenergetic wave-vector space. As a result, the variable is still a distribution function, but depending on the position-energy space (x, ε) (and time) only and not on the position-wave-vector space (x, k) (and time). Thus, we are able to reduce the seven-dimensional Boltzmann equation to a five-dimensional problem.

One may argue that the SHE model does not fit in the hierarchy of diffusive moment models, presented in this part, which have the property that the moments only depend on the position and time variables. However, the SHE equations contain a diffusive term and their derivation is similar to that of the energy-transport equations in Chap. 6 such that it seems appropriate to present the SHE model at this place.

7.1 Derivation from the Boltzmann Equation

As in Chaps. 5 and 6, we start with the semi-classical Boltzmann equation

$$\partial_t f + v(k) \cdot \nabla_x f + \frac{q}{\hbar} \nabla_x V \cdot \nabla_k f = Q_{\text{el}}(f) + Q_{\text{in}}(f), \quad x \in \mathbb{R}^3, \ k \in B, \ t > 0,$$

with initial condition $f(x,k,0) = f_I(x,k)$, $x \in \mathbb{R}^3$, $k \in B$, and the velocity $v(k) = \nabla_k E(k)$ defined by the energy band $E(k)$. The elastic collision operator reads as

$$(Q_{\text{el}}(f))(x,k,t) = \int_B \sigma_{\text{el}}(x,k,k')\delta(E'-E)(f'-f)\,\mathrm{d}k', \qquad (7.1)$$

where the scattering rate $\sigma_{\text{el}}(x,k,k')$ is symmetric in k and k', and Q_{in} denotes the inelastic collision operator modeling, for instance, inelastic phonon scattering and electron–electron collisions. First we scale the Boltzmann equation. The scaling is

Jüngel, A.: *Spherical Harmonics Expansion Equations.* Lect. Notes Phys. **773**, 157–170 (2009)
DOI 10.1007/978-3-540-89526-8_7

as in Sect. 6.1, i.e., with the mean free path of elastic and inelastic collisions λ_{el} and λ_{in}, respectively, we introduce the reference length $\lambda_0 = \sqrt{\lambda_{el}\lambda_{in}}$ and the parameter $\alpha = \sqrt{\lambda_{el}/\lambda_{in}}$. Then the scaled equation reads as follows:

$$\alpha^2 \partial_t f + \alpha \left(v(k) \cdot \nabla_x f + \nabla_x V \cdot \nabla_k f \right) = Q_{el}(f) + \alpha^2 Q_{in}(f), \qquad (7.2)$$

employing the same notation for the scaled and unscaled variables. We suppose that α is a small (positive) parameter such that it makes sense to study the limit $\alpha \to 0$ in (7.2).

The derivation of the SHE model is performed in three steps following [1]. Let (f_α, V_α) be a (smooth) solution of (7.2) and the Poisson equation (5.3). First we perform the limit $\alpha \to 0$ in (7.2). Then $Q_{el}(f) = 0$, where $f = \lim_{\alpha \to 0} f_\alpha$. By Proposition 4.5 (3) in Sect. 4.2, it follows that f is a function of the energy only, i.e., $f(x,k,t) = F(x,E(k),t)$ for some function F. Second, inserting the Chapman–Enskog expansion $f_\alpha = F + \alpha g_\alpha$ in (7.2) and dividing by α, we infer that

$$\alpha \partial_t f_\alpha + (v(k) \cdot \nabla_x F + \nabla_x V \cdot \nabla_k F)$$
$$+ \alpha \left(v(k) \cdot \nabla_x g_\alpha + \nabla_x V \cdot \nabla_k g_\alpha \right) = Q_{el}(g_\alpha) + \alpha Q_{in}(f_\alpha).$$

Here we have used that Q_{el} is linear and $Q_{el}(F) = 0$. The (formal) limit $\alpha \to 0$ yields

$$Q_{el}(g) = v(k) \cdot \nabla_x F + \nabla_x V \cdot \nabla_k F = v(k) \cdot (\nabla_x F + \nabla_x V \partial_\varepsilon F),$$

where $g = \lim_{\alpha \to 0} g_\alpha$ and $\partial_\varepsilon F = \partial F / \partial \varepsilon$. By Lemma 6.1, this equation is solvable if and only if

$$\int_B v(k) \cdot (\nabla_x F + \nabla_x V \partial_\varepsilon F) \delta(E(k) - \varepsilon) \, dk = 0 \quad \text{for all } \varepsilon.$$

This is the case since the above integral is equal to

$$(\nabla_x F + \nabla_x V \partial_\varepsilon F)(x, \varepsilon, t) \cdot \int_B \nabla_k E(k) \delta(E(k) - \varepsilon) \, dk,$$

and the integral vanishes (see (6.6)). Let $d_0(x,k,t)$ be the unique solution in $N(Q_{el})^\perp$ of

$$Q_{el}(d_0) = -v(k). \qquad (7.3)$$

We claim that $g = -(\nabla_x F + \nabla_x V \partial_\varepsilon F) \cdot d_0$ (up to the addition of an element of $N(Q_{el})$). Indeed, since F depends on $E(k)$ only and the collision integral is taken over the isoenergetic surface, we have

$$Q_{el}(g) = -(\nabla_x F + \nabla_x V \partial_\varepsilon F) \cdot Q_{el}(d_0) = (\nabla_x F + \nabla_x V \partial_\varepsilon F) \cdot v(k). \qquad (7.4)$$

The third step is to integrate the Boltzmann equation (7.2) over the isoenergetic surface,

$$\int_B \left(\partial_t F + \alpha \partial_t g \alpha + \alpha^{-1} (v(k) \cdot \nabla_x F + \nabla_x V \cdot \nabla_k F) \right.$$

$$+ (v(k) \cdot \nabla_x g \alpha + \nabla_x V \cdot \nabla_k g \alpha) - \alpha^{-1} Q_{el}(g\alpha) - Q_{in}(f\alpha) \big)$$

$$\times \delta(E(k) - \varepsilon) \frac{dk}{4\pi^3} = 0.$$

The isoenergetic surface integral over $Q_{el}(f)$ vanishes for all functions f. Therefore, by (7.4),

$$\int_B (v(k) \cdot \nabla_x F + \nabla_x V \cdot \nabla_k F) \delta(E(k) - \varepsilon) \frac{dk}{4\pi^3} = \int_B Q_{el}(g)\delta(E(k) - \varepsilon) \frac{dk}{4\pi^3} = 0.$$

We conclude that all terms of order α^{-1} vanish, and the limit $\alpha \to 0$ leads to

$$\int_B (\partial_t F + v(k) \cdot \nabla_x g + \nabla_x V \cdot \nabla_k g - Q_{in}(F)) \, \delta(E(k) - \varepsilon) \frac{dk}{4\pi^3} = 0.$$

With the density of states of energy ε,

$$N(\varepsilon) = \int_B \delta(E(k) - \varepsilon) \frac{dk}{4\pi^3}, \quad \varepsilon \in \mathbb{R}, \tag{7.5}$$

the first summand in the above integral becomes

$$\int_B \partial_t F \delta(E(k) - \varepsilon) \frac{dk}{4\pi^3} = N(\varepsilon)\partial_t F.$$

The second summand is defined as the divergence of the electron current density and it follows, employing $g = -(\nabla_x F + \nabla_x V \partial_\varepsilon F) \cdot d_0$, that

$$J(x, \varepsilon, t) = -\int_B v(k) g \delta(E(k) - \varepsilon) \frac{dk}{4\pi^3}$$

$$= D(x, \varepsilon)(\nabla_x F + \nabla_x V \partial_\varepsilon F)(x, \varepsilon, t), \tag{7.6}$$

where

$$D(x, \varepsilon) = \int_B \nabla_k E(k) \otimes d_0 \delta(E(k) - \varepsilon) \frac{dk}{4\pi^3} \in \mathbb{R}^{3 \times 3} \tag{7.7}$$

is the diffusion matrix. For the computation of the third summand, we choose a smooth test function ψ with compact support in the range $R(E)$ of E and use the definition of the delta distribution to obtain

$$\int_{\mathbb{R}} \psi(\varepsilon) \int_B \nabla_k g \delta(E(k) - \varepsilon) \, dk \, d\varepsilon = \int_B \psi(E(k)) \nabla_k g(E(k)) \, dk$$

$$= -\int_B g(E(k)) \psi'(E(k)) \nabla_k E(k) \, dk$$

$$= -\int_{\mathbb{R}} \psi'(\varepsilon) \int_B v(k) g(\varepsilon) \delta(E(k) - \varepsilon) \, dk \, d\varepsilon$$

$$= \int_{\mathbb{R}} \psi'(\varepsilon) J(\varepsilon) \, d\varepsilon = -\int_{\mathbb{R}} \psi(\varepsilon) \frac{\partial J}{\partial \varepsilon}(\varepsilon) \, d\varepsilon.$$

Since ψ is arbitrary, we conclude that

$$\int_B \nabla_k g \delta(E(k) - \varepsilon) \frac{dk}{4\pi^3} = -\frac{\partial J}{\partial \varepsilon}(x, \varepsilon, t).$$

Finally, the fourth summand defines the averaged inelastic collision term. We have shown the following theorem.

Theorem 7.1 (Spherical harmonics expansion equations). *Let the scattering rate of the elastic collision operator* (7.1) *be symmetric and let* (f_α, V_α) *be a solution of the Boltzmann–Poisson system* (7.2) *and* (5.3)*. Then the (formal) limit functions* $F = \lim_{\alpha \to 0} f_\alpha$ *and* $V = \lim_{\alpha \to 0} V_\alpha$ *satisfy the* spherical harmonics expansion (SHE) *equations*

$$N(\varepsilon)\partial_t F - \operatorname{div} J - \nabla V \cdot \frac{\partial J}{\partial \varepsilon} = S(F), \tag{7.8}$$

$$\lambda_D^2 \Delta V = n - C(x), \quad x \in \mathbb{R}^3,\ \varepsilon \in \mathbb{R},\ t > 0,$$

where the density of states $N(\varepsilon)$ *of energy* ε *is defined in* (7.5) *and the particle current density* $J(x, \varepsilon, t)$ *is introduced in* (7.6)*. With the diffusion matrix* (7.7)*, the particle density is given by* $n(x,t) = \int_{\mathbb{R}} F(x, \varepsilon, t)N(\varepsilon)\,d\varepsilon$*, and the averaged collision operator equals*

$$(S(F))(x, \varepsilon, t) = \int_B (Q_{\mathrm{in}}(F))(x, k, t)\delta(E(k) - \varepsilon)\frac{dk}{4\pi^3}.$$

Finally, the initial condition reads as

$$F(x, \varepsilon, t) = \int_B f_I(x, k, t)\delta(E(k) - \varepsilon)\frac{dk}{4\pi^3}, \quad x \in \mathbb{R}^3,\ \varepsilon \in \mathbb{R},\ t > 0.$$

Originally, the SHE model was derived in the physics literature from a truncated expansion of the Boltzmann equation in spherical harmonics (therefore its name). In 1956, an approximation of the Boltzmann equation was given by Herring and Vogt [2], truncating the expansion after the first term. In the following decades, the method was applied to the Boltzmann equation under various assumptions on the collision processes, for instance, including acoustic phonon and intervalley scattering for small electric fields [3], also including impact ionization scattering [4], or modeling multi-band effects [5]. The SHE model also appears in the studies of Stratton (see, e.g., [6, 7]). The first derivation of this model from a diffusion approximation of the Boltzmann equation is due to Dmitruk, Saul, and Reyna [8]. Often, spherically symmetric band diagrams were assumed for the derivation [9, 10]. A general band structure was employed in the works of Vecchi et al. [4, 11]. A more mathematical derivation was performed by Ben Abdallah, Degond [1], and co-workers; see, for instance, [12–14]. Schmeiser and Zwirchmayr gave a rigorous justification of the diffusion approximation for parabolic bands and electron–phonon scattering [15]. Numerical simulations of the SHE model can be found, for instance, in [4, 5, 16, 17]. An existence and uniqueness result for transient solutions was shown in [18].

The SHE equations were also employed in the context of collisions of electrons with the isolator surface in MOS transistors [19] and to model semiconductor superlattices [20]. Moreover, it appears in gas discharge [21] and plasma physics [22, 23].

The advantages of the SHE model are twofold. First, as already mentioned, the equations have to be solved in four dimensions of the position-energy space instead of the six-dimensional phase space of the Boltzmann equation. Second, the SHE equation (7.8) is mathematically of parabolic type which simplifies the analysis and numerical solution. In order to see the diffusive structure, we introduce the total energy variable $u = \varepsilon - V(x,t)$ and the transformed functions

$$f(x, \varepsilon - V(x,t), t) = F(x, \varepsilon, t),$$
$$d(x, \varepsilon - V(x,t), t) = D(x, \varepsilon),$$
$$\rho(x, \varepsilon - V(x,t), t) = N(\varepsilon).$$

Then $\nabla_x F = \nabla_x f - \nabla_x V \partial_u f = \nabla_x f - \nabla_x V \partial_\varepsilon F$ and $\partial_t F = \partial_t f - \partial_t V \partial_u f$. Therefore, with the notation

$$\nabla_* = \nabla_x + \nabla_x V \frac{\partial}{\partial \varepsilon},$$

we can write (7.8) in symmetric form as

$$N(\varepsilon) \partial_t F - \nabla_* \cdot J = S(F), \quad J = D \nabla_* F.$$

In the total energy variable, we obtain, since $\nabla_* F = \nabla_x F + \nabla_x V \partial_\varepsilon F = \nabla_x f$,

$$\rho(x, u, t) \partial_t f - \mathrm{div}_x (d \nabla_x f)(x, u, t) = S(f) + \rho(x, u, t) \partial_t V \frac{\partial f}{\partial u}.$$

The parabolicity property of the differential operator now follows from the positive definiteness of the diffusion matrix d or D, as stated in the following proposition.

Proposition 7.2 (Properties of the diffusion matrix). *The diffusion matrix (7.7) is symmetric and positive semi-definite.*

Proof. We show first the symmetry. We compute, for any (smooth) test function $\psi(\varepsilon)$, employing (7.3),

$$\int_{\mathbb{R}} D_{ij}(x, \varepsilon) \psi(\varepsilon) \, d\varepsilon = \int_{\mathbb{R}} \int_B \frac{\partial E}{\partial k_i} d_{0,j} \delta(E(k) - \varepsilon) \psi(\varepsilon) \frac{dk}{4\pi^3} \, d\varepsilon$$

$$= -\int_B Q_{\mathrm{el}}(d_{0,i}) d_{0,j} \psi(E(k)) \frac{dk}{4\pi^3}$$

$$= -\int_B Q_{\mathrm{el}}(\psi(E(k)) d_{0,i}) d_{0,j} \frac{dk}{4\pi^3}.$$

The last equality holds since $\psi(E(k))$ only depends on the energy. By Proposition 4.5, Q_{el} is symmetric on $L^2(B)$ and hence,

$$\int_{\mathbb{R}} D_{ij}(x,\varepsilon)\psi(\varepsilon)\,d\varepsilon = -\int_B \psi(E(k))d_{0,i}Q_{\mathrm{el}}(d_{0,j})\frac{dk}{4\pi^3} = \int_{\mathbb{R}} D_{ji}(x,\varepsilon)\psi(\varepsilon)\,d\varepsilon.$$

Since ψ is arbitrary, $D_{ij} = D_{ji}$.

Next, we show that (D_{ij}) is positive semi-definite. Let $z \in \mathbb{R}^3$. Then, by the definition of Q_{el},

$$z^\top D(x,\varepsilon)z = -\sum_{i,j=1}^3 \int_B z_i Q_{\mathrm{el}}(d_{0,i})d_{0,j}z_j\delta(E(k)-\varepsilon)\frac{dk}{4\pi^3}$$

$$= -\sum_{i,j=1}^3 \int_{B^2} \sigma_{\mathrm{el}}(x,k,k')\delta(E'-E)\delta(E-\varepsilon)z_i(d'_{0,i}-d_{0,i})z_jd_{0,j}\frac{dk'\,dk}{4\pi^3}.$$

Similar as in the proof of Proposition 4.5 (2), we can write

$$z^\top D(x,\varepsilon)z = \frac{1}{2}\sum_{i,j=1}^3 \int_{B^2} \sigma_{\mathrm{el}}(x,k,k')\delta(E'-E)\delta(E-\varepsilon)$$

$$\times z_i(d'_{0,i}-d_{0,i})z_j(d'_{0,j}-d_{0,j})\frac{dk'\,dk}{4\pi^3}$$

$$= \frac{1}{2}\int_{B^2} \sigma_{\mathrm{el}}\delta(E'-E)\delta(E-\varepsilon)\left|z\cdot(d'_0-d_0)\right|^2\frac{dk'\,dk}{4\pi^3} \geq 0.$$

This finishes the proof.   □

Ben Abdallah and Degond showed in [1, Prop. 3.6] a stronger property of the diffusion matrix: There exists a constant $K > 0$ such that

$$D_{ij}(x,\varepsilon) \geq \frac{K}{N(\varepsilon)}\int_B \frac{\partial E}{\partial k_i}\frac{\partial E}{\partial k_j}\delta(E(k)-\varepsilon)\frac{dk}{4\pi^3}, \quad i,j = 1,2,3.$$

The right-hand side defines a symmetric matrix which is degenerate at the critical points of E (i.e., $\nabla_k E(k) = 0$ at $\varepsilon = E(k)$). We show in Sect. 7.2 that this inequality is sharp (see (7.10)). The proof of the above inequality is based on the property

$$-\int_B Q_{\mathrm{el}}(f)f\frac{dk}{4\pi^3} \geq K_0 \int_B (f-[f])^2 N(E(k))\frac{dk}{4\pi^3},$$

where $K_0 > 0$ is a constant and $[f]$ the orthogonal projection on $N(Q_{\mathrm{el}})$,

$$[f](k) = \frac{1}{N(E(k))}\int_B f(k')\delta(E(k')-E(k))\frac{dk'}{4\pi^3}. \tag{7.9}$$

For a proof of the above property, we refer to Proposition 3.1 in [1].

Finally, we state the SHE equations in physical variables:

$$N(\varepsilon)\partial_t F - \frac{1}{q}\mathrm{div}_x J - \nabla_x V \cdot \frac{\partial J}{\partial\varepsilon} = S(F),$$

$$J = D(x,\varepsilon)\left(\nabla_x F + q\nabla_x V\frac{\partial F}{\partial\varepsilon}\right),$$

where $N(\varepsilon)$ is given by (7.5) and

$$D(x,\varepsilon) = \frac{q}{\hbar} \int_B \nabla_k E(k) \otimes d_0(x,k) \delta(E(k) - \varepsilon) \frac{dk}{4\pi^3}.$$

7.2 Some Explicit Models

The SHE model can be made more explicit for spherical symmetric energy bands. Therefore, we suppose similar as in Sect. 6.2 (and in [1]):

1. The scattering rate depends only on the energy,

$$\sigma_{el}(x,k,k') = s(x,E(k)) \quad \text{for all } k, k' \text{ with } E(k) = E(k').$$

2. The energy is spherical symmetric, i.e., $E = E(|k|)$, and strictly monotone in $|k|$. Thus, there exists a function γ such that $|k|^2 = \gamma(E(|k|))$ for all $k \in \mathbb{R}^3$.

We refer to the beginning of Sect. 6.2 for some comments on these assumptions. The SHE model becomes more explicit when we specify the density of states $N(\varepsilon)$ and the diffusion matrix $D(x,\varepsilon)$.

Proposition 7.3. *Under the above hypotheses on the scattering rate and the energy band, we have*

$$N(\varepsilon) = \frac{1}{2\pi^2} \sqrt{\gamma(\varepsilon)} \gamma'(\varepsilon), \quad D(x,\varepsilon) = \frac{4}{3} \frac{\gamma(\varepsilon)}{s(x,\varepsilon)\gamma'(\varepsilon)^2} \text{Id},$$

where Id *denotes the identity matrix in* $\mathbb{R}^{3\times3}$.

The formula for the diffusion matrix was also obtained by Ventura et al. [10] up to a multiplicative constant.

Proof. The expression for the density of states was already proved in Sect. 6.2; see the proof of Proposition 6.6. In that proof we have also shown that the hypothesis on the scattering rate allows us to simplify the elastic collision operator,

$$Q_{el}(f) = \frac{[f] - f}{\tau(x,E(k))}, \quad \text{where } \tau(x,\varepsilon) = \frac{1}{4\pi^3 s(x,\varepsilon)N(\varepsilon)},$$

and $[f]$ is defined in (7.9). Then, the solution of $Q_{el}(d_0) = -v(k) = -\nabla_k E(k)$ can be written explicitly as

$$d_0 = \tau(x,E(k))\nabla_k E(k),$$

which follows from the fact that $[d_0] = 0$. Hence,

$$D(x,\varepsilon) = \tau(x,\varepsilon) \int_{\mathbb{R}^3} \nabla_k E(k) \otimes \nabla_k E(k)\delta(E(k) - \varepsilon)\,dk. \tag{7.10}$$

Now, the proof proceeds as for Proposition 6.6. $\square$

Finally, we consider two examples.

Example 7.4 (Parabolic band approximation). In addition to the assumptions at the beginning of this section, we suppose that

- the energy band is parabolic, $E(k) = \frac{1}{2}|k|^2$, $k \in \mathbb{R}^3$, and
- the scattering rate is given by $s(x, \varepsilon) = s_1(x)\varepsilon^\beta$, $\beta \geq 0$.

The first assumption implies that $\gamma(\varepsilon) = 2\varepsilon$ and

$$N(\varepsilon) = \frac{\sqrt{2\varepsilon}}{\pi^2}, \quad D(x, \varepsilon) = \frac{2\varepsilon}{3s(x, \varepsilon)} \, \text{Id}.$$

With the second hypothesis, we obtain $D(x, \varepsilon) = (2/3s_1(x))\varepsilon^{1-\beta} \, \text{Id}$. Thus, the SHE equations read as

$$\frac{\sqrt{2\varepsilon}}{\pi^2} \partial_t F - \text{div}_x J - \nabla_x V \cdot \frac{\partial J}{\partial \varepsilon} = S(F), \quad x \in \mathbb{R}^3, \ \varepsilon > 0, \ t > 0,$$

$$J = \frac{2}{3s_1(x)} \varepsilon^{1-\beta} \left(\nabla_x F + \nabla_x V \frac{\partial F}{\partial \varepsilon} \right).$$

We notice that this equation is mathematically of degenerate type since the diffusion coefficient vanishes at $\varepsilon = 0$ if $\beta < 1$. Another mathematical difficulty arises from the fact that at critical points of the electric potential (i.e., $\nabla_x V(x, t) = 0$), we do not obtain information on the derivative of F with respect to ε. We refer to [18] for details.

It remains to determine the averaged collision operator $S(F)$. The precise structure depends on the assumptions on the inelastic collision integral. Here, we propose a simplified expression, the Fokker–Planck approximation

$$S(F) = s_0 \frac{\partial}{\partial \varepsilon} \left(\varepsilon^\beta N(\varepsilon)^2 \left(F + \frac{\partial F}{\partial \varepsilon} \right) \right),$$

similar to Proposition 6.7 in Sect. 6.2. $\square$

Example 7.5 (Nonparabolic band approximation). In the nonparabolic band approximation of Kane $\frac{1}{2}|k|^2 = E(1 + \alpha E)$ or $\gamma(\varepsilon) = 2\varepsilon(1 + \alpha\varepsilon)$, we obtain from Proposition 7.3 the formulas

$$N(\varepsilon) = \frac{1}{\pi^2} \sqrt{2\varepsilon(1 + \alpha\varepsilon)}(1 + 2\alpha\varepsilon), \quad D(x, \varepsilon) = \frac{2}{3} \frac{\varepsilon^{1-\beta}(1 + \alpha\varepsilon)}{s_1(x)(1 + 2\alpha\varepsilon)^2} \, \text{Id}.$$

These expressions have to be employed in (7.8) and (7.6). $\square$

7.3 Diffusion Approximation

The energy-transport equations can be derived from the SHE model in the diffusion approximation. We proceed in the following as in [1]. We recall the SHE equations from Sect. 7.1,

$$N(\varepsilon)\partial_t F - \mathrm{div}_x J - \nabla_x V \cdot \frac{\partial J}{\partial \varepsilon} = S(F), \quad x \in \mathbb{R}^3, \ \varepsilon \in \mathbb{R}, \ t > 0, \tag{7.11}$$

$$J = D(x,\varepsilon)\left(\nabla_x F + \nabla_x V \frac{\partial F}{\partial \varepsilon}\right). \tag{7.12}$$

Our main assumption is that the inelastic collision operator is the sum of electron–electron scattering and inelastic phonon collisions and that the carrier–carrier scattering is dominant:

$$Q_{\mathrm{in}}(f) = \frac{1}{\alpha}Q_{\mathrm{ee}}(f) + Q_{\mathrm{ph}}(f),$$

where $\alpha > 0$ is a small parameter and Q_{ee} and Q_{ph} are defined in (6.2) and (6.16), respectively. This assumption means that the energy loss due to inelastic phonon scattering occurs on a longer time scale than carrier–carrier collisions. Then, by the definition of the averaged collision term $S(F)$, we can write

$$S(F) = \frac{1}{\alpha}S_{\mathrm{ee}}(F) + S_{\mathrm{ph}}(F), \tag{7.13}$$

where S_{ee} and S_{ph} are defined by

$$(S_{\mathrm{ee}}(F))(x,\varepsilon,t) = \int_B (Q_{\mathrm{ee}}(F))(x,k,t)\delta(E(k) - \varepsilon)\,dk,$$

$$(S_{\mathrm{ph}}(F))(x,\varepsilon,t) = \int_B (Q_{\mathrm{ph}}(F))(x,k,t)\delta(E(k) - \varepsilon)\,dk.$$

The averaged collision operators have the following properties.

Lemma 7.6. (1) *The operator S_{ee} conserves mass and energy and S_{ph} conserves mass in the sense*

$$\int_{\mathbb{R}} S_{\mathrm{ee}}(F)\,d\varepsilon = \int_{\mathbb{R}} S_{\mathrm{ee}}(F)\varepsilon\,d\varepsilon = \int_{\mathbb{R}} S_{\mathrm{ph}}(F)\,d\varepsilon = 0$$

for all functions F.

(2) *The kernel of S_{ee} consists of Fermi–Dirac distributions*

$$F_{\mu,T} = \frac{1}{1 + e^{(\varepsilon - \mu)/T}},$$

where $\mu \in \mathbb{R}$ and $T > 0$.

(3) *For given $F = F(\varepsilon)$, let $L_F = DS_{\mathrm{ee}}(F)$ denote the (Fréchet) derivative of S_{ee} at the point F. Then the operator equation $L_F(G) = H$ is solvable if and only if*

$$\int_{\mathbb{R}} H(\varepsilon)\,d\varepsilon = \int_{\mathbb{R}} H(\varepsilon)\varepsilon\,d\varepsilon = 0. \tag{7.14}$$

Proof. The first statement for S_{ee} is proved in Lemma 6.2. The statement for S_{ph} follows directly from the definition of Q_{ph}. The second statement is shown similarly as in the proof of Proposition 4.6. For the last assertion, we observe that the derivative $DQ_{ee}(f)$ reads as

$$
(DQ_{ee}(f))(g) = \int_{B^3} \sigma_{ee}(k,k',k_1,k_1')\delta(E'+E_1'-E-E_1)
$$
$$
\times f'f_1'(1-f)(1-f_1)(\tilde{g}'+\tilde{g}_1'-\tilde{g}-\tilde{g}_1)\,dk\,dk'\,dk_1\,dk_1',
$$

where $\tilde{g} = g/f(1-f)$. Similarly as in the proof of Proposition 4.6, the following equation can be verified:

$$
\int_{\mathbb{R}} L_F(G)H\,d\varepsilon = -\frac{1}{4}\int_{\mathbb{R}}\int_{B^4}\sigma_{ee}(k,k',k_1,k_1')\delta(E'+E_1'-E-E_1)
$$
$$
\times F'F_1'(1-F)(1-F_1)(\tilde{G}'+\tilde{G}_1'-\tilde{G}-\tilde{G}_1)
$$
$$
\times (H'+H_1'-H-H_1)\,dk_1\,dk'\,dk_1\,dk_1'\,d\varepsilon,
$$

where $\tilde{G} = G/F(1-F)$. This expression shows that the operator L_F is self-adjoint with respect to the scalar product

$$
(G,H)_F = \int_B G(\varepsilon)H(\varepsilon)\frac{d\varepsilon}{F(1-F)}.
$$

Similarly as in the proof of Proposition 4.6, it can be shown that the kernel $N(L_F)$ of L_F is spanned by $F(1-F)$ and $\varepsilon F(1-F)$. By the Fredholm alternative (Lemma 5.1), the equation $L_F(G) = H$ is solvable if and only if H is an element of the orthogonal complement of $N(L_F)$ (with respect to the above scalar product), i.e., if and only if (7.14) holds. $\square$

Now, we turn to the derivation of the energy-transport model. Let $(F_\alpha, J_\alpha, V_\alpha)$ be a solution of the SHE equations (7.11) and (7.12) with collision operator (7.13) and of the Poisson equation (5.3), respectively. The (formal) limit $\alpha \to 0$ in (7.11) gives $S_{ee}(F) = 0$, where $F = \lim_{\alpha\to 0} F_\alpha$. Thus, by Lemma 7.6, F is given by $F_{\mu,T}$ for some μ and T. Inserting the Chapman–Enskog expansion $F_\alpha = F + \alpha G_\alpha$ into (7.11) and (7.12) leads to

$$
N(\varepsilon)\partial_t F + \alpha N(\varepsilon)\partial_t G_\alpha - \mathrm{div}_x J_\alpha - \nabla V_\alpha \cdot \frac{\partial J_\alpha}{\partial \varepsilon} = \frac{1}{\alpha}S_{ee}(F_\alpha) + S_{ph}(F_\alpha), \quad (7.15)
$$

$$
J_\alpha = D(x,\varepsilon)\left(\nabla_x F + \nabla_x V_\alpha \frac{\partial F}{\partial \varepsilon}\right) + \alpha D(x,\varepsilon)\left(\nabla_x G_\alpha + \nabla_x V_\alpha \frac{\partial G_\alpha}{\partial \varepsilon}\right). \quad (7.16)
$$

We develop the right-hand side of (7.15) employing Taylor expansion:

$$
\frac{1}{\alpha}S_{ee}(F_\alpha) + S_{ph}(F_\alpha) = \frac{1}{\alpha}\left(S_{ee}(F) + \alpha(DS_{ee}(F))(G_\alpha)\right) + S_{ph}(F) + \mathscr{O}(\alpha)
$$
$$
= (DS_{ee}(F))(G_\alpha) + S_{ph}(F) + \mathscr{O}(\alpha),
$$

since $S_{ee}(F) = 0$. Thus, performing the limit $\alpha \to 0$ in (7.15) and (7.16) gives

$$N(\varepsilon)\partial_t F - \mathrm{div}_x J - \nabla_x V \cdot \frac{\partial J}{\partial \varepsilon} - S_{\mathrm{ph}}(F) = (DS_{\mathrm{ee}}(F))(G), \tag{7.17}$$

$$J = D(x,\varepsilon)\left(\nabla_x F + \nabla_x V \frac{\partial F}{\partial \varepsilon}\right), \tag{7.18}$$

where $V = \lim_{\alpha \to 0} V_\alpha$ and $G = \lim_{\alpha \to 0} G_\alpha$. By Lemma 7.6 (3), the operator equation (7.17) is solvable if and only if

$$\int_{\mathbb{R}} \left(N(\varepsilon)\partial_t F - \mathrm{div}_x J - \nabla_x V \cdot \frac{\partial J}{\partial \varepsilon} - S_{\mathrm{ph}}(F)\right)\varepsilon^j \, d\varepsilon = 0, \quad j = 0, 1, \tag{7.19}$$

is satisfied. Defining the macroscopic particle and energy densities

$$n(x,t) = \int_{\mathbb{R}} F_{\mu,T}(x,\varepsilon,t)N(\varepsilon)\,d\varepsilon, \quad ne(x,t) = \int_{\mathbb{R}} F_{\mu,T}(x,\varepsilon,t)N(\varepsilon)\varepsilon\,d\varepsilon \tag{7.20}$$

and the macroscopic particle and energy current densities

$$J_n(x,t) = \int_{\mathbb{R}} J(x,\varepsilon,t)\,d\varepsilon, \quad J_e(x,t) = \int_{\mathbb{R}} J(x,\varepsilon,t)\varepsilon\,d\varepsilon,$$

we can reformulate the solvability condition as

$$\partial_t n - \mathrm{div} J_n = \nabla V \cdot \int_{\mathbb{R}} \frac{\partial J}{\partial \varepsilon}\,d\varepsilon + \int_{\mathbb{R}} S_{\mathrm{ph}}(F)\,d\varepsilon = 0,$$

$$\partial_t (ne) - \mathrm{div} J_e = \nabla V \cdot \int_{\mathbb{R}} \frac{\partial J}{\partial \varepsilon}\varepsilon\,d\varepsilon + \int_{\mathbb{R}} S_{\mathrm{ph}}(F)\varepsilon\,d\varepsilon = -\nabla V \cdot J_n + W(\mu,T),$$

employing Lemma 7.6 (1), where

$$W(\mu,T) = \int_{\mathbb{R}} S_{\mathrm{ph}}(F)\varepsilon\,d\varepsilon \tag{7.21}$$

is the energy relaxation term.

It remains to compute the fluxes J_n and J_e. Taking the derivative of $F = F_{\mu,T}$ with respect to x, we obtain from (7.18), employing the abbreviations $u_0 = \mu/T$ and $u_1 = -1/T$,

$$J = D(x,\varepsilon)\left(\frac{\partial F}{\partial u_0}\nabla_x u_0 + \frac{\partial F}{\partial u_1}\nabla_x u_1 + \nabla_x V \frac{\partial F}{\partial \varepsilon}\right)$$

$$= D(x,\varepsilon)F(1-F)\left(\nabla_x u_0 + \varepsilon\nabla_x u_1 - \frac{\nabla V_x}{T}\right).$$

Then, integrating J with respect to ε, we have shown the following theorem.

Theorem 7.7 (Energy-transport equations). *Let (F_α, V_α) be a solution of the SHE equations (7.11) and (7.12) and the Poisson equation (5.3). Then the limit functions $F = F_{\mu,T} = \lim_{\alpha \to 0} F_\alpha$ and $V = \lim_{\alpha \to 0} V_\alpha$ satisfy the energy-transport equations*

$$\partial_t n - \operatorname{div} J_n = 0, \quad \partial_t (ne) - \operatorname{div} J_e + J_n \cdot \nabla V = W(\mu, t),$$

$$J_n = D_{00} \left(\nabla \left(\frac{\mu}{T} \right) - \frac{\nabla V}{T} \right) - D_{01} \nabla \left(\frac{1}{T} \right),$$

$$J_e = D_{10} \left(\nabla \left(\frac{\mu}{T} \right) - \frac{\nabla V}{T} \right) - D_{11} \nabla \left(\frac{1}{T} \right),$$

$$\lambda_D^2 \Delta V = n - C(x),$$

where the diffusion coefficients are given by

$$D_{ij}(x, \mu, T) = \int_{\mathbb{R}} D(x, \varepsilon) F_{\mu, T} (1 - F_{\mu, T}) \varepsilon^{i+j} \, d\varepsilon, \quad i, j = 0, 1, \tag{7.22}$$

and the electron density n, the energy density ne, and the energy relaxation term $W(\mu, T)$ are defined in (7.20) and (7.21), respectively.

To be precise, the SHE equations for n and ne are obtained from the solvability condition (7.19) for the operator equation (7.17). As in Sect. 7.1, the diffusion coefficients are matrices in $\mathbb{R}^{3 \times 3}$.

Proposition 7.8 (Properties of the diffusion matrix and the relaxation term).
The following properties hold:
(1) The matrix $\mathscr{D} = (D_{ij})$ is symmetric and $D_{01} = D_{10}$.
(2) If the six functions $\{\partial E / \partial k_i, E \partial E / \partial k_i : i = 1, 2, 3\}$ are linearly independent, then $\mathscr{D}$ is symmetric and positive definite for any $\mu \in \mathbb{R}$ and $T > 0$.
(3) The relaxation term $W(\mu, T)$, defined through the inelastic collision operator (6.16), is monotone with respect to T,

$$W(\mu, T)(T - 1) \leq 0 \quad \text{for all } T > 0.$$

The proof of this proposition is very similar to the proofs of Propositions 6.4 and 6.5 and is therefore omitted.

Remark 7.9. The diffusion coefficients (7.22) are the same as those derived in Sect. 6.1 (see (6.14)). Indeed, with (7.7) we can write (7.22) as

$$D_{ij} = \int_B v(k) \otimes d_0 F (1 - F) \left(\int_{\mathbb{R}} \delta(E(k) - \varepsilon) \varepsilon^{i+j} \, d\varepsilon \right) \frac{dk}{4\pi^3}$$

$$= \int_B v(k) \otimes d_0 F (1 - F) E^{i+j} \frac{dk}{4\pi^3}.$$

The function d_0 is a solution of $Q_{\text{el}}(d_0) = -v(k)$. Since F is a function of the energy and not of the wave vector, $\tilde{d}_0 = F(1 - F) d_0$ solves $Q_{\text{el}}(\tilde{d}_0) = -vF(1 - F)$. Thus, the above expression becomes

$$D_{ij} = \int_B v(k) \otimes \tilde{d}_0 E^{i+j} \frac{dk}{4\pi^3},$$

which is equal to (6.14) after identifying d_0 and $\tilde{d}_0$. $\square$

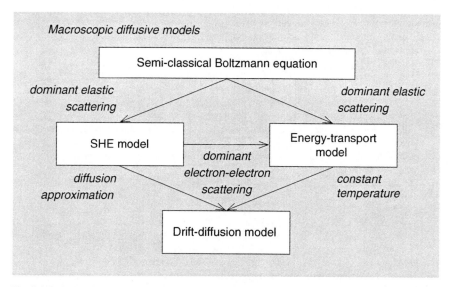

Fig. 7.1 Relations between the drift-diffusion, energy-transport, and SHE equations. A (high-field) drift-diffusion model is derived from the SHE model in [24]. The drift-diffusion equations can be directly derived from the Boltzmann equation in the diffusion approximation (see Sect. 5.1)

Figure 7.1 summarizes the relations between the diffusive models derived in this and the previous chapters.

References

1. N. Ben Abdallah and P. Degond. On a hierarchy of macroscopic models for semiconductors. *J. Math. Phys.* 37 (1996), 3308–3333.
2. C. Herring and E. Vogt. Transport and deformation-potential theory for many-valley semiconductors with anisotropic scattering. *Phys. Rev.* 101 (1956), 944–961.
3. C. Gray and H. Ralph. Solution of Boltzmann's equation for semiconductors using a spherical harmonic expansion. *J. Phys. C: Solid State Phys.* 5 (1972), 55–62.
4. M. Vecchi and M. Rudan. Modeling electron and hole transport with full-band structure effects by means of the spherical-harmonics expansion of the BTE. *IEEE Trans. Electr. Devices* 45 (1998), 230–238.
5. S. Singh, N. Goldsman, and I. Mayergoyz. Modeling multi-band effects of hot electron transport in silicon by self-consistent solution of the Boltzmann transport and Poisson equations. *Solid State Electr.* 39 (1996), 1695–1700.
6. R. Stratton. The influence of interelectronic collisions on conduction and breakdown in covalent semiconductors. *Proc. Royal Soc. London, Ser. A* 242 (1957), 355–373.
7. R. Stratton. Diffusion of hot and cold electrons in semiconductor barriers. *Phys. Rev.* 126 (1962), 2002–2014.
8. P. Dmitruk, A. Saul, and L. Reyna. High-electric field approximation to charge transport in semiconductor devices. *Appl. Math. Letters* 5 (1992), 99–102.

9. A. Abramo, F. Venturi, E. Sangiorgi, J. Highman, and B. Riccò. A numerical method to compute isotropic band models from anisotropic semiconductor band structures. *IEEE Trans. Computer Aided Design* 12 (1993), 1327–1336.

10. D. Ventura, A. Gnudi, G. Baccarani, and F. Odeh. Multidimensional spherical harmonics expansion of Boltzmann equation for transport in semiconductors. *Appl. Math. Letters* 5 (1992), 85–90.

11. M. Vecchi, D. Ventura, A. Gnudi, and G. Baccarani. Incorporating full band-structure effects in the spherical-harmonics expansion of the Boltzmann transport equation. In: *Numerical Modeling of Processes and Devices for Integrated Circuits, Workshop NUPAD V*, 55–58. IEEE Press, 1994.

12. J.-P. Bourgade. On spherical harmonics expansion type models for electron–phonon collisions. *Math. Meth. Appl. Sci.* 26 (2003), 247–271.

13. P. Degond and S. Mancini. Diffusion driven by collisions with the boundary. *Asympt. Anal.* 27 (2001), 47–73.

14. A. Mellet. Diffusion limit of a nonlinear kinetic model without the detailed balance principle. *Monatsh. Math.* 134 (2002), 305–329.

15. C. Schmeiser and A. Zwirchmayr. Elastic and drift-diffusion limits of electron–phonon interaction in semiconductors. *Math. Models Meth. Appl. Sci.* 8 (1998), 37–53.

16. A. Gnudi, D. Ventura, G. Baccarani, and F. Odeh. Two-dimensional MOSFET simulation by means of a multidimensional spherical harmonics expansion of the Boltzmann transport equation. *Solid State Electr.* 36 (1993), 575–581.

17. N. Goldsman, L. Henrickson, and J. Frey. A physics-based analytical/numerical solution to the Boltzmann transport equation for use in device simulation. *Solid State Electr.* 34 (1991), 389–396.

18. O. Hansen and A. Jüngel. Analysis of a spherical harmonics expansion model of plasma physics. *Math. Models Meth. Appl. Sci.* 14 (2004), 759–774.

19. P. Degond. Transport of trapped particles in a surface potential. In: C. Cioranescu et al. (eds.), *Nonlinear Partial Differential Equations and their Applications*, Collège de France Seminar, Studies Math. Appl. 14, 273–296. Elsevier, Amsterdam, 2002.

20. P. Degond and K. Zhang. Diffusion approximation of a scattering matrix model of a semiconductor superlattice. *SIAM J. Appl. Math.* 63 (2002), 279–298.

21. Y. Raiser. *Gas Discharge Physics*. Springer, Berlin, 1991.

22. M. Cho and D. Hasting. Dielectric processes and arcing rates of high voltage solar arrays. *J. Spacecraft Rockets* 28 (1991), 698–706.

23. T. Morrone. Time harmonic, spherical harmonic, and power series expansion of the Boltzmann equation. *Phys. Fluids* 11 (1969), 2617–2620.

24. P. Degond and A. Jüngel. High-field approximations of the energy-transport model for semiconductors with non-parabolic band structure. *Z. Angew. Math. Phys.* 52 (2001), 1053–1070.

Chapter 8
Diffusive Higher-Order Moment Equations

The drift-diffusion and energy-transport equations of Chaps. 5 and 6 are derived from the Boltzmann equation by considering the moments

$$n = \int_B F \frac{dk}{4\pi^3}, \quad ne = \int_B FE(k) \frac{dk}{4\pi^3},$$

where F is the distribution function. We have already indicated in Sect. 2.4 that this strategy can be generalized. In this chapter, we detail the derivation of a hierarchy of diffusive moment models.

We consider the semiconductor Boltzmann equation in the diffusive scaling

$$\alpha^2 \partial_t f + \alpha \left(v(k) \cdot \nabla_x f + \nabla_x V \cdot \nabla_k f \right) = Q(f), \quad x \in \mathbb{R}^3, \, k \in B, \, t > 0, \qquad (8.1)$$

together with periodic boundary conditions with respect to k and the initial condition $f(x,k,0) = f_I(x,k)$. The scaling means that we have changed the space and time scales according to $x \to x/\alpha$ and $t \to t/\alpha^2$, where α is the ratio of the mean free path between two consecutive collisions to some reference length (see Sects. 5.1 and 6.1 for details). The function $v(k) = \nabla_k E(k)$ is the mean velocity defined by the energy band $E(k)$. We suppose that α is small compared to one, and below we will perform the formal limit $\alpha \to 0$.

We further assume that the collision operator can be decomposed into two parts, a dominant part and a small part:

$$Q(f) = Q_0(f) + \alpha^2 Q_1(f).$$

In order to specify the assumptions on the collision terms Q_0 and Q_1, we need the so-called generalized Maxwellians which are introduced in the next section. In the following, we proceed as in [1].

Jüngel, A.: *Diffusive Higher-Order Moment Equations.* Lect. Notes Phys. **773**, 171–194 (2009)
DOI 10.1007/978-3-540-89526-8_8 © Springer-Verlag Berlin Heidelberg 2009

8.1 Derivation from the Boltzmann Equation

Entropy maximization and generalized Maxwellians. For a given distribution function $f(x,k,t)$, we introduce the scaled relative *entropy* (or free energy, since we include $E(k)$) by

$$(S(f))(x,t) = - \int_B f \left(\log f - 1 + E(k) \right) \frac{dk}{4\pi^3}.$$

The generalized Maxwellians are defined like in Sect. 2.2 as the maximizers of certain constrained extremal problems. In order to define this problem, let some *moments* $m(x,t) = (m_0(x,t), \ldots, m_N(x,t))$ be given. We assume that $N \geq 1$,

$$E(k) \text{ is even and } \kappa_i(k) = E(k)^i, \quad i = 0, \ldots, N, \tag{8.2}$$

where $\kappa(k) = (\kappa_0(k), \ldots, \kappa_N(k))$ are scalar *weight functions*. It is possible to employ more general (even) weight functions, but then the formulas below become more complicated. We refer to [1] for details. Also vector-valued weight functions can be considered, see [2].

As in the previous chapters, we set $\langle g \rangle = \int_B g(k) \, dk/4\pi^3$ for functions $g(k)$. Then we call the expression

$$\langle \kappa_i f \rangle = \int_B \kappa_i(k) f(k) \frac{dk}{4\pi^3}$$

the *i-th moment* of f. For given moments $m = (m_0, \ldots, m_N)$, we consider the following constrained maximization problem:

$$S(f^*) = \max \left\{ S(f) : \langle \kappa f(x, \cdot, t) \rangle = m(x,t) \text{ for } x \in \mathbb{R}^3, \, t > 0 \right\}. \tag{8.3}$$

Lemma 8.1. *The formal solution of* (8.3) *(if it exists) is*

$$f^*(x,k,t) = e^{\lambda^*(x,t) \cdot \kappa(k)},$$

where $\lambda^* = (\lambda_0^*, \ldots, \lambda_N^*)$ *are some Lagrange multipliers defined by the relation* $\langle \kappa f^* \rangle = m$.

Proof. With the Lagrange multipliers λ_i, we have to analyze the functional

$$G(f, \lambda) = \int_B f (\log f - 1 - E(k)) \frac{dk}{4\pi^3} - \lambda \cdot (\langle \kappa f \rangle - m).$$

The extremal condition for a function f^*,

$$0 = \frac{\partial G(f^*, \lambda)}{\partial f}(g) = \int_B g(\log f^* - E(k)) \frac{dk}{4\pi^3} - \lambda \cdot \langle \kappa g \rangle$$

$$= \int_B g \left(\log f^* - E(k) - \lambda \cdot \kappa \right) \frac{dk}{4\pi^3},$$

for all functions g implies the identity $\log f^* = E(k) + \lambda \cdot \kappa$. Then, by setting $\lambda_1^* = \lambda_1 + 1$ and $\lambda_i^* = \lambda_i$ for all $i \neq 1$, the lemma is proved. $\square$

Remark 8.2. We notice that the mathematical solution of (8.3) is quite delicate. In [3], it was shown that (8.3) can be uniquely solved whenever the multipliers $\widetilde{\lambda} = \widetilde{\lambda}(m)$ can be found. However, there are situations for which problem (8.3) has no solution. This is the case if the momentum space is unbounded and the polynomial weight functions grow superquadratically at infinity [4, 5]. When the constraint of the highest degree is relaxed (as an inequality instead of an equality), the constrained maximization problem is always uniquely solvable [6]. In particular, the maximization problem can be uniquely solved if one of the following conditions holds:

1. General band structure: B is a bounded set.
2. Kane's nonparabolic band approximation: $B = \mathbb{R}^3$ and $\kappa = (1, E, E^2)$, where

$$E(k) = \frac{|k|^2}{1 + \sqrt{1 + 2\alpha|k|^2}}. \tag{8.4}$$

Notice that $E(k)$ grows linearly with k at infinity such that $\kappa_i(k)$ is at most quadratic.
3. Parabolic band approximation: $B = \mathbb{R}^3$ and $\kappa = (1, |k|^2/2)$. $\square$

Given a function $f(x, k, t)$ with moments $m_i = \langle \kappa_i f \rangle$, we call the maximizer of (8.3) the *generalized Maxwellian* with respect to f and write $f^* = M[f]$. We infer from Lemma 8.1 that there exist Lagrange multipliers λ_i such that

$$M[f](x, k, t) = e^{\lambda(x,t) \cdot \kappa(k)}.$$

By definition, $M[f]$ and f have the same moments, $\langle \kappa_i M[f] \rangle = \langle \kappa_i f \rangle = m_i$.

The use of the above entropy functional implicitly assumes nondegenerate Maxwell–Boltzmann statistics. Degenerate Fermi–Dirac statistics can be also considered. Then the entropy functional reads as

$$S(f) = -\int_B \left(f \log f + \frac{1}{\eta}(1 - \eta f) \log(1 - \eta f) + E(k)f \right) \frac{dk}{4\pi^3}.$$

The (scaled) parameter η measures the semiconductor degeneracy. Fermi–Dirac statistics means that $\eta = 1$; if $\eta = 0$, we recover Maxwell–Boltzmann statistics since $\lim_{\eta \to 0} \eta^{-1}(1 - \eta f) \log(1 - \eta f) = -f$. Solving the constrained maximization problem with Fermi–Dirac statistics, a computation shows that the formal maximizer is

$$f^* = \frac{1}{\eta + e^{-\lambda \cdot \kappa}}. \tag{8.5}$$

Assumptions on the collision operators. Let (f_α, V_α) be a (smooth) solution of the Boltzmann equation (8.1) and the Poisson equation (5.3). We assume that f_α

converges to some function $F = \lim_{\alpha \to 0} f_\alpha$. We introduce the Hilbert space $L_F^2(B)$ with the scalar product

$$(f,g)_F = \int_B fgM[F]^{-1} \frac{dk}{4\pi^3} \tag{8.6}$$

and the corresponding norm $\| \cdot \|_F$, where $M[F]$ is the generalized Maxwellian with respect to F. We impose the following hypotheses on the collision operators. Instead of considering specific scattering models, we assume only abstract properties:

1. For all functions $f(k)$ and $i = 0, \dots, N$, $\langle \kappa_i Q_0(f) \rangle = 0$. The kernel of Q_0 consists of generalized Maxwellians, $N(Q_0) = \{f : f = \text{const.} M[f]\}$.
2. For all functions $f(k)$, $\langle Q_1(f) \rangle = 0$.
3. The derivative $L = DQ_0(M[F])$ is continuous, closed, and symmetric on $L_F^2(B)$ and its kernel is spanned by $M[F]$.

The first two hypotheses express the collisional invariants. Since $\kappa_0 = 1$ and $\kappa_1 = E$ by (8.2), the first assumption expresses in particular mass and energy conservation. Additionally, we suppose for Q_0 conservation for all moments with respect to the weight functions. This assumption is rather strong. However, it is satisfied, for instance, for the (simple) relaxation-time operator

$$Q_0(f) = \frac{1}{\tau}(M[f] - f), \tag{8.7}$$

where $\tau > 0$ is a (possibly space- and time-dependent) relaxation time. Since $M[f]$ and f have the same moments and the kernel consists of generalized Maxwellians, the first assertion is satisfied. Operators describing impurity scattering (4.11), acoustic phonon collisions in the elastic approximation (4.10), or electron–electron scattering (4.12) satisfy the first condition with $N = 1$.

The second hypothesis simply expresses mass conservation for the collision operator Q_1, which is physically reasonable. For instance, the phonon collision operator with transition rate (4.7) satisfies this condition.

Examples of collision operators, which fulfill the third hypothesis, are the low-density operator (4.22) and the relaxation-time operator (8.7). The assumption is quite natural in order to apply the Fredholm alternative for the Chapman–Enskog correction.

The above approach has the advantage that only abstract properties of the scattering terms are assumed, including many examples. On the other hand, there are some drawbacks too. In contrast to the diffusion approximation of Chap. 6, there are no scattering terms of first order in the Knudsen number α. For realistic semiconductors the order of magnitudes of the various scattering mechanisms can be quite complicated and the decomposition into two operators $Q_0(f)$ and $\alpha^2 Q_1(f)$ may be too simple. Furthermore, the generalized Maxwellian may not be a good approximation of the Boltzmann distribution function in certain situations, for instance in the drain region of an n^+nn^+ diode [7]. A possible way out could be the use of the generalized Fermi–Dirac distributions (8.5). However, the resulting expressions for the diffusion coefficients and densities become quite complicated since some integrals cannot be simplified as in the case of Maxwell–Boltzmann statistics. Another possibility could be the use of the superposition of two Maxwellian-type distribution functions modeling hot and cold electron populations in the semiconductor (see, for instance, [7, 8]).

Derivation of the diffusion moment model. The derivation consists of three steps. First we perform the (formal) limit $\alpha \to 0$ in the Boltzmann equation (8.1). Then $Q_0(F) = 0$. By the first hypothesis, F is a generalized Maxwellian, $F = M[F]$. In the second step, we insert the Chapman–Enskog expansion $f_\alpha = M[f_\alpha] + \alpha g_\alpha$ in the Boltzmann equation (8.1). Observing the (formal) expansion of the collision operator,

$$Q_0(f_\alpha) = Q_0(M[f_\alpha]) + \alpha D Q_0(M[f_\alpha]) g_\alpha + \mathcal{O}(\alpha^2),$$

we obtain

$$\alpha \partial_t (M[f_\alpha] + \alpha g_\alpha) + (v \cdot M[f_\alpha] + \nabla_x V \cdot \nabla_k M[f_\alpha]) + \alpha (v \cdot g_\alpha + \nabla_x V \cdot \nabla_k g_\alpha)$$
$$= \alpha^{-1} Q_0(M[f_\alpha]) + D Q_0(M[f_\alpha]) g_\alpha + \mathcal{O}(\alpha).$$

By the first hypothesis, we have $Q_0(M[f_\alpha]) = 0$. Hence, the formal limit $\alpha \to 0$ gives

$$v \cdot \nabla_x M[F] + \nabla_x V \cdot \nabla_k M[F] = D Q_0(M[F]) G = LG, \qquad (8.8)$$

where $G = \lim_{\alpha \to 0} g_\alpha$.

We claim that this operator equation is solvable. The third assumption and the Fredholm alternative (Lemma 5.1) imply that the equation $LG = H$ is solvable if and only if $H \in N(L^*)^\perp$ and its solution is unique in the space $N(L^*)^\perp$. Since L is assumed to be symmetric and its kernel consists of generalized Maxwellians only, we conclude that $LG = H$ is solvable if and only if $0 = (H, M[F])_F = \int_B H \, dk/4\pi^3$. Now, $v_i \kappa_j M[F]$ and $(\partial \kappa_i / \partial k_j) M[F]$ are odd functions in k, and hence their integrals vanish over B. Thus, the integral of the left-hand side of (8.8) over B vanishes too, and this operator equation is uniquely solvable in $N(L)^\perp$. We claim further that the solution is given, up to the addition of a function in $N(L)$, by

$$G = -\sum_{i=0}^{N} (\phi_i \cdot \nabla_x \lambda_i + i \lambda_i \phi_{i-1} \cdot \nabla_x V), \qquad (8.9)$$

where $\phi_i = (\phi_{i1}, \phi_{i2}, \phi_{i3})$ is the unique (vector-valued) solution in $N(L)^\perp$ to

$$L\phi_{ij} = -v_j \kappa_i M[F] = -\frac{\partial E}{\partial k_j} E^i M[F], \quad i = 0, \ldots, N, \; j = 1, 2, 3, \qquad (8.10)$$

and we have set $\phi_{-1} = 0$. Indeed, observing that

$$\nabla_x M[F] = \sum_{i=0}^{N} \kappa_i M[F] \nabla_x \lambda_i, \quad \nabla_k M[F] = \sum_{i=0}^{N} \lambda_i M[F] \nabla_k \kappa_i,$$

and $\nabla_k \kappa_i = iv\kappa_{i-1}$ (by (8.2)), we have

$$
\begin{aligned}
LG &= -\sum_{i=0}^{N} (L\phi_i \cdot \nabla_x \lambda_i + i\lambda_i \phi_{i-1} \cdot \nabla_x V) \\
&= \sum_{i=0}^{N} (\kappa_i v \cdot \nabla_x \lambda_i + \lambda_i \nabla_x V \cdot \nabla_k \kappa_i) M[F] = v \cdot \nabla_x M[F] + \nabla_x V \cdot \nabla_k M[F].
\end{aligned}
$$

The third step is concerned with the limit $\alpha \to 0$ in the moment equations. For this, we multiply the Boltzmann equation (8.1) by κ_i, integrate over B, and integrate by parts in the term involving the electric potential:

$$
\begin{aligned}
\alpha^2 \partial_t \langle \kappa_i f_\alpha \rangle + \alpha (\operatorname{div}_x \langle v\kappa_i f_\alpha \rangle - i\nabla_x V \cdot \langle v\kappa_{i-1} f_\alpha \rangle) \\
= \langle \kappa_i Q_0(f_\alpha) \rangle + \alpha^2 \langle \kappa_i Q_1(f_\alpha) \rangle, \quad i = 0, \dots, N,
\end{aligned}
$$

where we have set $\kappa_{-1} = 0$. By the first assumption on Q_0, the moments of $Q_0(f_\alpha)$ vanish. Inserting the Chapman–Enskog expansion and dividing the resulting equation by α^2 then lead to

$$
\partial_t \langle \kappa_i M[f_\alpha] \rangle + \alpha \partial_t \langle \kappa_i g_\alpha \rangle + \operatorname{div}_x \langle v\kappa_i g_\alpha \rangle - i\nabla_x V \cdot \langle v\kappa_{i-1} g_\alpha \rangle = \langle \kappa_i Q_1(f_\alpha) \rangle.
$$

We have used that $v\kappa_i M[f_\alpha]$ is an odd function in k and hence its integral over B vanishes. We perform the limit $\alpha \to 0$ to obtain

$$
\partial_t \langle \kappa_i M[F] \rangle + \operatorname{div}_x \langle v\kappa_i G \rangle - i\nabla_x V \cdot \langle v\kappa_{i-1} G \rangle = \langle \kappa_i Q_1(M[F]) \rangle.
$$

These are the evolution equations for the moments $m_i = \langle \kappa_i M[F] \rangle$. The fluxes $J_i = -\langle v\kappa_i G \rangle$ are determined by employing the expression (8.9) for G. This gives the following theorem.

Theorem 8.3 (Diffusive moment equations). *Let $N \geq 1$, assume (8.2), and let the hypotheses on the collision operators on page 174 hold. Furthermore, let (f_α, V_α) be a (smooth) solution of the Boltzmann–Poisson system (8.1) and (5.3). Then the limit functions $F = \lim_{\alpha \to 0} f_\alpha$ and $V = \lim_{\alpha \to 0} V_\alpha$ satisfy $F = M[F] = e^{\lambda \cdot \kappa}$ and the higher-order diffusive moment equations*

$$
\partial_t m_i - \operatorname{div} J_i + iJ_{i-1} \cdot \nabla V = W_i, \quad x \in \mathbb{R}^3, \, t > 0, \tag{8.11}
$$

$$
J_i = \sum_{j=0}^{N} \left(D_{ij} \nabla \lambda_j + jD_{i,j-1} \lambda_j \nabla V \right), \quad i = 0, \dots, N, \tag{8.12}
$$

$$
\lambda_D^2 \Delta V = m_0 - C(x).
$$

The moments m_i and the right-hand side W_i are defined by

$$
m_i = \int_B FE(k)^i \frac{dk}{4\pi^3}, \quad W_i = \int_B Q_1(F)E(k)^i \frac{dk}{4\pi^3}, \tag{8.13}
$$

the matrices $D_{ij} \in \mathbb{R}^{3 \times 3}$ are given by

$$D_{ij} = \int_B v(k) \otimes \phi_j E(k)^i \frac{dk}{4\pi^3}, \quad i, j = 0, \ldots, N, \tag{8.14}$$

where we have set $D_{i,-1} = 0$, and the functions $\phi_j = (\phi_{j1}, \phi_{j2}, \phi_{j3})$ are the (unique) solutions in $N(L)^\perp$ of the operator equations (8.10). Furthermore, m and λ are related through the first equation in (8.13). The initial conditions read as

$$m_i(\cdot, 0) = \int_B f_I E(k)^i \frac{dk}{4\pi^3}, \quad i = 0, \ldots, N.$$

Diffusive higher-order moment models were already derived in the engineering literature in the 1990s. For instance, Sonoda et al. [9] derived heuristically a system for the moments $\langle M[F] \rangle$, $\langle E(k)M[F] \rangle$, and $\langle E(k)^2 M[F] \rangle$. Grasser et al. [8, 10] employ the same moments and give a more systematic derivation, even for nonparabolic bands.

Some properties of the diffusion matrices. Next, we show that the matrices D_{ij} are symmetric and, under some additional assumptions, positive definite. This indicates that the system (8.11) and (8.12) may be well posed, at least locally in time.

Proposition 8.4 (Properties of the diffusion matrix). *The following properties hold:*

(1) The matrix $\mathscr{D} \in \mathbb{R}^{3(N+1) \times 3(N+1)}$, defined by $\mathscr{D} = (D_{ij})$ with $D_{ij} \in \mathbb{R}^{3 \times 3}$, is symmetric.

(2) Let the operator $-L = -DQ_0(M[F])$ be coercive on $N(L)^\perp$, i.e., there exists a constant $\beta > 0$ such that for all $g \in N(L)^\perp$, $(-Lg, g)_F \geq \beta \|g\|_F^2$ (see (8.6) for the definition of $(\cdot, \cdot)_F$). Furthermore, let $\{v_i E^j : i = 1, 2, 3, \ j = 0, \ldots, N\}$ be linearly independent functions in k. Then the diffusion matrix $\mathscr{D} = (D_{ij})$ is positive definite, i.e., for all $\xi \in \mathbb{R}^{3(N+1)}$ with $\xi \neq 0$,

$$\xi^\top \mathscr{D} \xi > 0.$$

Proof. (1) We have to show that $D_{ij}^\top = D_{ji}$. We write $D_{ij} = (D_{ij}^{\ell m}) \in \mathbb{R}^{3 \times 3}$. Since L is symmetric on $L^2(B)$, by the third assumption on page 174, we have

$$D_{ij}^{\ell m} = (v_\ell \kappa_i M[F], \phi_{jm})_F = -(L\phi_{i\ell}, \phi_{jm})_F = -(\phi_{i\ell}, L\phi_{jm})_F$$
$$= (\phi_{i\ell}, v_m \kappa_j M[F])_F = D_{ji}^{m\ell}.$$

(2) The proof is inspired by the proof of Proposition IV.6 in [11]. We write $D_{ij} = (D_{ij}^{\ell m})$ and $\xi = (\xi_0, \ldots, \xi_N) \in \mathbb{R}^{3(N+1)}$ with $\xi_i = (\xi_{i\ell}) \in \mathbb{R}^3$. Let $(\xi_0, \ldots, \xi_N) \neq 0$. Then, by the definition of the matrices D_{ij},

$$\xi^\top \mathscr{D} \xi = \sum_{i,j=0}^N \xi_i^\top D_{ij} \xi_j = \sum_{i,j=0}^N \sum_{\ell,m=1}^3 \xi_{i\ell} D_{ij}^{\ell m} \xi_{jm} = \sum_{i,j=0}^N \sum_{\ell,m=1}^3 \int_B \xi_{i\ell} \kappa_i v_\ell \phi_{jm} \xi_{jm} \frac{dk}{4\pi^3}.$$

Since $v_\ell \kappa_i M[F] = -L\phi_{i\ell}$, we obtain

$$\sum_{i,j=0}^{N} \xi_i^\top D_{ij} \xi_j = -\sum_{i,j=0}^{N} \sum_{\ell,m=1}^{3} \int_B \xi_{i\ell} L\phi_{i\ell} \phi_{jm} \xi_{jm} M[F]^{-1} \frac{dk}{4\pi^3}$$

$$= \sum_{i,j=0}^{N} \sum_{\ell,m=1}^{3} \left(-L(\xi_{i\ell}\phi_{i\ell}), \xi_{jm}\phi_{jm} \right)_F$$

$$= \left(-L \left(\sum_{i=0}^{N} \sum_{\ell=1}^{3} \xi_{i\ell}\phi_{i\ell} \right), \sum_{i=0}^{N} \sum_{\ell=1}^{3} \xi_{i\ell}\phi_{i\ell} \right)_F.$$

As $\phi_{i\ell} \in N(L)^\perp$, the coercivity and boundedness of $-L$ (with bound $c_L > 0$) give

$$\sum_{i,j=0}^{N} \xi_i^\top D_{ij} \xi_j \geq \beta \left\| \sum_{i=0}^{N} \sum_{\ell=1}^{3} \xi_{i\ell}\phi_{i\ell} \right\|_F^2 \geq \frac{\beta}{c_L^2} \left\| -L \left(\sum_{i=0}^{N} \sum_{\ell=1}^{3} \xi_{i\ell}\phi_{i\ell} \right) \right\|_F^2$$

$$= \frac{\beta}{c_L^2} \left\| \sum_{i=0}^{N} \sum_{\ell=1}^{3} \xi_{i\ell} v_\ell \kappa_i M[F] \right\|_F^2 = \frac{\beta}{c_L^2} \int_B \left| \sum_{i=0}^{N} \sum_{\ell=1}^{3} \xi_{i\ell} v_\ell \kappa_i \right|^2 M[F] \frac{dk}{4\pi^3} > 0,$$

since the functions $v_\ell \kappa_i$ are supposed to be linearly independent.   □

Remark 8.5. We discuss in the following the assumptions of the above proposition. The functions $v_\ell E^i$ ($\ell = 1, 2, 3$, $i = 0, \ldots, N$) are linearly independent, for instance, in the case of the parabolic and nonparabolic band approximation. The operator $-L = -DQ_0(M[F])$ is coercive on $N(L)^\perp$ if Q_0 is the relaxation-time operator (8.7) and if E^i ($i = 0, \ldots, N$) are linearly independent. For a proof of this statement, let $g \in N(L)^\perp$. We show first that $M[g] \in N(L)$. This follows if we have shown that $M[M[g]] = M[g]$. Thus, let $M[g] = e^{\lambda \cdot \kappa}$ and $M[M[g]] = e^{\mu \cdot \kappa}$. Since the moments of $M[g]$ and $M[M[g]]$ coincide by construction, we have

$$\int_B \kappa \left(e^{\lambda \cdot w} - e^{\mu \cdot \kappa} \right) dk = 0 \quad \text{and} \quad \int_B (\lambda \cdot \kappa - \mu \cdot \kappa) \left(e^{\lambda \cdot \kappa} - e^{\mu \cdot \kappa} \right) dk = 0.$$

By the strict monotonicity of $x \mapsto e^x$, the integrand of the second integral vanishes and therefore $(\lambda - \mu) \cdot \kappa = 0$. Since $\kappa_i = E^i$ are assumed to be linearly independent, $\lambda = \mu$. Hence, $M[M[g]] = M[g]$ and $M[g] \in N(L)$. This property shows that $(M[g], g)_F = 0$ and

$$(-Lg, g)_F = -(Q_0(g), g)_F = -\frac{1}{\tau} (M[g] - g, g)_F = \frac{1}{\tau} \|g\|_F^2.$$

This proves the coercivity of $-L$ on $N(L)^\perp$.   □

The diffusion matrices can be simplified if the operator Q_0 is of relaxation-time type.

Proposition 8.6 (Simplified diffusion matrix). *Let $Q_0(f) = (M[f] - f)/\tau$, where $\tau > 0$ is some (scaled) relaxation time, and set $E(k) = e(\frac{1}{2}|k|^2)$ for some scalar function e. Then*

$$D_{ij} = \frac{\tau}{3} \int_B e\left(\frac{1}{2}|k|^2\right) e'\left(\frac{1}{2}|k|^2\right)^2 |k|^2 \exp\left(\sum_{\ell=0}^{N} \lambda_\ell e\left(\frac{1}{2}|k|^2\right)^\ell\right) \frac{dk}{4\pi^3} \text{Id},$$

where Id *is the unit matrix in* $\mathbb{R}^{3\times3}$.

Under the assumption of the proposition, we can identify D_{ij} with its diagonal element and simply write $\mathscr{D} = (D_{ij}) \in \mathbb{R}^{(N+1)\times(N+1)}$.

Proof. First, we observe that in the case of the relaxation-time operator, the function G, which is the limit of the Chapman–Enskog correction, can be written explicitly, enabling us to solve the operator Eq. (8.10) explicitly. Indeed, from the Chapman–Enskog expansion and the Boltzmann equation (8.1), we infer that

$$g_\alpha = \frac{1}{\alpha}(f_\alpha - M[f_\alpha]) = -\frac{\tau}{\alpha}Q_0(f_\alpha)$$
$$= -\tau\alpha(\partial_t f_\alpha - Q_1(f_\alpha)) - \tau(v \cdot \nabla_x f_\alpha + \nabla_x V \cdot \nabla_k f_\alpha),$$

and the formal limit $\alpha \to 0$ gives

$$G = -\tau(v \cdot \nabla_x M[F] + \nabla_x V \cdot \nabla_k M[F]) = -\tau\sum_{j=0}^{N}(\kappa_j v \cdot \nabla_x \lambda_j + \nabla_x V \cdot \nabla_k \kappa_j \lambda_j)M[F].$$

Comparing with (8.9), the solution ϕ_j of (8.10) reads as

$$\phi_j = \tau v \kappa_j M[F] = \tau\nabla_k EE^j M[F].$$

By definition (8.14), this implies

$$D_{ij} = \int_B E^i \nabla_k E \otimes \phi_j \frac{dk}{4\pi^3} = \tau\int_B E^{i+j}\nabla_k E \otimes \nabla_k EM[F]\frac{dk}{4\pi^3}.$$

Since $\nabla_k E(k) = ke'(\frac{1}{2}|k|^2)$, we obtain

$$D_{ij} = \tau\int_B e\left(\frac{1}{2}|k|^2\right)^{i+j} e'\left(\frac{1}{2}|k|^2\right)^2 k \otimes kM[F]\frac{dk}{4\pi^3}.$$

The function $k \mapsto k \otimes k$ is odd in every off-diagonal element such that the above integral vanishes except in the diagonal elements. Since each diagonal element has the same value and $M[F] = e^{\lambda \cdot \kappa}$, the expression for D_{ij} follows.   □

The diffusion coefficients can be further simplified under additional assumptions on the energy band structure. We consider some examples.

Example 8.7 (Monotone energy band). Let the assumption of Proposition 8.6 hold. We suppose additionally that the energy band $e(\frac{1}{2}|k|^2)$ is strictly monotone in $|k|$ and that $e(0) = 0$ and $\lim_{z\to\infty} e(z) = \infty$. This allows us to choose $B = \mathbb{R}^3$. Then, with spherical coordinates (ρ, θ, ϕ), for $i, j = 0, \ldots, N$, we have

$$D_{ij} = \frac{\tau}{12\pi^3} \int_0^{2\pi} \int_0^\pi \int_0^\infty e\left(\frac{1}{2}\rho^2\right)^{i+j} e'\left(\frac{1}{2}\rho^2\right)^2 \rho^4 \exp\left(\sum_{\ell=0}^N \lambda_\ell e\left(\frac{1}{2}\rho^2\right)^\ell\right)$$
$$\times \sin\theta\, d\rho\, d\theta\, d\phi.$$

Now we perform the change of variables $\varepsilon = e(\frac{1}{2}\rho^2)$, setting $\gamma(\varepsilon) = \rho^2$. Therefore, $d\varepsilon = e'(\frac{1}{2}\rho^2)\rho\, d\rho$ and $e'(\frac{1}{2}\rho^2) = 2/\gamma'(\varepsilon)$ such that

$$D_{ij} = \frac{2\tau}{3\pi^2} \int_0^\infty \varepsilon^{i+j} \frac{\gamma(\varepsilon)^{3/2}}{\gamma'(\varepsilon)} \exp\left(\sum_{\ell=0}^N \lambda_\ell \varepsilon^\ell\right) d\varepsilon. \tag{8.15}$$

In the special case $N = 1$ and for constant relaxation times, the same diffusion coefficients were derived in [11, formulas (3.36), (4.17)]. Notice that the above transformation allows us to simplify the expression for the moments:

$$m_i = \int_B e\left(\frac{1}{2}|k|^2\right)^i \exp\left(\sum_{\ell=0}^N \lambda_\ell e\left(\frac{1}{2}|k|^2\right)^\ell\right) \frac{dk}{4\pi^3}$$
$$= \frac{1}{\pi^2} \int_0^\infty e\left(\frac{1}{2}\rho^2\right)^i \exp\left(\sum_{\ell=0}^N \lambda_\ell e\left(\frac{1}{2}\rho^2\right)^\ell\right) \rho^2\, d\rho$$
$$= \frac{1}{2\pi^2} \int_0^\infty \varepsilon^i \sqrt{\gamma(\varepsilon)}\gamma'(\varepsilon) \exp\left(\sum_{\ell=0}^N \lambda_\ell \varepsilon^\ell\right) d\varepsilon, \tag{8.16}$$

where $i = 0,\ldots,N$. Expressions (8.15) and (8.16) coincide with those for the energy-transport model if $\tau = 2\pi\sqrt{\gamma(\varepsilon)}\gamma'(\varepsilon)$, see (6.18), (6.19) and (6.20). □

Example 8.8 (Nonparabolic and parabolic band approximation). In the case of Kane's nonparabolic band approximation (8.4), we can further simplify the integrals (8.15) and (8.16). Since $\gamma(\varepsilon) = |k|^2 = 2\varepsilon(1 + \alpha\varepsilon)$ and $\gamma'(\varepsilon) = 2(1 + 2\alpha\varepsilon)$, with the nonparabolicity parameter $\alpha > 0$, we compute

$$D_{ij} = \frac{\sqrt{8}\tau}{3\pi^2} \int_0^\infty \varepsilon^{i+j+3/2} \frac{(1 + \alpha\varepsilon)^{3/2}}{1 + 2\alpha\varepsilon} \exp\left(\sum_{\ell=0}^N \lambda_\ell \varepsilon^\ell\right) d\varepsilon,$$
$$m_i = \frac{\sqrt{2}}{\pi^2} \int_0^\infty \varepsilon^{i+1/2}(1 + \alpha\varepsilon)^{1/2}(1 + 2\alpha\varepsilon) \exp\left(\sum_{\ell=0}^N \lambda_\ell \varepsilon^\ell\right) d\varepsilon,$$

where $i = 0,\ldots,N$. The relations for the parabolic band approximation are obtained by setting $\alpha = 0$ in the above formulas:

$$D_{ij} = \frac{\sqrt{8}\tau}{3\pi^2} \int_0^\infty \varepsilon^{i+j+3/2} \exp\left(\sum_{\ell=0}^N \lambda_\ell \varepsilon^\ell\right) d\varepsilon, \tag{8.17}$$

$$m_i = \frac{\sqrt{2}}{\pi^2} \int_0^\infty \varepsilon^{i+1/2} \exp\left(\sum_{\ell=0}^N \lambda_\ell \varepsilon^\ell\right) d\varepsilon, \tag{8.18}$$

where $i, j = 0,\ldots,N$. □

8.2 Some Explicit Examples

In this section we verify that the above strategy leads to the drift-diffusion and energy-transport models of Chaps. 5 and 6, if one or two moments, respectively, are chosen. This provides an alternative derivation, already indicated in Sect. 2.4, but for simpler collision operators than those considered in Chaps. 5 and 6. Moreover, a higher-order model is presented.

Drift-diffusion equations. The Maxwellian for the drift-diffusion model is obtained by choosing $N = 0$ in the entropy maximization problem (8.3). The only weight function is $\kappa_0 = 1$, and the Maxwellian equals $M[F] = e^{\lambda_0 - E(k)}$. The balance equation is given by (8.11), $\partial_t m_0 + \operatorname{div} J_0 = 0$, since the mass conservation leads to $W_0 = 0$. We need to compute the current density J_0 since the case $N = 0$ was excluded in Theorem 8.3. We have to solve

$$LG = v \cdot \nabla_x M[F] + \nabla_x V \cdot \nabla_k M[F] = v \cdot \nabla_x (\lambda_0 - V) M[F] \qquad (8.19)$$

(see (8.8)). Let ϕ_0 be the unique solution in $N(L)^{\perp}$ of $L\phi_0 = -vM[F]$. It is not difficult to verify that $G = -\nabla_x (\lambda_0 - V) \cdot \phi_0$ solves (8.19). Thus,

$$J_0 = -\langle vG \rangle = \langle v \otimes \phi_0 \rangle \nabla_x (\lambda_0 - V).$$

Notice that Theorem 8.3 would only give $J_i = \langle v \otimes \phi_0 \rangle \nabla_x \lambda_0$ since we have modified the definition of the Lagrange multipliers. The flux can be written in terms of the particle density m_0. Indeed, since

$$m_0 = \int_B M[F] \frac{dk}{4\pi^3} = Ae^{\lambda_0}, \quad \text{where } A = \int_B e^{-E(k)} \frac{dk}{4\pi^3} > 0,$$

we obtain $\nabla_x \lambda_0 = (\nabla_x m_0)/m_0$ and hence

$$J_0 = \mu_0 (\nabla_x m_0 - m_0 \nabla_x V), \quad \text{where } \mu_0 = \frac{1}{m_0} \int_B v \otimes \phi_0 \frac{dk}{4\pi^3}.$$

This gives the drift-diffusion equations for the particle density $n = m_0$ and the current density $J_n = J_0$ (see Sect. 5.1):

$$\partial_t n - \operatorname{div} J_n = 0, \quad J_n = \mu_0 (\nabla n - n \nabla V).$$

The model of Sect. 5.1 is obtained in the parabolic band approximation. By (8.17) and (8.18), we infer that

$$m_0 = \frac{\sqrt{2}}{\pi^2} e^{\lambda_0} \int_0^{\infty} \varepsilon^{1/2} e^{-\varepsilon} d\varepsilon = \frac{\sqrt{2}}{\pi^2} e^{\lambda_0} \Gamma\left(\frac{3}{2}\right) = \frac{2}{(2\pi)^{3/2}} e^{\lambda_0},$$

$$\mu_0 = \frac{\sqrt{8}\tau}{3\pi^2 m_0} e^{\lambda_0} \int_0^{\infty} \varepsilon^{3/2} e^{-\varepsilon} d\varepsilon = \frac{\sqrt{8}\tau e^{\lambda_0}}{3\pi^2 m_0} \Gamma\left(\frac{5}{2}\right) = \frac{\sqrt{8}\tau}{3\pi^2} \frac{(2\pi)^{3/2}}{2} \Gamma\left(\frac{5}{2}\right) = \tau,$$

where Γ is the Gamma function defined in (1.60).

Energy-transport equations. The energy-transport model is obtained from the choice $N = 1$. Then $M[F] = e^{\lambda_0 + \lambda_1 E}$. By Theorem 8.3, the balance equations are

$$\partial_t m_0 - \operatorname{div} J_0 = 0, \quad \partial_t m_1 - \operatorname{div} J_1 + \nabla V \cdot J_0 = W_1,$$

where the particle current density J_0 and the energy current density J_1 are given by

$$J_0 = D_{00}(\nabla \lambda_0 + \lambda_1 \nabla V) + D_{01} \nabla \lambda_1,$$
$$J_1 = D_{10}(\nabla \lambda_0 + \lambda_1 \nabla V) + D_{11} \nabla \lambda_1,$$

the diffusion coefficients are defined by $D_{ij} = \langle E^i v \otimes \phi_j \rangle$, and the moments are functions of the Lagrange multipliers λ_0 and λ_1:

$$m_0 = e^{\lambda_0} \int_B e^{\lambda_1 E(k)} \frac{dk}{4\pi^3}, \quad m_1 = e^{\lambda_0} \int_B e^{\lambda_1 E(k)} E(k) \frac{dk}{4\pi^3}.$$

The above equations correspond to the energy-transport model derived in Sect. 6.1.

The energy-transport equations can be made more explicit in the parabolic energy band approximation, for instance. We set $\lambda_1 = -1/T$ and interpret T as the electron temperature. By Example 8.8, we have

$$m_i = \frac{\sqrt{2}}{\pi^2} e^{\lambda_0} \int_0^\infty \varepsilon^{i+1/2} e^{-\varepsilon/T} \, d\varepsilon = \frac{\sqrt{2}}{\pi^2} e^{\lambda_0} T^{i+3/2} \Gamma\left(i + \frac{3}{2}\right), \quad i = 0, 1,$$

and thus, the electron density $n = m_0$ and the energy density $ne = m_1$ are

$$n = \frac{2}{(2\pi)^{3/2}} T^{3/2} e^{\lambda_0}, \quad ne = \frac{3}{(2\pi)^{3/2}} T^{5/2} e^{\lambda_0} = \frac{3}{2} nT, \tag{8.20}$$

which coincides with the expressions in Example 6.8. The diffusion coefficients become, using (8.17),

$$D_{ij} = \frac{\sqrt{8}\tau}{3\pi^2} e^{\lambda_0} \int_0^\infty \varepsilon^{i+j+3/2} e^{-\varepsilon/T} \, d\varepsilon = \frac{\sqrt{8}\tau}{3\pi^2} e^{\lambda_0} T^{i+j+5/2} \Gamma\left(i + j + \frac{5}{2}\right),$$

and computing the gamma functions, we obtain the diffusion matrix

$$\mathscr{D} = (D_{ij}) = \tau n T \begin{pmatrix} 1 & \dfrac{5}{2}T \\[2mm] \dfrac{5}{2}T & \dfrac{35}{4}T^2 \end{pmatrix}. \tag{8.21}$$

Remark 8.9. Energy-transport models with diffusion matrices similar to (6.26) can be derived by taking a relaxation time which depends on the macroscopic energy:

$$\tau = \tau_0 \left(\frac{\langle M[F] \rangle}{\langle \varepsilon M[F] \rangle} \right)^\beta,$$

where $\tau_0 > 0$ and $\beta \in \mathbb{R}$. Then $\tau = \tau_0(m_0/m_1)^\beta = (\frac{2}{3})^\beta \tau_0 T^{-\beta}$, and the diffusion matrix (8.21) becomes

$$\mathscr{D} = \left(\frac{2}{3}\right)^\beta \tau_0 n T^{1-\beta} \begin{pmatrix} 1 & \dfrac{5}{2}T \\ \dfrac{5}{2}T & \dfrac{35}{4}T^2 \end{pmatrix}.$$

This matrix is very similar to (6.26) for $\beta = 1$ except the coefficients $5/2$ and $35/4$. The matrix (6.26) is obtained if the relaxation time depends on the *microscopic* energy, i.e., $\tau = \tau(\varepsilon) = \varepsilon_0/\varepsilon$ for some $\varepsilon_0 > 0$. This is not surprising since the relaxation time used to derive (6.26) also depends on the energy ε, see (6.24). We infer that

$$D_{ij} = \frac{\sqrt{8}}{3\pi^2}e^{\lambda_0}\int_0^\infty \tau(\varepsilon)\varepsilon^{i+j+3/2}e^{-\varepsilon/T}\,d\varepsilon = \frac{\sqrt{8}\varepsilon_0}{3\pi^2}e^{\lambda_0}T^{i+j+3/2}\Gamma\left(i+j+\frac{3}{2}\right),$$

giving

$$\mathscr{D} = \frac{2}{3}\varepsilon_0 n \begin{pmatrix} 1 & \dfrac{3}{2}T \\ \dfrac{3}{2}T & \dfrac{15}{4}T^2 \end{pmatrix}, \tag{8.22}$$

which is the diffusion matrix of the Chen model. $\qquad\square$

Fourth-order moment equations. Finally, we consider the case $N = 2$ and $\kappa = (1, E, E^2)$. In the parabolic band approximation, the weight functions are at most of fourth order in k, which explains the name of the model. The balance equations are given by (8.11),

$$\partial_t m_0 - \text{div} J_0 = 0, \tag{8.23}$$

$$\partial_t m_1 - \text{div} J_1 + \nabla V \cdot J_0 = W_1, \tag{8.24}$$

$$\partial_t m_2 - \text{div} J_2 + 2\nabla V \cdot J_1 = W_2, \tag{8.25}$$

where W_1 and W_2 are the averaged collision terms, defined in Theorem 8.3, and the fluxes read as

$$J_i = D_{i0}(\nabla\lambda_0 + \lambda_1\nabla V) + D_{i1}(\nabla\lambda_1 + 2\lambda_2\nabla V) + D_{i2}\nabla\lambda_2, \quad i = 0, 1, 2,$$

with the diffusion coefficients (8.14). The derivation of the model is based on the solvability of the constrained maximization problem (8.3). According to Remark 8.2, this problem with weight functions $(1, E, E^2)$ is solvable if the nonparabolic band approximation (8.4) with nonparabolicity parameter $\alpha > 0$ is assumed. A more explicit formulation is obtained in the parabolic band approximation. Strictly speaking, the extremal problem may be unsolvable in this situation. However, we may consider the parabolic band case as the limiting situation when the nonparabolicity parameter in the nonparabolic band approximation tends to zero. Then (8.17) and (8.18) lead to the expressions

$$m_i = \frac{\sqrt{2}}{\pi^2} e^{\lambda_0} \int_0^\infty \varepsilon^{i+1/2} e^{\lambda_1 \varepsilon + \lambda_2 \varepsilon^2} \, d\varepsilon, \quad i = 0, 1, 2, \tag{8.26}$$

$$D_{ij} = \frac{\sqrt{8}\tau}{3\pi^2} \int_0^\infty \varepsilon^{i+j+3/2} e^{\lambda_1 \varepsilon + \lambda_2 \varepsilon^2} \, d\varepsilon, \quad i, j = 0, 1, 2.$$

Unfortunately, the above integrals cannot be further simplified. Moreover, the interpretation of the Lagrange multipliers λ_i becomes more difficult. In the next section, we discuss a reformulation of the fourth-order moment model and compare it with higher-order models in the literature.

8.3 Drift-Diffusion Formulation

We show that the fluxes of the higher-order moment model can be written in a drift-diffusion-type form, which allows for a numerical decoupling of the stationary model.

Proposition 8.10. *Let the assumptions of Theorem 8.3 and Proposition 8.4 (2) hold. Then we can write*

$$J_i = \nabla d_i + F_i(d) d_i \nabla V,$$

where $d_i = D_{i0}$, $d = (d_0, \dots, d_N)$, and

$$F_i(d) = \sum_{j=1}^N j \frac{D_{i,j-1}}{D_{i0}} \lambda_j, \quad i = 0, \dots, N.$$

The Lagrange multipliers λ_j are implicitly given by the values of $d_i = \langle E^i v \otimes \phi_0 \rangle$ and $L\phi_0 = -vM[F] = -ve^{\lambda \cdot \kappa}$. The mapping $d = d(\lambda)$ can be inverted since $\det d'(\lambda) = \det \mathscr{D} > 0$.

Proof. We claim that the first sum in the flux formulation (8.12) equals ∇D_{i0}. Indeed, from

$$L(\nabla \phi_{jm}) = \nabla(L\phi_{jm}) = -v_m E^j \sum_{\ell=0}^N \nabla \lambda_\ell E^\ell M[F] = -\sum_{\ell=0}^N \nabla \lambda_\ell v_m E^{j+\ell} M[F]$$

$$= L\left(\sum_{\ell=0}^N \nabla \lambda_\ell \phi_{j+\ell,m} \right)$$

and the unique solvability in $N(L)^\perp$, we obtain the relation

$$\nabla \phi_j = \sum_{\ell=0}^{N} \nabla \lambda_\ell \phi_{j+\ell} + cM[F],$$

where c is a constant vector. Hence, by (8.14), setting $j = 0$,

$$\nabla D_{i0} = \langle E^i v \otimes \nabla \phi_0 \rangle = \sum_{\ell=0}^{N} \nabla \lambda_\ell \langle E^i v \otimes \phi_\ell \rangle = \sum_{\ell=0}^{N} \nabla \lambda_\ell D_{i\ell}.$$

Then (8.12) becomes

$$J_i = \nabla D_{i0} + D_{i0} \nabla V \sum_{j=0}^{N} j \frac{D_{i,j-1}}{D_{i0}} \lambda_j,$$

proving the first assertion.

It remains to show that the determinant of the matrix $d'(\lambda)$ is positive. Since

$$L\left(\frac{\partial \phi_{jm}}{\partial \lambda_\ell}\right) = -v_m E^j \frac{\partial M[F]}{\partial \lambda_\ell} = -v_m E^{j+\ell} M[F] = L\phi_{j+\ell,m},$$

which gives $\partial \phi_0 / \partial \lambda_\ell = \phi_\ell + cM[F]$ for some and thus,

$$\frac{\partial D_{i0}}{\partial \lambda_\ell} = \left\langle E^i v \otimes \frac{\partial \phi_0}{\partial \lambda_\ell} \right\rangle = \langle E^i v \otimes \phi_\ell \rangle = D_{i\ell},$$

the Jacobian of $d(\lambda)$ consists of the elements $\partial d_i / \partial \lambda_j = \partial D_{i0} / \partial \lambda_j = D_{ij}$. The matrix $\mathscr{D} = (D_{ij})$ is positive definite by Proposition 8.4, and we have $\det d'(\lambda) = \det \mathscr{D} > 0$. □

Remark 8.11. The numerical decoupling of the higher-order moment model can be done as follows. Under the assumptions of the above proposition, the stationary model reads as

$$-\mathrm{div}\, J_i = -i\nabla V \cdot J_{i-1} + W_i, \quad J_i = \nabla d_i + F_i(d)d_i \nabla V, \quad i = 0,\dots,N.$$

We assume that V is given, and $W_i = W_i(d,V)$ may depend on d and V. We also write $J_i = J_i(d,V)$. During the iteration procedure, we may "freeze" the nonlinearities: Let $\tilde{d}$ be given (e.g., from the previous iteration step) and consider the system

$$-\mathrm{div}\, J_i(d,V) = -i\nabla V \cdot J_{i-1}(\tilde{d},V) + W_i(\tilde{d},V), \quad J_i(d,V) = \nabla d_i + F_i(\tilde{d})d_i \nabla V.$$

This system is decoupled since each equation is a scalar elliptic differential equation for d_i. Furthermore, the linear equations can by "symmetrized" by local Slotboom variables as described, for instance, in [12], to treat the convective part $F_i(\tilde{d})d_i \nabla V$. Finally, the "symmetrized" equations can be numerically discretized by mixed finite elements [12, 13]. □

We come back to the fourth-order moment model discussed in the previous section. Assuming the parabolic band approximation, the model can be rewritten as follows. We rewrite the functions $F_i(d)$:

$$F_i(d) = \lambda_1 + 2\frac{d_{i+1}}{d_i}\lambda_2, \quad i = 0, 1, 2.$$

Observing that $d_i = (2\tau/3)m_{i+1}$ and integrating by parts, we obtain from (8.26)

$$m_i = -\frac{\sqrt{2}}{\pi^2}e^{\lambda_0}\int_0^\infty \frac{2}{2i+3}\varepsilon^{i+3/2}(\lambda_1 + 2\lambda_2\varepsilon)e^{\lambda_1\varepsilon+\lambda_2\varepsilon^2}\,d\varepsilon \tag{8.27}$$

$$= -\frac{2}{2i+3}(\lambda_1 m_{i+1} + 2\lambda_2 m_{i+2}) = -\frac{3}{(2i+3)\tau}(\lambda_1 d_i + 2\lambda_2 d_{i+1}).$$

Hence,

$$F_i(d) = \frac{1}{d_i}(\lambda_1 d_i + 2\lambda_2 d_{i+1}) = -\frac{(2i+3)\tau}{3}\frac{m_i}{d_i},$$

and the fluxes become, for constant relaxation time,

$$J_i = \nabla d_i + F_i(d)d_i\nabla V = \frac{2}{3}\tau\left(\nabla m_{i+1} - \frac{2i+3}{2}m_i\nabla V\right), \quad i = 0, 1, 2. \tag{8.28}$$

Together with the balance equations (8.23), (8.24), and (8.25), we obtain a system of three equations for the unknowns m_0, m_1, and m_2. If τ depends on x or t, the variables are τm_0, τm_1, and τm_2. In the expression for J_2, the moment m_3 is needed. It can be computed from m_0, m_1, and m_2 using the relation

$$m_3 = -\frac{1}{2\lambda_2}\left(\frac{5}{2}m_1 + \lambda_1 m_2\right), \tag{8.29}$$

which comes from (8.27), where λ_1, λ_2 are functions of $m = (m_0, m_1, m_2)$. The fourth-order model with the above current relations can also be interpreted as a system of parabolic equations in the variables m_1, m_2, and m_3; the particle density m_0 is then a function of m_1, m_2, and m_3.

It remains to show that the function $m(\lambda)$ with $\lambda = (\lambda_0, \lambda_1, \lambda_2)$ can be inverted. This comes from the fact that the matrix $dm/d\lambda = (m_{i+j})_{i,j} \in \mathbb{R}^{3\times3}$ is positive definite (and hence its determinant is positive) since it is equal to the Hessian of the strictly convex function

$$\lambda \mapsto m_0 = \frac{\sqrt{2}}{\pi^2}\int_0^\infty \varepsilon^{1/2}e^{\lambda_0+\lambda_1\varepsilon+\lambda_2\varepsilon^2}\,d\varepsilon.$$

The final fourth-order model consists of the balance equations (8.23), (8.24), and (8.25) and the current relations (8.28) in the variables m_0, m_1, and m_2.

Remark 8.12. Grasser et al. have derived a related fourth-order model, called the *six-moments transport equations* (see formulas (124)–(129) in [14]). The model equations are given by (8.23), (8.24), (8.25), and (8.28) together with

$$m_0 = n, \quad m_1 = \frac{3}{2}nT, \quad m_2 = \frac{5 \cdot 3}{4}nT^2\beta_n. \tag{8.30}$$

Here, the variables are the particle density n, the electron temperature T, and the kurtosis β_n. This notion is inspired from the energy-transport model in the parabolic band approximation, where $m_2 = \frac{15}{4}nT^2$ (see the coefficient D_{22} in (8.22)). In this sense, β_n measures the deviation from the heated Maxwellian $M[F] = e^{\lambda_0 - \varepsilon/T}$. More generally, the kurtosis is defined by

$$\beta_n = \frac{3}{5}\frac{m_0 m_2}{m_1^2}.$$

By the Cauchy–Schwarz inequality,

$$m_1^2 = \frac{2}{\pi^4}e^{2\lambda_0}\left(\int_0^\infty \varepsilon^{1/4}\varepsilon^{5/4}e^{\lambda_1\varepsilon+\lambda_2\varepsilon^2}\,d\varepsilon\right)^2$$
$$\leq \frac{2}{\pi^4}e^{2\lambda_0}\int_0^\infty \varepsilon^{1/2}e^{\lambda_1\varepsilon+\lambda_2\varepsilon^2}\,d\varepsilon\int_0^\infty \varepsilon^{5/2}e^{\lambda_1\varepsilon+\lambda_2\varepsilon^2}\,d\varepsilon = m_0 m_2,$$

we obtain the restriction $\beta_n \geq 3/5$.

Grasser et al. [14] define heuristically m_3 in terms of the lower-order moments by setting

$$m_3 = \frac{7 \cdot 5 \cdot 3}{8}nT^2\beta_n^c, \tag{8.31}$$

where the constant exponent c is fitted from Monte Carlo simulations of the Boltzmann equation, computing the numerical moment m_3^{MC}. It has been found that the choice $c = 3$ gives the smallest deviation of the ratio m_3^{MC}/m_3 from the desired value one [14].

In the model derived above, m_3 is implicitly defined in terms of the lower-order moments, see (8.29). Using the notation (8.30) and setting $\lambda_1 = -1/T$ as in the energy-transport equations, we obtain from (8.29)

$$m_3 = -\frac{15}{8}\frac{(1 - \beta_n)nT}{\lambda_2}.$$

The expression (8.31) is obtained by setting $\lambda_2 = -(1 - \beta_n)/7T\beta_n^c$. Since it should hold $\lambda_2 < 0$ (in order to have integrability of $e^{\lambda_1\varepsilon+\lambda_2\varepsilon^2}$ for $\varepsilon \geq 0$), we conclude the restriction $\beta_n \leq 1$. Together with the above condition, the kurtosis has to satisfy the bounds $3/5 \leq \beta_n \leq 1$ [15]. Clearly, $\beta_n = 1$ corresponds to the energy-transport case for which $\lambda_2 = 0$.

Thus, the model of Grasser et al. is contained in our model hierarchy with the heuristic choice $\lambda_2 = -(1 - \beta_n)/7T\beta_n^c$. $\square$

8.4 Symmetrization and Entropy

In Sect. 6.3, we have shown that the electric force terms can be eliminated in the energy-transport equations by employing dual entropy variables (see (6.31)). It is possible to extend this methodology to the higher-order moment models derived above. To this end, we define *generalized dual entropy variables* $v = (v_0, \ldots, v_N)^\top$ by

$$\lambda = Pv,$$

where $\lambda = (\lambda_0, \ldots, \lambda_N)^\top$ are the Lagrange multipliers (or the primal entropy variables), and the transformation matrix $P = (P_{ij}) \in \mathbb{R}^{(N+1) \times (N+1)}$ is the following upper triangular matrix:

$$P_{ij} = (-1)^{i+j} \binom{j}{i} a_{ij} V^{j-i} \quad \text{with} \quad a_{ij} = \begin{cases} 1 & \text{if } i \le j, \\ 0 & \text{if } i > j, \end{cases}$$

where $i, j = 0, \ldots, N$. The dual entropy formulation allows us to "symmetrize" the equations in the sense of the following proposition.

Proposition 8.13 (Dual entropy formulation). *Define the dual entropy variables* $v = (v_0, \ldots, v_N)^\top$, *the transformed moments* $\rho = (\rho_0, \ldots, \rho_N)^\top$, *and the thermodynamic fluxes* $F = (F_0, \ldots, F_N)^\top$ *by*

$$\lambda = Pv, \quad \rho = P^\top m, \quad \text{and} \quad F = P^\top J.$$

Then the model equations (8.11) *and* (8.12) *can be equivalently written as*

$$\partial_t \rho_i - \operatorname{div} F_i = (P^\top W + V^{-1} \partial_t V Rm)_i, \quad F_i = \sum_{j=0}^{N} C_{ij} \nabla v_j,$$

where $W = (0, W_1, \ldots, W_N)^\top$, $R = (R_{ij})$ *is given by* $R_{ij} = (i - j) P_{ji}$, *and the new diffusion matrix* $C = (C_{ij})$ *is defined by* $C = P^\top DP$.

The proof of the proposition is quite technical and is based on some properties of the transformation matrix P, which are shown first.

Lemma 8.14. *The following properties hold:*
   (1) *The matrix* $Q = (Q_{ij})$ *given by* $Q_{ij} = \binom{j}{i} a_{ij} V^{j-i}$ *is the inverse of* P.
   (2) *We set* $j \delta_{i,j-1} = 0$ *for* $j = 0$. *Then, for all* $i, j = 0, \ldots, N$,

$$\sum_{k=0}^{N} (j-k) P_{ik} Q_{kj} = -\sum_{k=0}^{N} (j-k) Q_{ik} P_{kj} = j \delta_{i,j-1} V.$$

(3) *For all* $i = 0, \ldots, N-1$, $j = 1, \ldots, N$,

$$-j P_{i,j-1} + (i+1) P_{i+1,j} = 0.$$

Proof. (1) By the definition of the coefficients a_{ij}, we have $\sum_\ell P_{i\ell}Q_{\ell j} = 0$ for all $i > j$. Let $i < j$. Then

$$\sum_{\ell=0}^N P_{i\ell}Q_{\ell j} = \sum_{\ell=i}^j (-1)^{i+\ell}\binom{\ell}{i}\binom{j}{\ell}V^{j-i} = V^{j-i}\sum_{\ell=i}^j (-1)^{i+\ell}\binom{j}{i}\binom{j-i}{\ell-i}$$

$$= V^{j-i}\binom{j}{i}\sum_{p=0}^{j-i}(-1)^p\binom{j-i}{p} = 0.$$

Furthermore, for $i = j$, we obtain

$$\sum_{\ell=0}^N P_{i\ell}Q_{\ell i} = \sum_{\ell=i}^i (-1)^{i+\ell}\binom{\ell}{i}\binom{i}{\ell} = 1.$$

Thus, PQ equals the identity matrix.

(2) The definition of a_{ij} yields $\sum_\ell (j-\ell)P_{i\ell}Q_{\ell j} = 0$ for $i \geq j$. Next, for $i < j-1$ it follows that

$$\sum_{\ell=0}^N (j-\ell)P_{i\ell}Q_{\ell j} = V^{j-i}\sum_{\ell=i}^{j-1}(j-\ell)(-1)^{i+\ell}\binom{\ell}{i}\binom{j}{\ell}$$

$$= V^{j-i}\sum_{\ell=i}^{j-1}(-1)^{i+\ell}j\binom{j-1}{i}\binom{j-1-i}{\ell-i}$$

$$= jV^{j-i}\binom{j-1}{i}\sum_{p=0}^{j-1-i}(-1)^p\binom{j-1-i}{p} = 0.$$

If $i = j-1$ then

$$\sum_{\ell=0}^N (j-\ell)P_{i\ell}Q_{\ell j} = V\sum_{\ell=j-1}^{j-1}(j-\ell)(-1)^{j-1+\ell}\binom{\ell}{j-1}\binom{j}{\ell}$$

$$= V\binom{j-1}{j-1}\binom{j}{j-1} = jV.$$

The second equality is shown in a similar way.

(3) For $i \geq j$ we have $P_{i,j-1} = 0$ and $P_{i+1,j} = 0$. If $i < j$ then

$$-jP_{i,j-1} + (i+1)P_{i+1,j} = (-1)^{i+j+1}V^{j-1-i}\left(-j\binom{j-1}{i} + (i+1)\binom{j}{i+1}\right) = 0.$$

This shows the lemma. □

Proof (of Proposition 8.13). First we show the relation for the new fluxes. Employing the definitions $C = P^\top DP$ and $v = Q\lambda$ and the property $QP = \mathrm{Id}$, we obtain

$$\sum_{j=0}^{N} C_{ij}\nabla v_j = \sum_{j,\ell,p,n=0}^{N} P_{\ell i}D_{\ell p}P_{pj}\nabla(Q_{jn}\lambda_n)$$

$$= \sum_{j,\ell,p,n=0}^{N} P_{\ell i}D_{\ell p}(P_{pj}Q_{jn}\nabla\lambda_n + P_{pj}\nabla Q_{jn}\lambda_n)$$

$$= \sum_{\ell,p=0}^{N} P_{\ell i}D_{\ell p}\nabla\lambda_p + \sum_{\ell,p,n=0}^{N} P_{\ell i}D_{\ell p}\left(\sum_{j=0}^{N}(n-j)P_{pj}Q_{jn}\right)V^{-1}\nabla V\lambda_n,$$

since $\nabla Q_{jn} = (n-j)V^{-1}\nabla V Q_{jn}$. Now, using Lemma 8.14 (2),

$$\sum_{j=0}^{N} C_{ij}\nabla v_j = \sum_{\ell,p=0}^{N} P_{\ell i}D_{\ell p}\nabla\lambda_p + \sum_{\ell,p,n=0}^{N} P_{\ell i}D_{\ell p}n\delta_{p,n-1}\nabla V\lambda_n$$

$$= \sum_{\ell,n=0}^{N} P_{\ell i}(D_{\ell n}\nabla\lambda_n + nD_{\ell,n-1}\nabla V\lambda_n) = \sum_{\ell=0}^{N} P_{\ell i}J_\ell = F_i.$$

Next, we compute the transformed balance equations. By the definition of F_i,

$$\text{div}\, F_i = \sum_{j=0}^{N} \text{div}\,(P_{ji}J_j) = \sum_{j=0}^{N}(P_{ji}\,\text{div}\,J_j + \nabla P_{ji}\cdot J_j) \tag{8.32}$$

$$= \sum_{j=0}^{N} P_{ji}(\text{div}\,J_j - jJ_{j-1}\cdot\nabla V) + \sum_{j=0}^{N}(\nabla P_{ji}\cdot J_j + jP_{ji}J_{j-1}\cdot\nabla V).$$

We show that the second sum vanishes. Observing that $\nabla P_{ji} = (i-j)V^{-1}\nabla V P_{ji}$, we find

$$A = \sum_{j=0}^{N}(\nabla P_{ji}\cdot J_j + jP_{ji}J_{j-1}\cdot\nabla V) = \sum_{j=0}^{N}\left((i-j)P_{ji}V^{-1}\nabla V\cdot J_j + jP_{ji}J_{j-1}\cdot\nabla V\right).$$

Since the first sum can be rewritten, by Lemma 8.14 (2), as

$$\sum_{j=0}^{N}(i-j)P_{ji}V^{-1}\nabla V\cdot J_j = \sum_{j,\ell=0}^{N}(i-\ell)\delta_{j\ell}P_{\ell i}V^{-1}J_j\cdot\nabla V$$

$$= \sum_{j,\ell,p=0}^{N}(i-\ell)P_{jp}Q_{p\ell}P_{\ell i}V^{-1}J_j\cdot\nabla V = \sum_{j,p=0}^{N}\left(\sum_{\ell=0}^{N}(i-\ell)Q_{p\ell}P_{\ell i}\right)P_{jp}V^{-1}J_j\cdot\nabla V$$

$$= -\sum_{j,p=0}^{N} i\delta_{p,i-1}P_{jp}J_j\cdot\nabla V = -\sum_{j=0}^{N} iP_{j,i-1}J_j\cdot\nabla V,$$

we obtain

$$A = \sum_{j=0}^{N-1}(-iP_{j,i-1} + (j+1)P_{j+1,i})J_j\cdot\nabla V = 0,$$

using Lemma 8.14 (3). Hence, with the balance equations (8.11), (8.32) becomes

$$\operatorname{div} F_i = \sum_{j=0}^{N} P_{ji}(\partial_t m_j - W_j). \tag{8.33}$$

We employ the definition $\rho = P^\top m$ to rewrite the first sum,

$$\sum_{j=0}^{N} P_{ji} \partial_t m_j = \sum_{j=0}^{N} \left(\partial_t (P_{ji} m_j) - \partial_t P_{ji} m_j \right)$$

$$= \partial_t \rho_i - V^{-1} \partial_t V \sum_{j=0}^{N} (i-j) P_{ji} m_j = \partial_t \rho_i - V^{-1} \partial_t V \sum_{j=0}^{N} R_{ij} m_j.$$

This finishes the proof. $\square$

We consider two examples.

Example 8.15. (Energy-transport model) The transformation matrix P and its inverse Q read in the case $N = 1$ as follows:

$$P = \begin{pmatrix} 1 & -V \\ 0 & 1 \end{pmatrix}, \quad Q = \begin{pmatrix} 1 & V \\ 0 & 1 \end{pmatrix}.$$

Defining the chemical potential μ by $\lambda_0 = \mu/T$, where $T = -1/\lambda_1 > 0$ is the particle temperature, the dual entropy variable $v = Q\lambda$ becomes (see Sect. 6.3)

$$v_0 = \lambda_0 + V\lambda_1 = \frac{\mu - V}{T}, \quad v_1 = \lambda_1 = -\frac{1}{T}.$$

The quantity $\mu - V$ is known as the *electro-chemical potential*. $\square$

Example 8.16. (Fourth-order model) For $N = 2$, the transformation matrix is given by

$$P = \begin{pmatrix} 1 & -V & V^2 \\ 0 & 1 & -2V \\ 0 & 0 & 1 \end{pmatrix}.$$

Introducing the chemical potential and the temperature as in the previous example and the *second-order temperature* θ as in [10] by $\lambda_2 = -1/\theta T$, the dual entropy variables are

$$v_0 = \frac{\mu - V}{T} - \frac{V^2}{\theta T}, \quad v_1 = -\frac{1}{T} - \frac{2V}{\theta T}, \quad v_2 = -\frac{1}{\theta T}. \quad \square$$

The dual entropy formulation allows us to prove entropy dissipation. We define the relative entropy S by

$$S(t) = -\int_{\mathbb{R}^3} (m \cdot (\lambda - \bar\lambda) - m_0 + \bar m_0) \, dx \leq 0,$$

where $\lambda = (\lambda_0, \dots, \lambda_N)^\top$ and $m = (m_0, \dots, m_N)^\top$. The vectors $\bar{\lambda} = (V, -1, 0, \dots, 0)^\top$ and $\bar{m}_0 = m_0(\bar{\lambda})$ are the equilibrium values (since $e^{\bar{\lambda}\cdot\kappa} = e^{V - E(k)}$ is the equilibrium distribution function in the presence of an electric field). Notice that for the energy-transport model (i.e., $N = 1$), the relative entropy becomes (see (8.20))

$$S = -\int_{\mathbb{R}^3} \left(n \left(\ln \left(\frac{n}{T^{3/2}} \right) - \frac{5}{2} - \log \frac{2}{(2\pi)^{3/2}} - V \right) + \frac{3}{2} nT + \frac{2}{(2\pi)^{3/2}} e^V \right) dx.$$

Proposition 8.17 (Entropy inequality). *Assume that the electric potential is time-independent and that*

$$\int_{\mathbb{R}^3} W \cdot (\lambda - \bar{\lambda}) \, dx \le 0. \tag{8.34}$$

Then any (smooth) solution λ of the higher-order moment equations (8.11) and (8.12) satisfies the entropy inequality

$$-\frac{dS}{dt} + \int_{\mathbb{R}^3} \sum_{i,j=0}^{N} C_{ij} \nabla v_i \cdot \nabla v_j \, dx \le 0.$$

The second integral on the left-hand side is called the *entropy production*. It is nonnegative since the diffusion matrix is positive (semi-) definite. Thus, the entropy is nondecreasing in time.

Proof. We introduce the relative entropy density $s(\lambda) = -m \cdot (\lambda - \bar{\lambda}) + m_0 - \bar{m}_0$. The moments are given by $m_i = \int_B E(k)^i e^{\lambda \cdot \kappa} dk/4\pi^3$. Then $\partial m_0/\partial \lambda_i = m_i$ and we obtain

$$\frac{\partial s}{\partial \lambda_i} = -\frac{\partial m}{\partial \lambda_i} \cdot (\lambda - \bar{\lambda}) - m_i + \frac{\partial m_0}{\partial \lambda_i} = -\frac{\partial m}{\partial \lambda_i} \cdot (\lambda - \bar{\lambda})$$

and

$$\partial_t m \cdot (\lambda - \bar{\lambda}) = \sum_{i=0}^{N} \frac{\partial m}{\partial \lambda_i} \cdot (\lambda - \bar{\lambda}) \partial_t \lambda_i = -\sum_{i=0}^{N} \frac{\partial s}{\partial \lambda_i} \partial_t \lambda_i = -\partial_t s(\lambda). \tag{8.35}$$

The balance equation (8.11) is formally equivalent to (8.33); multiplying the latter equations by $v_i - \bar{v}_i$, where $\bar{v} = Q\bar{\lambda}$, and summing over $i = 0, \dots, N$, it follows that

$$(P^\top \partial_t m)^\top (v - \bar{v}) - (\operatorname{div} F)^\top (v - \bar{v}) = (P^\top W)^\top (v - \bar{v}).$$

Integrating over x and employing the definition $v = Q\lambda$ give

$$\int_{\mathbb{R}^3} \partial_t m^\top PQ(\lambda - \bar{\lambda}) \, dx - \int_{\mathbb{R}^3} \sum_{i,j=0}^{N} \operatorname{div}(C_{ij} \nabla v_j)(v_i - \bar{v}_i) \, dx$$
$$= \int_{\mathbb{R}^3} W^\top PQ(\lambda - \bar{\lambda}) \, dx.$$

Finally, integrating by parts in the second integral, taking into account that $\nabla \bar{v} = 0$, and using (8.35) yield

$$-\int_{\mathbb{R}^3} \partial_t s(\lambda)\,dx + \int_{\mathbb{R}^3} \sum_{i,j=0}^{N} C_{ij}\nabla v_i \cdot \nabla v_j\,dx = \int_{\mathbb{R}^3} W^{\top}(\lambda - \bar{\lambda})\,dx \le 0,$$

which proves the lemma. $\quad\square$

In [16, Lemma 4.11], it was shown that assumption (8.34) on W holds for an inelastic phonon collision operator in the case of the energy-transport model. This hypothesis also holds if

$$W_i = -\frac{1}{\tau}(m_i - \bar{m}_i), \quad \text{where } \bar{m}_i = m_i(\bar{\lambda}),$$

since

$$W \cdot (\lambda - \bar{\lambda}) = -\frac{1}{\tau}\int_B (e^{\kappa \cdot \lambda} - e^{\kappa \cdot \bar{\lambda}})(\kappa \cdot \lambda - \kappa \cdot \bar{\lambda})\frac{dk}{4\pi^3} \le 0.$$

Figure 8.1 summarizes the relations between the diffusive models discussed in this chapter.

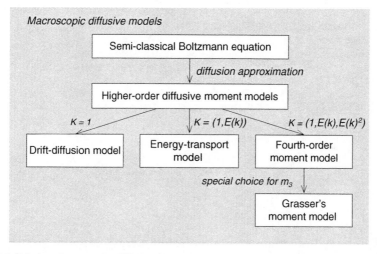

Fig. 8.1 Relations between the diffusive moment models. From the diffusive moment hierarchy, explicit models can be derived by choosing the weight functions κ. The moment m_3 is discussed in Remark 8.12

References

1. A. Jüngel, S. Krause, and P. Pietra. A hierarchy of diffusive higher-order moment equations for semiconductors. *SIAM J. Appl. Math.* 68 (2007), 171–198.
2. H. Struchtrup. Derivation of 13 moment equations for rarefied gas flow to second order accuracy for arbitrary interaction potentials. *SIAM Multiscale Model. Simul.* 3 (2005), 221–243.

3. S. Ihara. *Information Theory for Continuous Systems*. World Scientific, Singapore, 1993.
4. W. Dreyer, M. Junk, and M. Kunik. On the approximation of kinetic equations by moment systems. *Nonlinearity* 14 (2001), 881–906.
5. M. Junk and V. Romano. Maximum entropy moment systems of the semiconductor Boltzmann equation using Kane's dispersion relation. *Continuum Mech. Thermodyn.* 17 (2004), 247–267.
6. J. Schneider. Entropic approximation in kinetic theory. *ESAIM: Math. Mod. Numer. Anal.* 38 (2004), 541–561.
7. T. Grasser, H. Kosina, C. Heitzinger, and S. Selberherr. Characterization of the hot electron distribution function using six moments. *J. Appl. Phys.* 91 (2002), 3869–3879.
8. T. Grasser. Non-parabolic macroscopic transport models for semiconductor device simulation. *Physica A* 349 (2005), 221–258.
9. K. Sonoda, M. Yamaji, K. Taniguchi, C. Hamaguchi, and S. Dunham. Moment expansion approach to calculate impact ionization rate in submicron silicon devices. *J. Appl. Phys.* 80 (1996), 5444–5448.
10. T. Grasser, H. Kosina, M. Gritsch, and S. Selberherr. Using six moments of Boltzmann's equation for device simulation. *J. Appl. Phys.* 90 (2001), 2389–2396.
11. N. Ben Abdallah and P. Degond. On a hierarchy of macroscopic models for semiconductors. *J. Math. Phys.* 37 (1996), 3308–3333.
12. P. Degond, A. Jüngel, and P. Pietra. Numerical discretization of energy-transport models for semiconductors with nonparabolic band structure. *SIAM J. Sci. Comput.* 22 (2000), 986–1007.
13. S. Holst, A. Jüngel, and P. Pietra. A mixed finite-element discretization of the energy-transport equations for semiconductors. *SIAM J. Sci. Comput.* 24 (2003), 2058–2075.
14. T. Grasser, H. Kosina, and S. Selberherr. Hot carrier effects within macroscopic transport models. *Internat. J. High Speed Electr. Sys.* 13 (2003), 873–901.
15. T. Grasser, R. Kosik, C. Jungemann, H. Kosina, and S. Selberherr. Nonparabolic macroscopic transport models for device simulation based on bulk Monte Carlo data. *J. Appl. Phys.* 97 (2005), 093710.
16. N. Ben Abdallah, P. Degond, and S. Génieys. An energy-transport model for semiconductors derived from the Boltzmann equation. *J. Stat. Phys.* 84 (1996), 205–231.

Chapter 9
Hydrodynamic Equations

In Sect. 2.1, we have considered two different time scalings. In the diffusion scaling, assumed in Chaps. 5, 6, 7, and 8, the typical time is of the order of the time between two consecutive collisions divided by the square of the Knudsen number α^2, which is supposed to be small compared to one. In this chapter, we consider a shorter time scale. More precisely, we suppose that the typical time is of the order of the time between two scattering events divided by α. We show that with this scaling hydrodynamic equations can be derived. In contrast to the models of the previous chapters, hydrodynamic models are mathematically not of parabolic but of hyperbolic type.

9.1 Derivation from the Boltzmann Equation

Like in the previous chapters, the starting point is the semiconductor Boltzmann equation for the distribution function $f = f(x,k,t)$,

$$\partial_t f + v(k) \cdot \nabla_x f + \frac{q}{\hbar} \nabla_x V \cdot \nabla_k f = Q(f), \quad x \in \mathbb{R}^3, \ k \in B, \ t > 0,$$

where $v(k) = \nabla_k E(k)/\hbar$ is the group velocity, $E(k)$ the energy band structure depending on the (pseudo) wave vector k, $V(x,t)$ the electric potential, computed from the Poisson equation (5.3), and B the Brillouin zone. The initial condition reads as $f(x,k,0) = f_I(x,k)$. We assume that the collision operator is given as the sum

$$Q(f) = Q_0(f) + Q_1(f)$$

and that the mean free path λ_0 of collisions described by Q_0 is much smaller than the mean free path λ given by Q_1.

First, we scale the Boltzmann equation. We proceed similarly as in Sect. 2.3. We introduce the reference length λ, which is the mean free path corresponding to Q_1, and the reference velocity $v_0 = \sqrt{k_B T_L/m^*}$. This velocity corresponds to a particle

Jüngel, A.: *Hydrodynamic Equations*. Lect. Notes Phys. **773**, 195–213 (2009)
DOI 10.1007/978-3-540-89526-8_9

with kinetic energy of the order of the thermal energy $k_B T_L$. Furthermore, we define the reference wave vector $m^* v_0 / \hbar$, the reference potential $U_T = k_B T_L / q$, and the reference times $\tau = \lambda / v_0$ and $\tau_0 = \lambda_0 / v_0$. This defines the dimensionless variables

$$x = \lambda x_s, \quad t = \tau t_s, \quad k = \frac{m^* v_0}{\hbar} k_s,$$

and the dimensionless functions

$$V = U_T V_s, \quad Q_0(f) = \frac{1}{\tau_0} Q_{s,0}(f), \quad Q_1(f) = \frac{1}{\tau} Q_{s,1}(f).$$

Inserting this scaling into the Boltzmann equation and multiplying the resulting equation by τ_0, we obtain, omitting the index "s",

$$\alpha \partial_t f + \alpha \left(v(k) \cdot \nabla_x f + \nabla_x V \cdot \nabla_k f \right) = Q_0(f) + \alpha Q_1(f), \tag{9.1}$$

where $\alpha = \lambda_0 / \lambda$ is the ratio between the mean free paths corresponding to the collision operators Q_0 and Q_1, respectively. We assume that $\alpha \ll 1$, i.e., there are much more scattering events described by Q_0 than by Q_1.

Derivation of the hydrodynamic equations. The hydrodynamic model is derived from the moment equations of (9.1). Let f_α be a solution of (9.1) and V_α be a solution of the Poisson equation (5.3). We introduce the following weight functions: $\kappa_0(k) = 1$, $\kappa_1(k) = v(k)$, and $\kappa_2(k) = E(k)$. The corresponding moments have a physical interpretation: $n = \langle f \rangle$ is the electron density, $J = -\langle v(k) f \rangle$ the electron current density, and $\langle E(k) f \rangle$ the energy density, relative to the distribution function f, where $\langle g \rangle = \int_B g(k) \, dk / 4\pi^3$. Then, the moment equations are obtained from (9.1) by multiplication by κ_i / α and integration over the Brillouin zone, as in Sect. 2.3:

$$\partial_t \langle \kappa_i f_\alpha \rangle + \mathrm{div}_x \langle v \kappa_i f_\alpha \rangle - \nabla_x V \cdot \langle \nabla_k \kappa_i f_\alpha \rangle = \alpha^{-1} \langle \kappa_i Q_0(f_\alpha) \rangle + \langle \kappa_i Q_1(f_\alpha) \rangle, \tag{9.2}$$

where $i = 0, 1, 2$. Unfortunately, this set of equations is not closed: The integrals $\langle v v f_\alpha \rangle = \langle v \otimes v f_\alpha \rangle$ and $\langle v E f_\alpha \rangle$ cannot be written in terms of the lower-order moments n, J, and ne. This is referred to as the *closure problem*. To solve this problem, we make some simplifying assumptions on the collision operators and perform the formal limit $\alpha \to 0$.

We impose the following assumptions:

1. The energy band is approximated by a parabolic band and $B = \mathbb{R}^3$. Then the weight functions become $\kappa_0(k) = 1$, $\kappa_1(k) = k$, and $\kappa_2(k) = \frac{1}{2} |k|^2$.
2. All moments up to second order of Q_0 vanish, i.e., $\langle \kappa_i Q_0(f) \rangle = 0$ for all functions f.
3. The kernel of Q_0 is spanned by the Maxwellians, $N(Q_0) = \{ f : f(k) = M[f](k) = M(k) = \exp(\lambda_0 + \lambda_1 \cdot v(k) + \frac{1}{2} \lambda_2 |k|^2) \}$, where λ_0 and λ_2 are real numbers and λ_1 is a vector.
4. The collisions described by Q_1 conserve mass, i.e., $\langle Q_1(f) \rangle = 0$ for all functions $f(k)$.

The first assumption simplifies the computations below. We will discuss the case of a general band structure at the end of this section. The second condition means that the collision operator Q_0 conserves mass, momentum, and energy. For instance, a relaxation-time operator satisfies this hypothesis. The third assumption signifies that the equilibrium state of the system given by Q_0 is given by Maxwellians. In view of the parabolic band approximation, the Maxwellians can be written in a more common form. Instead of λ_0, λ_1, and λ_2, we introduce the electron density n, the mean velocity u, and the electron temperature T by $n = \langle M \rangle$, $\lambda_2 = -1/T$, and $\lambda_1 = u/T$. Then a computation shows that

$$n = \int_{\mathbb{R}^3} \exp\left(\lambda_0 + \lambda_1 \cdot k + \frac{1}{2}\lambda_2 |k|^2\right) \frac{dk}{4\pi^3} = \frac{2}{(2\pi)^{3/2}} T^{3/2} e^{\lambda_0 + |u|^2/2T},$$

and hence,

$$M(k) = \frac{1}{2}\left(\frac{2\pi}{T}\right)^{3/2} n e^{-|u-k|^2/2T}. \tag{9.3}$$

Finally, the last condition implies mass conservation for the total collision operator $Q_0 + \alpha Q_1$ which is physically reasonable.

The derivation of the hydrodynamic equations is based on two steps, as explained in Sect. 2.3. In the first step, the (formal) limit $\alpha \to 0$ in the Boltzmann equation (9.1) leads to

$$Q_0(f) = 0, \quad \text{where } f = \lim_{\alpha \to 0} f_\alpha.$$

Then, by the third assumption, $f = M$ for some n, u, and T. In the second step, we perform the limit $\alpha \to 0$ in the moment equations (9.2), employing the first assumption on Q_0:

$$\partial_t \langle \kappa_i M \rangle + \mathrm{div}_x \langle \kappa_i k M \rangle - \nabla_x V \cdot \langle \nabla_k \kappa_i M \rangle = \langle \kappa_i Q_1(M) \rangle, \quad i = 0, 1, 2. \tag{9.4}$$

We need to compute the higher-order moments.

Lemma 9.1. *Let the Maxwellian M be given by (9.3). Then*

$$\langle kM \rangle = nu, \quad \langle k \otimes kM \rangle = n(u \otimes u) + nT\,\mathrm{Id}, \quad \left\langle \frac{1}{2}k|k|^2 M \right\rangle = \frac{1}{2}nu(|u|^2 + 5T).$$

Proof. We employ the following identities:

$$\left\langle e^{-|z|^2/2} \right\rangle = 2(2\pi)^{-3/2}, \quad \left\langle z_i z_j e^{-|z|^2/2} \right\rangle = 2(2\pi)^{-3/2}\delta_{ij}.$$

Then, with the transformation $z = (k - u)/\sqrt{T}$, we obtain

$$\langle kM \rangle = \frac{1}{2}(2\pi)^{3/2} n \left\langle (u + \sqrt{T}z)e^{-|z|^2/2} \right\rangle = nu,$$

since $z \mapsto z e^{-|z|^2/2}$ is an odd function and hence, its integral vanishes. Furthermore,

$$\langle k \otimes kM \rangle = \frac{1}{2}(2\pi)^{3/2} n \left\langle (u + \sqrt{T}z) \otimes (u + \sqrt{T}z) \mathrm{e}^{-|z|^2/2} \right\rangle$$

$$= n(u \otimes u) + \frac{1}{2}(2\pi)^{3/2} nT \langle z \otimes z \mathrm{e}^{-|z|^2/2} \rangle = n(u \otimes u) + nT \,\mathrm{Id}.$$

Finally, we have for $i = 1, 2, 3$

$$\left\langle \frac{1}{2} k_i |k|^2 M \right\rangle = \frac{1}{4}(2\pi)^{3/2} n \sum_{j=1}^{3} \left\langle (u_i + \sqrt{T}z_i)(u_j + \sqrt{T}z_j)^2 \mathrm{e}^{-|z|^2/2} \right\rangle.$$

A straightforward computation shows that

$$\left\langle \frac{1}{2} k_i |k|^2 M \right\rangle = \frac{1}{2} n u_i |u|^2 + \frac{1}{2} nT \sum_{j=1}^{3} (u_i + 2u_j \delta_{ij}) = \frac{1}{2} n u_i |u|^2 + \frac{5}{2} nT u_i.$$

This finishes the proof. □

Inserting the expressions of Lemma 9.1 into the moment equations (9.4) leads to the following result.

Theorem 9.2 (Hydrodynamic equations). *Let the above assumptions on the energy band and the collision operators hold and let (f_α, V_α) be a solution of the Boltzmann–Poisson system (9.1) and (5.3). Then the limit function $f = \lim_{\alpha \to 0} f_\alpha$ equals the Maxwellian (9.3), where the functions n, $J_n = -nu$, and ne are solutions of the* hydrodynamic equations

$$\partial_t n - \operatorname{div} J_n = 0, \tag{9.5}$$

$$\partial_t J_n - \operatorname{div}\left(\frac{J_n \otimes J_n}{n}\right) - \nabla(nT) + n\nabla V = -\langle kQ_1(M) \rangle, \tag{9.6}$$

$$\partial_t(ne) - \operatorname{div}\left(J_n(e + T)\right) + J_n \cdot \nabla V = \left\langle \frac{1}{2}|k|^2 Q_1(M) \right\rangle, \tag{9.7}$$

$$\lambda_D^2 \Delta V = n - C(x),$$

where $e = \frac{1}{2}|u|^2 + \frac{3}{2}T$ is the sum of the kinetic and thermal energies, and $V = \lim_{\alpha \to 0} V_\alpha$. The initial conditions are

$$n(\cdot, 0) = \int_B f_I \frac{dk}{4\pi^3}, \quad J_n(\cdot, 0) = -\int_B k f_I \frac{dk}{4\pi^3}, \quad (ne)(\cdot, 0) = \int_B \frac{1}{2}|k|^2 f_I \frac{dk}{4\pi^3}.$$

Discussion of the equations. The production terms $\langle kQ_1(M) \rangle$ and $\langle \frac{1}{2}|k|^2 Q_1(M) \rangle$ can be specified if Q_1 is given, for instance, by the low-density operator

$$(Q_1(f))(x, k, t) = \int_{\mathbb{R}^3} \sigma\left(x, k, k'\right) \left(M_{\mathrm{eq}} f' - M'_{\mathrm{eq}} f\right) dk', \tag{9.8}$$

where the collision cross-section σ is assumed to be symmetric in k and k' and $M_{\mathrm{eq}}(k) = \frac{1}{2}(2\pi)^{3/2}e^{-|k|^2/2}$ is the Maxwellian of the thermal equilibrium state. Notice that M_{eq} is normalized, i.e., $\langle M_{\mathrm{eq}}\rangle = 1$.

Lemma 9.3. *The low-density collision operator* (9.8) *satisfies, for all functions* $f(k)$,

$$\langle Q_1(f)\rangle = 0,$$

$$\langle kQ_1(f)\rangle = -\int_{\mathbb{R}^3} \frac{kf}{\tau_p(x,k)}\frac{dk}{4\pi^3},$$

$$\left\langle \frac{1}{2}|k|^2 Q_1(f)\right\rangle = \int_{\mathbb{R}^3} \frac{f(e_0(x,k)-e\frac{1}{2}|k|^2)}{\tau_e(x,k)}\frac{dk}{4\pi^3},$$

where the momentum relaxation time τ_p, *the energy relaxation time* τ_e, *and the averaged energy* e_0 *are given by, respectively,*

$$\frac{1}{\tau_p(x,k)} = \int_{\mathbb{R}^3}\sigma(x,k,k')M'_{\mathrm{eq}}\left(1-\frac{k'}{k}\right)dk', \tag{9.9}$$

$$\frac{1}{\tau_e(x,p)} = \int_{\mathbb{R}^3}\sigma(x,k,k')M'_{\mathrm{eq}}\,dk',$$

$$e_0(x,k) = \left(\int_{\mathbb{R}^3}\sigma(x,k,k')M'_{\mathrm{eq}}\,dk'\right)^{-1}\int_{\mathbb{R}^3}\sigma(x,k,k')\frac{1}{2}|k'|^2 M'_{\mathrm{eq}}\,dk'.$$

Furthermore, if the scattering rate σ *does not depend on* k *and* k', *then*

$$\langle kQ_1(M)\rangle = \frac{J_n}{\tau_0},\quad \left\langle \frac{1}{2}|k|^2 Q_1(M)\right\rangle = -\frac{n}{\tau_0}\left(e-\frac{3}{2}\right), \tag{9.10}$$

where $\tau_0 = 1/\sigma$.

Expression (9.9) for the momentum relaxation time can be also found in [1, Sect. 5.2.1].

Proof. The first moment is given by

$$\langle kQ_1(f)\rangle = \frac{1}{4\pi^3}\int_{\mathbb{R}^6}\sigma(k,k')(M_{\mathrm{eq}}f' - M'_{\mathrm{eq}}f)k\,dk'\,dk,$$

and exchanging k and k' in the first sum gives

$$\langle kQ_1(f)\rangle = \frac{1}{4\pi^3}\int_{\mathbb{R}^6}\sigma(k,k')M'_{\mathrm{eq}}f(k'-k)\,dk'\,dk$$

$$= -\int_{\mathbb{R}^3}kf\left(\int_{\mathbb{R}^3}\sigma(k,k')M'_{\mathrm{eq}}\left(1-\frac{k'}{k}\right)dk'\right)\frac{dk}{4\pi^3},$$

showing the second equation involving τ_p. The third equation follows from

$$\left\langle \frac{1}{2}|k|^2 Q_1(f) \right\rangle = \frac{1}{2}\int_{\mathbb{R}^3} f\left(\int_{\mathbb{R}^3} \sigma(k,k')|k'|^2 M'_{\text{eq}}\,dk' - |k|^2 \int_{\mathbb{R}^3} \sigma(k,k')M'_{\text{eq}}\,dk' \right) \frac{dk}{4\pi^3}.$$

Furthermore, if $\sigma = \sigma(x)$, we compute

$$Q_1(f) = \sigma(x)\left(M_{\text{eq}} \int_{\mathbb{R}^3} f'\,dk' - f \int_{\mathbb{R}^3} M'_{\text{eq}}\,dk' \right) = 4\pi^3 \sigma(x)(n M_{\text{eq}} - f),$$

and hence, by the definition of J_n,

$$\langle k Q_1(M) \rangle = \sigma(n\langle k M_{\text{eq}}\rangle - \langle kM \rangle) = \sigma J_n.$$

By the second identity in Lemma 9.1, we infer that $\langle \frac{1}{2}|k|^2 M \rangle = \frac{1}{2}n|u|^2 + \frac{3}{2}nT = ne$, such that

$$\left\langle \frac{1}{2}|k|^2 Q_1(M) \right\rangle = \sigma\left(n\left\langle \frac{1}{2}|k|^2 M_{\text{eq}} \right\rangle - \left\langle \frac{1}{2}|k|^2 M \right\rangle \right) = \sigma\left(\frac{3}{2}n - ne \right),$$

which proves the lemma. $\square$

For wave-vector-independent scattering rates, the above lemma shows that the averaged collision integrals are of relaxation-time type. Indeed, in the absence of external forces, in the homogeneous case, and for constant scattering rate, we conclude from (9.6) and (9.7) that

$$\partial_t \int_{\mathbb{R}^3} J_n\,dx = -\frac{1}{\tau_0}\int_{\mathbb{R}^3} J_n\,dx, \quad \partial_t \int_{\mathbb{R}^3} ne\,dx = -\frac{1}{\tau_0}\int_{\mathbb{R}^3} \left(ne - \frac{3}{2}n \right)\,dx.$$

These differential equations can be solved explicitly. Since the total mass $\rho = \int_{\mathbb{R}^3} n(x,t)\,dx$ is constant in time, it follows that

$$\int_{\mathbb{R}^3} J_n(x,t)\,dx = e^{-t/\tau_0}\int_{\mathbb{R}^3} J_n(x,0)\,dx,$$

$$\int_{\mathbb{R}^3} (ne)(x,t)\,dx = \left(\int_{\mathbb{R}^3} (ne)(x,0)\,dx - \frac{3}{2}\rho \right)e^{-t/\tau_0} + \frac{3}{2}\rho.$$

Thus, as $t \to \infty$, $\int_{\mathbb{R}^3} J_n\,dx$ converges exponentially fast to zero with rate $1/\tau_0$ and $\int_{\mathbb{R}^3} ne\,dx$ converges exponentially fast to $\frac{3}{2}\rho$ with the same rate. Notice that $\frac{3}{2}\rho$ is the particle energy at equilibrium with velocity $u = 0$ and temperature $T = 1$.

The low-density collision operator (9.8) is a very simplified scattering model. We refer to Jacoboni and Lugli [2] for scattering integrals including acoustic deformation potential scattering, intravalley collisions, and impurity scattering. Often, the production integrals are expressed in terms of the deviation of the associated moment $\langle w_i M \rangle$ from its equilibrium value $\langle w_i M \rangle_{\text{eq}}$:

$$\langle w_i Q_1(M) \rangle = -\frac{\langle w_i M \rangle - \langle w_i M \rangle_{\text{eq}}}{\tau_i(M)},$$

where the macroscopic relaxation time $\tau_i(M)$ is a functional of the Maxwellian. This relaxation time is frequently modeled in an ad hoc manner, either by fitting to bulk Monte Carlo data or by empirical models and sometimes, the presentation (9.10) is assumed [3]. A relaxation-time approximation, separating the effects of interband and intraband collisions and assuming that the intravalley scattering occurs much more frequently than intervalley collisions, was suggested in Rudan et al. [4]. We refer to the work of Grasser [5, Sect. 5] for a discussion and more references.

We remark that the unscaled hydrodynamic equations with the collisional moments (9.10) are written as follows:

$$\partial_t n - \frac{1}{q} \operatorname{div} J_n = 0, \tag{9.11}$$

$$\partial_t J_n - \frac{1}{q} \operatorname{div} \left(\frac{J_n \otimes J_n}{n} \right) - \frac{q k_B}{m^*} \nabla(nT) + \frac{q^2}{m^*} n \nabla V = -\frac{J_n}{\tau_0}, \tag{9.12}$$

$$\partial_t (ne) - \frac{1}{q} \operatorname{div} \left(J_n(e + k_B T) \right) + J_n \cdot \nabla V = -\frac{n}{\tau_0} \left(e - \frac{3}{2} k_B T_L \right), \tag{9.13}$$

where the energy is given by

$$e = \frac{m^*}{2q^2} \frac{|J_n|^2}{n^2} + \frac{3}{2} k_B T. \tag{9.14}$$

In the absence of electric fields and scattering integrals, Eqs. (9.5), (9.6), and (9.7) correspond to the *Euler equations* of gas dynamics. Mathematically, they constitute a quasilinear hyperbolic system of conservation laws since mass, momentum, and energy are conserved. In this context, the expression nT in (9.6) can be interpreted as the *gas pressure*, which is generalized to the *stress tensor* $P = \langle (k-u) \otimes (k-u)M \rangle$. The *heat flux* is defined by $q = \langle \frac{1}{2}(k-u)|k-u|^2 M \rangle$. Since the Maxwellian M is even in $k-u$, this integral vanishes here and thus, $q = 0$.

The hydrodynamic model for semiconductors was first introduced by Bløtekjær [6] and Baccarani and Wordeman [3]. Bløtekjær derived the equations from the semiconductor Boltzmann equation by the moment method with a heuristic closure. In particular, he allowed for a nonvanishing heat flux $q = -\kappa \nabla T$, inspired from the Fourier law, where κ is the heat conductivity of the electron gas, and the expression $-\operatorname{div}(\kappa \nabla T)$ is added to the left-hand side of the energy equation (9.7). This choice was criticized in [7], since $\kappa \nabla T$ only approximates the diffusive component of the heat flux. For a uniform temperature, $\nabla T = 0$, and hence $q = 0$. However, the convective part of the heat flux needs to be included in some situations. Under some simplifying assumptions (like isotropy of the distribution function), another (scaled) expression for the heat flux was derived by Anile and Romano [8]:

$$q = -\frac{5}{2} T \nabla T + \frac{5}{2} n T u \left(\frac{1}{\tau_p} - \frac{1}{\tau_e} \right) \tau_e,$$

where τ_p and τ_e are the momentum and energy relaxation times, respectively. When $\tau_p = \tau_e$, one obtains the usual Fourier law. A nonvanishing heat flux was employed in Rudan et al. [4] as a closure in the moment equations. We remark that a nonzero heat flux can be derived by a Chapman–Enskog expansion method in the gas dynamics context, leading to the fluid dynamical Navier–Stokes equations [9].

There is a huge literature about hydrodynamic limits of the Boltzmann equation, in particular to derive the Euler and Navier–Stokes equations. The first rigorous result for the hydrodynamical limit was carried out by Caflisch [10] and generalized by Lachowicz [11]. The compressible Euler equations were derived by Nishida [12] and Ukai and Asano [13] based on the work by Grad [14]. The Navier–Stokes equations were derived from the Boltzmann equation by De Masi, Esposito, and Lebowitz [15] and simultaneously by Bardos, Golse, and Levermore [9, 16]. The above (formal) derivation is based on Bardos et al. [9]. For more references we refer to the books of Cercignani [17, 18].

When the electron temperature is assumed to be constant, Eqs. (9.5) and (9.6) are referred to as the *isothermal hydrodynamic model*. In gas dynamics, also *isentropic hydrodynamic equations* are considered, where the pressure $P = nT$ is replaced by $P = P_0 n^{\beta - 1}$ with $P_0 > 0$ and $\beta > 1$. Lions, Perthame, and co-workers proved the existence of global weak solutions of the transient isentropic model in one space dimension with vanishing electric field for all $\beta > 1$ [19, 20], whereas Matsumura and Nishida considered the isothermal case $\beta = 1$ [21]. The existence of global solutions of the one-dimensional isothermal equations including electric forces and the coupling to the Poisson equation was shown in [22, 23]. The isentropic case was treated in [24, 25]. The full hydrodynamic system (9.5), (9.6), and (9.7) (including a nonvanishing heat flux) was studied in [26, 27]. Finally, Gamba and Morawetz have analyzed the steady-state system [28, 29].

The solutions of the hydrodynamic equations may develop shock waves and contact discontinuities. They occur, for instance, for velocities exceeding the sound speed which is related to the electron temperature. As the hydrodynamic model is hyperbolic in the supersonic regions, special numerical methods have to be employed. Many numerical discretizations were proposed, for instance Godunov or Nessyahu–Tadmor schemes [25, 30, 31], ENO (essentially non-oscillatory) shock-capturing algorithms [32], second-order upwind-type finite-difference discretizations [33], stabilized finite-element approximations [34], high-resolution-centered schemes [35], and relaxation methods [36].

General energy band structure. The hydrodynamic model can be extended to general energy bands. Let the weight functions be given by $\kappa_0(k) = 1$, $\kappa_1(k) = k$, and $\kappa_2(k) = E(k)$. Performing the limit $\alpha \to 0$ in the moment equations (9.2) and assuming that all κ-moments of $Q_0(f_\alpha)$ vanish, we obtain

$$\partial_t \langle \kappa_i M \rangle + \mathrm{div}_x \langle v \kappa_i M \rangle - \nabla_x V \cdot \langle \nabla_k \kappa_i M \rangle = \langle \kappa_i Q_1(M) \rangle, \quad i = 0, 1, 2, \qquad (9.15)$$

where $M = \lim_{\alpha \to 0} f_\alpha$ is the Maxwellian $M(k) = \exp(\lambda_0 + \lambda_1 \cdot k + \lambda_2 E(k))$. We introduce similar as above the electron density $n = \langle M \rangle$, the current density $J_n = -nu = -\langle kM \rangle$ with the averaged crystal momentum u, and the energy density $ne = \langle E(k)M \rangle$. Then we can reformulate the above moment equations as

$$\partial_t n - \operatorname{div} J_n = 0,$$
$$\partial_t (nu) + \operatorname{div} (nu \otimes u + P) - n\nabla V = C_1,$$
$$\partial_t (ne) + \operatorname{div} ((ne)u + R) - nU\nabla V = C_2,$$

where $P = \langle (k-u) \otimes (k-u)M \rangle$ is the stress tensor, $R = \langle (k-u)E(k)M \rangle$ a part of the energy flux (which is defined by $\langle kE(k)M \rangle$), $U = \langle v(k)M \rangle / n$ is the averaged electron velocity, and $C_1 = \langle kQ_1(M) \rangle$ and $C_2 = \langle E(k)Q_1(M) \rangle$ are the moments of the collision term. Clearly, in the case of the parabolic band approximation, the averaged crystal momentum u and the electron velocity U coincide.

As discussed by Anile and Romano [37], the choice $\kappa_1(k) = v(k)$ instead of $\kappa_1(k) = k$ is also possible. Using this choice, the term $n\nabla V$ in the momentum equation has to be replaced by

$$\langle \nabla_k \otimes v(k)M \rangle \nabla V = \langle (m^*)^{-1} M \rangle \nabla V,$$

where $(m^*)^{-1} = (\nabla_k \otimes \nabla_k)E$ is the inverse of the (scaled) effective mass tensor. If the effective mass tensor can be approximated by the scalar value m^*, we recover the expression $(m^*)^{-1} n\nabla V$.

The difficulty now is to express the functions P, R, and U in terms of the lower-order moments n, nu, and ne. Usually, this can be done in an approximate way only. For instance, the Maxwellian may be approximated by

$$M_{\mathrm{approx}}(k) = e^{\lambda_0 + \lambda_2 E(k)} g(\lambda_1, \lambda_2, k, E(k)),$$

where g is a function of the Lagrange multipliers and weight functions (see, for instance, [38] for some choices for g), such that the integrals involving M_{approx} can be evaluated analytically.

The moment equations (9.15) form a symmetric hyperbolic system of conservation laws. In order to see this, we differentiate the moments with respect to the Lagrange multipliers, leading to

$$\sum_{j=0}^{2} \langle \kappa_i \kappa_j M \rangle \partial_t \lambda_j + \sum_{j=0}^{2} \langle v\kappa_i \kappa_j M \rangle \nabla_x \lambda_j - \nabla_x V \cdot \langle \nabla_k \kappa_i M \rangle = \langle \kappa_i Q_1(M) \rangle. \qquad (9.16)$$

The matrix $(\langle \kappa_i \kappa_j M \rangle)$ is symmetric and positive definite since it is the Hessian of the convex function $\lambda \mapsto \langle M \rangle = \langle e^{\kappa \cdot \lambda} \rangle$. Furthermore, the matrix $(\langle v\kappa_i \kappa_j M \rangle)$ is symmetric. Thus, it can be diagonalized, and the eigenvalues are real numbers. Therefore, the system (9.15), considered as a system of linearized equations with lower-order terms, is of symmetric hyperbolic type [31].

9.2 Extended Hydrodynamic Equations

The derivation of the hydrodynamic models presented in the previous section can be generalized to an arbitrary number of weight functions. Similar as in Chap. 8, this leads to a hierarchy of models. Employing more moments than in the previous section gives the so-called *extended hydrodynamic models*. Engineers started to pay attention to extended moment models from the end of the 1980s. Nekovee et al. [39] derived moment models from the Boltzmann equation, based on the works [40, 41], by expanding the distribution function around a Maxwellian involving Hermite polynomials. An arbitrary number of moments was considered by Struchtrup in the context of gas dynamics [42] and semiconductor theory [43]. He employed the entropy maximization closure to close the system of moment equations, similar to the procedure presented in Sect. 2.2. This closure was employed by Anile and Romano for the transport in semiconductors [37], first for silicon devices [44] and later for GaAs devices [45, 46].

In the following we detail the general approach for deriving higher-order hydrodynamic moment models, parallel to the higher-order diffusive moment models of Chap. 8, and discuss the extended hydrodynamic model of Anile and Romano [37].

Higher-order hydrodynamic moment models. Let $\kappa = \kappa(k) = (\kappa_0, \ldots, \kappa_N)$ be weight functions, $m = m(x,t) = (m_0, \ldots, m_N)$ be moments, and let, for a given distribution function $f(x,k,t)$,

$$(S(f))(x,t) = -\int_B f\,(\log f - 1 + E(k))\,\frac{dk}{4\pi^3}$$

be the relative entropy or energy using Maxwell–Boltzmann statistics (see Sect. 2.2). As in the previous section, we set $\langle g \rangle = \int_B g(k)\,dk/4\pi^3$. It is shown in Lemma 8.1 that the constrained maximization problem

$$S(f^*) = \max\left\{S(f) : \langle \kappa f(x,\cdot,t) \rangle = m(x,t) \text{ for } x \in \mathbb{R}^3,\ t > 0\right\} \qquad (9.17)$$

possesses the formal solution $f^*(x,k,t) = e^{\lambda(x,t)\cdot\kappa(k)}$ (if it exists), where $\lambda = (\lambda_0, \ldots, \lambda_N)$ are some Lagrange multipliers which are given implicitly by the relation $\langle \kappa f^* \rangle = m$. Clearly, in order to obtain an integrable solution if the Brillouin zone B is unbounded, the weight functions have to be chosen appropriately. For a given function f with moments $m = \langle \kappa f \rangle$, we call the maximizer of (9.17) the *generalized Maxwellian* of f, $f^* = M[f]$.

We consider the Boltzmann equation in the hydrodynamic scaling (see (9.1)):

$$\alpha \partial_t f_\alpha + \alpha\left(v(k)\cdot\nabla_x f_\alpha + \nabla_x V\cdot\nabla_k f_\alpha\right) = Q(f_\alpha). \qquad (9.18)$$

We assume for simplicity that the collision operator is given by the relaxation-time approximation,

$$Q(f) = \frac{1}{\tau}(M[f] - f). \qquad (9.19)$$

Then the kernel of Q consists of all functions satisfying $f = M[f]$ and, by construction, all moments of Q vanish, $\langle \kappa_i Q(f) \rangle = 0$ for all functions f and all $i = 0, \ldots, N$. First, we perform the formal limit $\alpha \to 0$ in (9.18) leading to $Q(f) = 0$ for the limit function $f = \lim_{\alpha \to 0} f_\alpha$. Hence, $f = M[f]$. Then, performing the limit $\alpha \to 0$ in the moment equations

$$\partial_t \langle \kappa_i f_\alpha \rangle + \mathrm{div}_x \langle v \kappa_i f_\alpha \rangle - \nabla_x V_\alpha \cdot \langle \nabla_k \kappa_i f_\alpha \rangle = \alpha^{-1} \langle \kappa_i Q(f_\alpha) \rangle = 0,$$

where V_α is a solution of the Poisson equation, gives the limit equations

$$\partial_t \langle \kappa_i M[f] \rangle + \mathrm{div} \langle v \kappa_i M[f] \rangle - \nabla_x V \cdot \langle \nabla_k \kappa_i M[f] \rangle = 0, \quad i = 0, \dots, N.$$

We have shown the following result.

Theorem 9.4 (Hydrodynamic moment equations). *Let (f_α, V_α) be a solution of the Boltzmann–Poisson system* (9.18), (5.3) *with collision operator* (9.19). *Then the formal limit functions $f = M[f] = \lim_{\alpha \to 0} f_\alpha$ and $V_\alpha = \lim_{\alpha \to 0} V_\alpha$ satisfy the hydrodynamic moment equations*

$$\partial_t m_i - \mathrm{div}\, J_i + \nabla_x V \cdot I_i = 0, \quad i = 0, \dots, N, \tag{9.20}$$

where $m_i = \langle \kappa_i M[f] \rangle$ are the moments, $J_i = -\langle v \kappa_i M[f] \rangle$ are the fluxes, and $I_i = -\langle \nabla_k \kappa_i M[f] \rangle$ are some auxiliary integrals. The initial conditions are given by

$$m_i(\cdot, 0) = \langle \kappa_i f_I \rangle, \quad i = 0, \dots, N.$$

The theorem is valid for more general collision operators if the distribution function $f = f_\alpha$ in the moment equations is assumed to be close to the Maxwellian such that f can be substituted approximately by $M[f]$. The right-hand side of the moment equations then changes to $\alpha^{-1} \langle \kappa_i Q(M[f]) \rangle$. With this closure, no limit $\alpha \to 0$ needs to be performed but the substitution of f by $M[f]$ may be criticized from a formal point of view. Equation (9.20) forms a system of symmetric hyperbolic equations. This can be proved as in the previous section (see (9.16)).

The macroscopic entropy of the system (9.20) is obtained by setting $f = M[f] = e^{\kappa \cdot \lambda}$ in the microscopic entropy $-\int_B f(\log f - 1) \, dk / 4\pi^3$, giving

$$S_0(t) = -\int_{\mathbb{R}^3} \int_B e^{\kappa \cdot \lambda} (\kappa \cdot \lambda - 1) \frac{dk}{4\pi^3} \, dx = -\int_{\mathbb{R}^3} (m \cdot \lambda - m_0) \, dx.$$

We claim that this function is constant in time. Indeed, by a straightforward computation and (9.20),

$$\frac{dS_0}{dt} = -\int_{\mathbb{R}^3} \partial_t m \cdot \lambda \, dx = -\int_{\mathbb{R}^3} \sum_{i=0}^{N} \left(\langle v \kappa_i M[f] \rangle \nabla_x \lambda_i + \nabla_x V \cdot \langle \nabla_k \kappa_i M[f] \rangle \lambda_i \right) dx$$

$$= -\int_{\mathbb{R}^3} \left(\mathrm{div}_x \langle v M[f] \rangle + \nabla_x V \cdot \langle \nabla_k M[f] \rangle \right) dx = 0.$$

Extended hydrodynamic model. In the following we discuss the moment model for $N = 3$, studied by Anile and Romano in [8, 37]. The weight functions are given by $\kappa_0(k) = 1$, $\kappa_1(k) = v(k)$, $\kappa_2(k) = E(k)$, and $\kappa_3(k) = v(k)E(k)$. The main task is to explicitly formulate the constitutive equations. The moments are defined in terms of the Maxwellian

$$M[f] = e^{\kappa \cdot \lambda} = \exp\left(\lambda_0 + \lambda_1 \cdot v(k) + \lambda_2 E(k) + \lambda_3 \cdot v(k)E(k)\right),$$

by the formula

$$m = \int_B \kappa e^{\kappa \cdot \lambda} \frac{dk}{4\pi^3}. \tag{9.21}$$

In order to obtain the dependence of the Lagrange multipliers λ_i on the moments m_i, we need to invert the constraints (9.21). This gives the closure relations for the fluxes. We will invert (9.21) under a physical condition on the distribution function.

We make two assumptions. First, we suppose that $E(K)$ is isotropic. Then, the distribution function at equilibrium,

$$M_{eq} = \exp\left(\lambda_{0,eq} - E(k)\right),$$

is isotropic too. Anile and Romano argued in [8, 37] as follows. Monte Carlo simulations for silicon semiconductor devices have shown that the anisotropy of the distribution function even far from equilibrium is small. This is due to the fact that the main collision mechanisms in silicon are interactions between electrons and acoustic and nonpolar optical phonons, which are isotropic. This motivates the second assumption: The anisotropy of the distribution function $M[f]$ is supposed to be small. Since at equilibrium the Lagrange multipliers are given by

$$\lambda_0 = \lambda_{0,eq}, \quad \lambda_1 = 0, \quad \lambda_2 = -1, \quad \lambda_3 = 0,$$

we expand λ_i up to second order around these equilibrium values, introducing the *anisotropy parameter δ*,

$$\lambda_0 = \lambda_0^{(0)} + \delta^2 \lambda_0^{(2)}, \qquad\qquad \lambda_1 = \delta \lambda_1^{(1)},$$
$$\lambda_2 = \lambda_2^{(0)} + \delta^2 \lambda_2^{(2)}, \qquad\qquad \lambda_3 = \delta \lambda_3^{(1)}.$$

Then, since $e^x = 1 + x + x^2/2 + \mathcal{O}(|x|^3)$ as $x \to 0$, we approximate $M[f]$, up to terms of order $\mathcal{O}(\delta^3)$, by

$$M[f] = \exp\left(\lambda_0^{(0)} + \lambda_2^{(0)} E(k)\right)\left(1 + \delta v(k) \cdot A_1 + \delta^2 A_2 + \frac{\delta^2}{2}|v(k) \cdot A_1|^2\right), \tag{9.22}$$

where $A_1 = \lambda_1^{(1)} + \lambda_3^{(1)} E(k)$ and $A_2 = \lambda_0^{(2)} + \lambda_2^{(2)} E(k)$. Inserting this expression into (9.21) shows that the moments m_0 and m_2 are of order one, whereas m_1 and m_3 are of order $\mathcal{O}(\delta)$.

Expressions for λ_i in terms of m_j are obtained by inverting (9.21). We insert (9.22) into (9.21) and equate equal powers of δ:

$$m_0 = e^{\lambda_0^{(0)}} \int_B e^{\lambda_2^{(0)} E} \frac{dk}{4\pi^3}, \tag{9.23}$$

$$m_2 = e^{\lambda_0^{(0)}} \int_B e^{\lambda_2^{(0)} E} E \frac{dk}{4\pi^3}, \tag{9.24}$$

$$m_1 = e^{\lambda_0^{(0)}} \int_B e^{\lambda_2^{(0)} E} v (v \cdot A_1) \frac{dk}{4\pi^3}, \tag{9.25}$$

$$m_3 = e^{\lambda_0^{(0)}} \int_B e^{\lambda_2^{(0)} E} v E (v \cdot A_1) \frac{dk}{4\pi^3}. \tag{9.26}$$

$$0 = \int_B e^{\lambda_2^{(0)} E} \left(A_2 + \frac{1}{2} |v \cdot A_1|^2 \right) \frac{dk}{4\pi^3}, \tag{9.27}$$

$$0 = \int_B e^{\lambda_2^{(0)} E} \left(A_2 + \frac{1}{2} |v \cdot A_1|^2 \right) E \frac{dk}{4\pi^3}. \tag{9.28}$$

The energy $e = m_2/m_0$ is a function of $\lambda_2^{(0)}$ only, by (9.23) and (9.24):

$$e = \frac{\langle e^{\lambda_2^{(0)} E} E \rangle}{\langle e^{\lambda_2^{(0)} E} \rangle},$$

and this relation can be inverted for given e, at least numerically [37, Sect. 5.1]. Then, for given m_0, (9.23) can be inverted for $\lambda_0^{(0)}$:

$$\lambda_0^{(0)} = \log m_0 - \log \int_B e^{\lambda_2^{(0)} E} \frac{dk}{4\pi^3}.$$

The parameters $\lambda_1^{(1)}$ and $\lambda_3^{(1)}$ can be computed by inverting equations (9.25) and (9.26), which form a linear system in these parameters. Finally, $\lambda_0^{(2)}$ and $\lambda_2^{(2)}$ are calculated from the system (9.27) and (9.28). The results are lengthy expressions, being of the form

$$\lambda_1^{(1)} = a_{11} m_1 + a_{12} m_3, \qquad \lambda_0^{(2)} = b_{01} m_1 \cdot m_1 + 2 b_{02} m_1 \cdot m_3 + b_{03} m_3 \cdot m_3,$$

$$\lambda_3^{(1)} = a_{31} m_1 + a_{32} m_3, \qquad \lambda_2^{(2)} = b_{21} m_1 \cdot m_1 + 2 b_{22} m_1 \cdot m_3 + b_{23} m_3 \cdot m_3,$$

where the coefficients a_{ij} and b_{ij} are functions of $\lambda_0^{(0)}$ and $\lambda_2^{(0)}$ or, equivalently, of m_0 and m_2. With these expressions, the Maxwellian (9.22) can be written as a function of the moments m_i, and thus, the fluxes $J_i = -\langle v \kappa_i M[f] \rangle$ and the integrals $I_i = -\langle \nabla_k \kappa_i M[f] \rangle$ can be expressed in terms of m_i only.

Anile and Romano have derived extended hydrodynamic models including the production terms

$$C_1 = \langle Q(M[f]) v \rangle, \quad C_2 = \langle Q(M[f]) E \rangle, \quad C_3 = \langle Q(M[f]) v E \rangle$$

on the right-hand sides of (9.20). They assumed that the collision operator Q describes acoustic and nonpolar phonon scattering and collisions with impurities. With the above approximation of $M[f]$, explicit but lengthy formulas for C_i can be derived. We refer to [8, Sect. 3.5] for the precise expressions.

9.3 Relaxation-Time Limits

By performing the so-called relaxation-time limits in the hydrodynamic model, it is possible to recover the drift-diffusion and energy-transport equations. We rewrite the hydrodynamic equations as follows:

$$\partial_t n - \frac{1}{q} \operatorname{div} J_n = 0, \tag{9.29}$$

$$\partial_t J_n - \frac{1}{q} \operatorname{div}\left(\frac{J_n \otimes J_n}{n}\right) - \frac{qk_B}{m^*}\nabla(nT) + \frac{q^2}{m^*}n\nabla V = -\frac{J_n}{\tau_p}, \tag{9.30}$$

$$\partial_t(ne) - \frac{1}{q}\operatorname{div}\left(J_n(e + k_B T)\right) + J_n \cdot \nabla V - \operatorname{div}\left(\kappa \nabla T\right) = -\frac{n}{\tau_e}\left(e - \frac{3}{2}k_B T_L\right), \tag{9.31}$$

where the energy e is given by (9.14) and the heat conductivity is defined by $\kappa = \kappa_0 \tau_p nk_B T/m^*$, with $\kappa_0 > 0$. These equations deviate from (9.5), (9.6), and (9.7) by the inclusion of the heat conduction term and by the introduction of two different relaxation mechanisms: momentum relaxation with rate τ_p and energy relaxation with rate τ_e. The reason is that we wish to consider two different time scales for momentum and energy relaxation. The coupling of the electric potential V to the electron density through the Poisson equation does not need to be considered in the subsequent considerations. Therefore, V will be treated as a given function.

First, we scale the above equations. We choose the reference length λ (for instance, the device diameter), the reference particle density C_m (for instance, the maximal value of the doping concentration), the reference temperature T_L (lattice temperature), the reference potential $U_T = k_B T_L/q$, and the reference electron current density $J_0 = qC_m\lambda/\tau$, where the reference time τ is given by the assumption that the thermal energy is of the same order as the geometric average of the kinetic energies needed to cross the semiconductor device in time τ and τ_p, respectively,

$$k_B T_L = \sqrt{m^*\left(\frac{\lambda}{\tau}\right)^2}\sqrt{m^*\left(\frac{\lambda}{\tau_p}\right)^2}.$$

With these reference values we can define the nondimensional variables

$$x = \lambda x_s, \qquad\qquad t = \tau t_s, \qquad\qquad n = C_m n_s,$$
$$J_n = J_0 J_{n,s}, \qquad\qquad V = U_T V_s, \qquad\qquad T = T_L T_s.$$

Replacing the dimensional variables in (9.29), (9.30), and (9.31) by the scaled ones, we obtain the scaled equations (omitting the index "s")

$$\partial_t n - \operatorname{div} J_n = 0, \tag{9.32}$$

$$\alpha \partial_t J_n - \alpha \operatorname{div}\left(\frac{J_n \otimes J_n}{n}\right) - \nabla(nT) + n\nabla V = -J_n, \tag{9.33}$$

$$\partial_t (ne) - \text{div} \, (J_n(e+T)) + J_n \cdot \nabla V - \text{div} \, (\kappa_0 n T \nabla T) = -\frac{n}{\beta} \left(e - \frac{3}{2} \right), \qquad (9.34)$$

with the energy

$$e = \alpha \frac{|J_n|^2}{2n^2} + \frac{3}{2} T \qquad (9.35)$$

and the nondimensional parameters

$$\alpha = \frac{\tau_p}{\tau}, \quad \beta = \frac{\tau_e}{\tau}.$$

We consider the following formal limits:

- $\alpha \to 0$ and $\beta \to 0$;
- $\alpha \to 0$ and β fixed;
- $\beta \to 0$ and α fixed.

The limit $\alpha \to 0$, $\beta \to 0$ corresponds to the physical situation when the kinetic energy needed to cross the domain in time τ is much smaller than the thermal energy. Then the hydrodynamic equations (9.32), (9.33), and (9.34) become in the limit $\alpha \to 0$ and $\beta \to 0$

$$\partial_t n - \text{div} J_n = 0, \quad J_n = \nabla(nT) - n\nabla V, \quad e = \frac{3}{2}.$$

Furthermore, by (9.35), $e = 3T/2$, and hence, the scaled temperature $T = 1$. The limit equations are the drift-diffusion model studied in Chap. 5.

The limit $\alpha \to 0$ with β fixed leads to another system of equations:

$$\partial_t n - \text{div} J_n = 0, \quad J_n = \nabla(nT) - n\nabla V, \qquad (9.36)$$

$$\partial_t \left(\frac{3}{2} nT \right) - \text{div} \left(\frac{5}{2} J_n T + \kappa_0 n T \nabla T \right) + J_n \cdot \nabla V = -\frac{3}{2} \frac{n}{\beta} (T - 1). \qquad (9.37)$$

We claim that this corresponds to an energy-transport model. In order to see this, we introduce the entropy variables μ/T and $-1/T$. Since $n = 2(2\pi)^{-3/2} T^{3/2} e^{\mu/T}$ (see (6.25)), we obtain after some computations

$$J_n = nT\nabla \left(\frac{\mu}{T} \right) - \frac{5}{2} nT^2 \nabla \left(\frac{1}{T} \right) - n\nabla V,$$

$$\frac{5}{2} J_n T + \kappa_0 n T \nabla T = \frac{5}{2} nT^2 \nabla \left(\frac{\mu}{T} \right) - \left(\frac{25}{4} + \kappa_0 \right) nT^3 \nabla \left(\frac{1}{T} \right) - \frac{5}{2} nT\nabla V.$$

Hence, we can write the above limit equations as an energy-transport model in the variables μ/T and $-1/T$ with the diffusion matrix

$$\mathcal{D} = nT \begin{pmatrix} 1 & \dfrac{5}{2}T \\ \dfrac{5}{2}T & \left(\dfrac{25}{4} + \kappa_0\right)T^2 \end{pmatrix}.$$

This matrix is positive definite for $n > 0$ and $T > 0$ if and only if κ_0 is positive. The heat conduction term is necessary to obtain well posedness of the energy-transport system (at least locally in time).

The limit $\beta \to 0$ with α fixed gives the equation $e = \frac{3}{2}$ which only means that the sum of the kinetic and thermal energies is constant in space and time. More interesting is the limit $\beta \to 0$ in the energy-transport model (9.36) and (9.37). This yields $T = 1$ and the drift-diffusion equations

$$\partial_t n - \mathrm{div}\, J_n = 0, \quad J_n = \nabla n - n\nabla V.$$

The limit $\alpha \to 0$, $\beta \to 0$ from the isentropic hydrodynamic to the drift-diffusion equations was first analyzed by Marcati and Natalini [47] and later by Lattenzio and Marcati [48, 49] in the presence of certain uniform bounds. A complete proof in one space dimension was given in [50, 51] and in several space dimensions in [52]. The isothermal model was treated by Junca and Rascle [53]. Chen et al. [54] proved the limit for the hydrodynamic model including the energy equation. The limit $\alpha \to 0$ was analyzed by Gasser and Natalini in [55].

The derivation of hydrodynamic models and the three relaxation-time limits are summarized in Fig. 9.1.

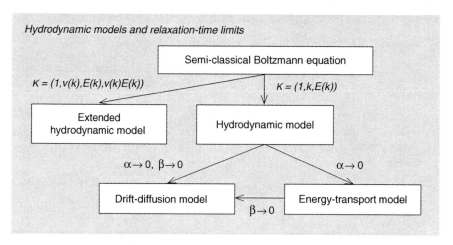

Fig. 9.1 Derivation of hydrodynamic models and relaxation-time limits in the hydrodynamic and energy-transport equations

References

1. M. Lundstrom. *Fundamentals of Carrier Transport*. 2nd edition, Cambridge University Press, Cambridge, 2000.
2. C. Jacoboni and P. Lugli. *The Monte Carlo Method for Semiconductor Device Simulation*. Springer, Vienna, 1989.
3. G. Baccarani and M. Wordeman. An investigation of steady state velocity overshoot effects in Si and GaAs devices. *Solid State Electr.* 28 (1985), 407–416.
4. M. Rudan, A. Gnudi, and W. Quade. A generalized approach to the hydrodynamic model of semiconductor equations. In: G. Baccarani (ed.), *Process and Device Modeling for Microelectronics*, 109–154. Elsevier, Amsterdam, 1993.
5. T. Grasser. Non-parabolic macroscopic transport models for semiconductor device simulation. *Physica A* 349 (2005), 221–258.
6. K. Bløtekjær. Transport equations for electrons in two-valley semiconductors. *IEEE Trans. Electr. Devices* 17 (1970), 38–47.
7. S.-C. Lee and T.-W. Tang. Transport coefficients for a silicon hydrodynamic model extracted from inhomogeneous Monte-Carlo calculations. *Solid State Electr.* 35 (1992), 561–569.
8. A. Anile and V. Romano. Hydrodynamic modeling of charge transport in semiconductors. *Meccanica* 35 (2000), 249–296.
9. C. Bardos, F. Golse, and C. Levermore. Fluid dynamical limits in kinetic equations. I. Formal derivations. *J. Stat. Phys.* 63 (1991), 323–344.
10. R. Caflisch. The fluid dynamical limit of the nonlinear Boltzmann equation. *Commun. Pure Appl. Math.* 33 (1980), 651–666.
11. M. Lachowicz. On the initial layer and the existence theorem for the nonlinear Boltzmann equation. *Math. Meth. Appl. Sci.* 9 (1987), 342–366.
12. T. Nishida. Fluid dynamical limit of the nonlinear Boltzmann equation to the level of the compressible Euler equation. *Commun. Math. Phys.* 61 (1978), 119–148.
13. S. Ukai and K. Asano. The Euler limit and the initial layer of the nonlinear Boltzmann equation. *Hokkaido Math. J.* 12 (1983), 311–332.
14. H. Grad. Asymptotic equivalence of the Navier-Stokes and non-linear Boltzmann equation. *Proc. Amer. Math. Soc.* 17 (1965), 154–183.
15. A. De Masi, R.Esposito, and J. Lebowitz. Incompressible Navier-Stokes and Euler limits of the Boltzmann equation. *Commun. Pure Appl. Math.* 42 (1989), 1189–1214.
16. C. Bardos, F. Golse, and C. Levermore. Fluid dynamical limits in kinetic equations. II. Convergence proofs for the Boltzmann equation. *Commun. Pure Appl. Math.* 46 (1993), 667–753.
17. C. Cercignani. *The Boltzmann Equation and Its Applications*. Springer, Berlin, 1988.
18. C. Cercignani, R. Illner, and M. Pulvirenti. *The Mathematical Theory of Dilute Gases*. Springer, New York, 1994.
19. P.-L. Lions, B. Perthame, and E. Souganidis. Existence of entropy solutions for the hyperbolic system of isentropic gas dynamics in Eulerian and Lagrangian coordinates. *Commun. Pure Appl. Math.* 44 (1996), 599–638.
20. P.-L. Lions, B. Perthame, and E. Tadmor. Kinetic formulation for the isentropic gas dynamics and p-system. *Commun. Math. Phys.* 163 (1994), 415–431.
21. A. Matsumura and T. Nishida. Initial boundary value problems for the equations of motion of compressible viscous and heat-conductive fluids. *Commun. Math. Phys.* 89 (1983), 445–464.
22. S. Cordier. Global solutions to the isothermal Euler-Poisson plasma model. *Appl. Math. Letters* 8 (1995), 19–24.
23. F. Poupaud, M. Rascle, and J. Vila. Global solutions to the isothermal Euler-Poisson system with arbitrarily large data. *J. Diff. Eqs.* 123 (1995), 93–121.
24. P. Marcati and R. Natalini. Weak solutions to a hydrodynamic model for semiconductors: The Cauchy problem. *Proc. Roy. Soc. Edinb., Sect. A* 125 (1995), 115–131.
25. B. Zhang. Convergence of the Gudonov scheme for a simplified one-dimensional hydrodynamic model for semiconductor devices. *Commun. Math. Phys.* 157 (1993), 1–22.

26. D. Wang and G.-Q. Chen. Formation of singularities in compressible Euler-Poisson fluids with heat diffusion and damping relaxation. *J. Diff. Eqs.* 144 (1998), 44–65.

27. L. Yeh. Well-posedness of the hydrodynamic model for semiconductors. *Math. Meth. Appl. Sci.* 19 (1996), 1489–1507.

28. I. Gamba. Stationary transonic solutions of a one-dimensional hydrodynamic model for semiconductors. *Commun. Part. Diff. Eqs.* 17 (1992), 553–577.

29. I. Gamba and C. Morawetz. A viscous approximation for a 2-D steady semiconductor or transonic gas dynamic flow: existence theorem for potential flow. *Commun. Pure Appl. Math.* 49 (1996), 999–1049.

30. A. Anile, V. Romano, and G. Russo. Extended hydrodynamical model of carrier transport in semiconductors. *SIAM J. Appl. Math.* 61 (2000), 74–101.

31. R. LeVeque. *Numerical Methods for Conservation Laws.* Birkhäuser, Basel, 1990.

32. E. Fatemi, J. Jerome, and S. Osher. Solution of the hydrodynamic device model using high-order nonoscillatory shock-capturing algorithms. *IEEE Trans. Computer-Aided Design* 10 (1991), 232–244.

33. L. Ballestra and R. Sacco. Numerical problems in semiconductor simulation using the hydrodynamic model: a second-order finite difference scheme. *J. Comput. Phys.* 195 (2004), 320–340.

34. M. Fortin and G. Yang. Simulation of the hydrodynamic model of semiconductor devices by a finite element method. *COMPEL* 15 (1996), 4–21.

35. A. Anile, N. Nikiforakis, and R. Pidatella. Assessment of a high resolution centered scheme for the solution of hydrodynamic semiconductor equations. *SIAM J. Sci. Comput.* 22 (2000), 1533–1548.

36. A. Jüngel and S. Tang. A relaxation scheme for the hydrodynamic equations for semiconductors. *Appl. Numer. Math.* 43 (2002), 229–252.

37. A. Anile and V. Romano. Non parabolic transport in semiconductors: closure of the moment equations. *Continuum Mech. Thermodyn.* 11 (1999), 307–325.

38. T. Grasser, H. Kosina, C. Heitzinger, and S. Selberherr. Characterization of the hot electron distribution function using six moments. *J. Appl. Phys.* 91 (2002), 3869–3879.

39. M. Nekovee, B. Guerts, H. Boots, and M. Schuurmans. Failure of extended-moment-equation approaches to describe ballistic transport in submicrometer structures. *Phys. Rev. B* 45 (1992), 6643–6651.

40. A. Bringer and G. Schön. Extended moment equations for electron transport in semiconducting submicron structures. *J. Appl. Phys.* 64 (1988), 2447–2455.

41. T. Portengen, M. Boots, and M. Schuurmans. A priori incorporation of ballistic and heating effects in a four-moment approach to the Boltzmann equation. *J. Appl. Phys.* 68 (1990), 2817–2823.

42. H. Struchtrup. Extended moments method for electrons in semiconductors. *Physica A* 275 (2000), 229–255.

43. S. Liotta and H. Struchtrup. Moment equations for electrons in semiconductors: comparison of spherical harmonics and full moments. *Solid State Electr.* 44 (2000), 95–103.

44. V. Romano. Non parabolic band transport in semiconductors: closure of the production terms in the moment equations. *Continuum Mech. Thermodyn.* 12 (2000), 31–51.

45. G. Mascali and V. Romano. Hydrodynamical model of charge transport in GaAs based on the maximum entropy principle. *Continuum Mech. Thermodyn.* 14 (2002), 405–423.

46. G. Mascali and V. Romano. Simulation of Gunn oscillations with a nonparabolic hydrodynamical model based on the maximum entropy principle. *COMPEL* 24 (2005), 35–54.

47. P. Marcati and R. Natalini. Weak solutions to a hydrodynamic model for semiconductors and relaxation to the drift-diffusion equation. *Arch. Rat. Mech. Anal.* 129 (1995), 129–145.

48. C. Lattenzio. On the 3-D bipolar isentropic Euler-Poisson model for semiconductors and the drift-diffusion limit. *Math. Models Meth. Appl. Sci.* 10 (2000), 351–360.

49. C. Lattanzio and P. Marcati. The relaxation to the drift-diffusion system for the 3-D isentropic Euler-Poisson model for semiconductors. *Discrete Contin. Dyn. Sys.* 5 (1999), 449–455.

50. A. Jüngel and Y.-J. Peng. A hierarchy of hydrodynamic models for plasmas: zero-relaxation-time limits. *Commun. Part. Diff. Eqs.* 24 (1999), 1007–1033.
51. A. Jüngel and Y.-J. Peng. Zero-relaxation-time limits in hydrodynamic models for plasmas revisited. *Z. Angew. Math. Phys.* 51 (2000), 385–396.
52. W.-A. Yong. Diffusive relaxation limit of multidimensional isentropic hydrodynamic models for semiconductors. *SIAM J. Appl. Math.* 64 (2004), 1737–1748.
53. S. Junca and M. Rascle. Relaxation of the isothermal Euler-Poisson system to the drift-diffusion equations. *Quart. Appl. Math.* 58 (2000), 511–521.
54. G.-Q. Chen, J. Jerome, and B. Zhang. Particle hydrodynamic moment models in biology and microelectronics: singular relaxation limits. *Nonlin. Anal.* 30 (1997), 233–244.
55. I. Gasser and R. Natalini. The energy transport and the drift diffusion equations as relaxation limits of the hydrodynamic model for semiconductors. *Quart. Appl. Math.* 57 (1999), 269–282.

Part IV
Microscopic Quantum Models

When the active region in a semiconductor device is smaller than about 100 nm, quantum mechanical effects, which go beyond the semi-classical description of the previous chapters, usually have to be included in the modeling of the transport phenomena. In fact, there are devices whose performance is based on quantum mechanical phenomena, such as laser diodes and resonant tunneling diodes. We present three different formulations of the evolution of quantum particles: the Schrödinger picture, the density-matrix formalism, and the quantum-kinetic Wigner formulation.

Chapter 10
The Schrödinger Equation

In this chapter, we reconsider the Schrödinger equation, which was already introduced in Sect. 1.2. Only two aspects of the modeling with the Schrödinger equation
are presented: the relation to the Liouville–von Neumann equation and the modeling
of transparent boundary conditions.

10.1 Density-Matrix Formulation

We consider three alternative formulations of the quantum mechanical motion of an
ensemble of electrons: the Schrödinger formulation, the density-matrix formulation,
and the kinetic Wigner formulation. The kinetic picture is introduced in the following chapter. Here, we consider the density-matrix representation and relate it to the
Schrödinger picture.

We assume that there exists an operator $\widehat{\rho}$, called the *density-matrix operator*,
satisfying the *Liouville–von Neumann equation* in the operator formulation

$$i\hbar\partial_t\widehat{\rho} = [H,\widehat{\rho}], \quad t > 0, \quad \widehat{\rho}(0) = \widehat{\rho}_I, \tag{10.1}$$

where H is the quantum mechanical Hamiltonian, for instance, $H = -(\hbar^2/2m)\Delta -
qV(x,t)$, and $[H,\widehat{\rho}] = H\widehat{\rho} - \widehat{\rho}H$ is the commutator. We suppose that the density-
matrix operator $\widehat{\rho}(t)$ is positive and self-adjoint for all $t \geq 0$. To be precise, some
additional properties (compactness, trace-class) are needed for the following functional analytical arguments; we refer to [1, 2] for details. The self-adjointness (and
compactness) of $\widehat{\rho}$ implies the existence of a complete orthonormal set of eigenfunctions (ψ_j) of $L^2(\mathbb{R}^3)$ with corresponding (real) eigenvalues (λ_j). The eigenfunctions
of $\widehat{\rho}_I$ are denoted by (ψ_j^0). We claim that the wave functions ψ_j are stationary solutions of the Schrödinger equation $i\hbar\partial_t\psi = H\psi$. In order to show this claim, we need
some more properties of the density-matrix operator.

Each density-matrix operator has the unique integral representation

Jüngel, A.: *The Schrödinger Equation*. Lect. Notes Phys. **773**, 217–230 (2009)
DOI 10.1007/978-3-540-89526-8_10   © Springer-Verlag Berlin Heidelberg 2009

$$(\hat{\rho}\psi)(x,t) = \int_{\mathbb{R}^3} \rho(x,y,t)\psi(y,t)\,dy, \quad t \geq 0, \tag{10.2}$$

where ρ is the density-matrix (function). The "diagonal" of the density-matrix can be interpreted as the *particle density*

$$n(x,t) = 2\rho(x,x,t). \tag{10.3}$$

The factor 2 takes into account the two possible states of the spin of the particles (see Sect. 1.6). Furthermore, the *particle current density* is defined by

$$J(x,t) = \frac{i\hbar q}{m}(\nabla_r - \nabla_s)\rho(x,x,t). \tag{10.4}$$

The notation $(\nabla_r - \nabla_s)\rho(x,x,t)$ means $(\nabla_r\rho(r,s,t) - \nabla_s\rho(r,s,t))|_{r=s=x}$. Moreover, the following properties hold.

Proposition 10.1 (Properties of the density-matrix). *The density-matrix solves the* Liouville–von Neumann equation *in the "matrix" formulation*

$$i\hbar\partial_t\rho(x,y,t) = (H_x - H_y)\rho(x,y,t), \quad t > 0, \quad \rho(x,y,0) = \rho_I(x,y), \quad x,y \in \mathbb{R}^3,$$

where H_x denotes the Hamiltonian only acting on the variable x (for instance, $H_x = -(\hbar^2/2m)\Delta_x - qV(x,t)$) and H_y only acts on the variable y. The initial datum ρ_I is computed from

$$(\hat{\rho}_I\psi)(x) = \int_{\mathbb{R}^3} \rho_I(x,y)\psi(y)\,dy.$$

Furthermore, the density-matrix can be expanded in terms of the eigenfunctions ψ_j,

$$\rho(x,y,t) = \sum_{j=1}^{\infty} \lambda_j\psi_j(x,t)\overline{\psi_j(y,t)}. \tag{10.5}$$

Proof. By the self-adjointness of H_y, we obtain for all functions $\psi(y,t)$:

$$\int_{\mathbb{R}^3} i\hbar\partial_t\rho(x,y,t)\psi(y,t)\,dy = i\hbar(\partial_t\hat{\rho})\psi(x,t) = (H\hat{\rho}\psi - \hat{\rho}H\psi)(x,t)$$

$$= \int_{\mathbb{R}^3} (H_x\rho(x,y,t)\psi(y,t) - \rho(x,y,t)H_y\psi(y,t))\,dy$$

$$= \int_{\mathbb{R}^3} (H_x\rho(x,y,t)\psi(y,t) - H_y\rho(x,y,t)\psi(y,t))\,dy$$

$$= \int_{\mathbb{R}^3} (H_x - H_y)\rho(x,y,t)\psi(y,t)\,dy.$$

This shows the first claim. In order to show the second one, we employ (10.2) for the eigenfunction $\psi = \psi_j$, multiply this equation by $\overline{\psi_\ell(x,t)}$, and integrate over $\mathbb{R}^3$. Then, in view of the orthonormality of (ψ_j),

$$\delta_{j\ell}\lambda_\ell = \int_{\mathbb{R}^3}\int_{\mathbb{R}^3} \rho(x,y,t)\psi_j(y,t)\overline{\psi_\ell(x,t)}\,dx\,dy. \tag{10.6}$$

The set $(\psi_j(x,t)\overline{\psi_\ell(y,t)})_{j,\ell}$ is a complete orthonormal set of $L^2(\mathbb{R}^3 \times \mathbb{R}^3)$. Therefore, the density-matrix can be expanded in this basis:

$$\rho(x,y,t) = \sum_{n,p=1}^{\infty} c_{np}(t)\psi_n(x,t)\overline{\psi_p(y,t)}.$$

Inserting this expansion into (10.6) and employing again the orthonormality of (ψ_j), it follows that the coefficients $c_{\ell j}(t)$ equal $\delta_{j\ell}\lambda_\ell$ such that (10.5) follows. $\square$

The density-matrix operators $\widehat{\rho}_I$ and $\widehat{\rho}$ can be expanded in the form

$$\widehat{\rho}_I = \sum_{j=1}^{\infty} \lambda_j |\psi_j^0\rangle\langle\psi_j^0|, \quad \widehat{\rho} = \sum_{j=1}^{\infty} \lambda_j |\psi_j\rangle\langle\psi_j|, \tag{10.7}$$

where the "bra-ket" notation $\langle\psi_j|, |\psi_j\rangle$ denotes the projection operator onto the jth eigenspace of $\widehat{\rho}$.

Now, we can state our main result.

Theorem 10.2 (Mixed-state Schrödinger equation). *Let $\widehat{\rho}$ be a density-matrix operator, satisfying the Liouville–von Neumann equation (10.1), with a complete orthonormal set of eigenfunctions (ψ_j) and eigenvalues (λ_j). The eigenfunctions of the initial-data operator $\widehat{\rho}_I$ are denoted by (ψ_j^0). Then ψ_j is the solution of the Schrödinger equation*

$$i\hbar\partial_t\psi_j = H\psi_j, \quad t > 0, \quad \psi_j(\cdot,0) = \psi_j^0 \quad in \ \mathbb{R}^3, \ j \in \mathbb{N}. \tag{10.8}$$

The particle density $n(x,t)$ can be written as

$$n(x,t) = \sum_{j=1}^{\infty} \lambda_j |\psi_j(x,t)|^2, \quad x \in \mathbb{R}^3, \ t > 0. \tag{10.9}$$

Conversely, let (ψ_j, λ_j) be a sequence of solutions of the Schrödinger equation (10.8) with numbers $\lambda_j \geq 0$. Then the density-matrix operator, defined by (10.7), solves the Liouville–von Neumann equation (10.1).

In view of (10.9), the number λ_j can be interpreted as the occupation probability of the jth state. The sequence of Schrödinger equations (10.8) together with the occupation probabilities (λ_j) is referred to as the *mixed-state Schrödinger equations*. The quantum system is called to be in a *mixed state*. The above proposition roughly states that the Liouville–von Neumann equation is equivalent to the mixed-state Schrödinger equations.

Proof. Let $\widehat{\rho}$ be a solution of the Liouville–von Neumann equation (10.1), represented as in (10.7). On the other hand, the solution of the Liouville–von Neumann equation can be written *formally* as

$$\widehat{\rho}(t) = e^{-iHt/\hbar}\widehat{\rho}_I e^{iHt/\hbar}, \quad t \geq 0,$$

since

$$\partial_t \widehat{\rho} = -\frac{i}{\hbar} H e^{-iHt/\hbar} \widehat{\rho}_I e^{iHt/\hbar} + \frac{i}{\hbar} e^{-iHt/\hbar} \widehat{\rho}_I H e^{iHt/\hbar} = -\frac{i}{\hbar} (H\widehat{\rho} - \widehat{\rho}H).$$

Here, we have used the fact that the Hamiltonian H and the operator $e^{iHt/\hbar}$ commute. Then, inserting the expansion (10.7) for $\widehat{\rho}_I$ in the above formula gives

$$\widehat{\rho}(t) = \sum_{j=1}^{\infty} \lambda_j |e^{-iHt/\hbar} \psi_j^0\rangle \langle e^{iHt/\hbar} \psi_j^0|.$$

Comparing this expression with the expansion (10.7) for $\widehat{\rho}$ shows that $\psi_j = e^{-iHt/\hbar} \psi_j^0$. Finally, differentiation with respect to time yields $\partial_t \psi_j = -(i/\hbar) H \psi_j$ which is equivalent to the Schrödinger equation (10.8).

Conversely, let ψ_j be the solution of the Schrödinger equation (10.8) and let $\widehat{\rho}$ be given by (10.7). Then

$$\partial_t \widehat{\rho} = \sum_{j=1}^{\infty} \lambda_j \left(|\partial_t \psi_j\rangle \langle \psi_j| + |\psi_j\rangle \langle \partial_t \psi_j| \right)$$

$$= \sum_{j=1}^{\infty} \lambda_j \left(-\frac{i}{\hbar} |H\psi_j\rangle \langle \psi_j| + \frac{i}{\hbar} |\psi_j\rangle \langle H\psi_j| \right) = -\frac{i}{\hbar}(H\widehat{\rho} - \widehat{\rho}H).$$

Thus, $\widehat{\rho}$ is a solution of the Liouville–von Neumann equation (10.1). $\square$

If the initial quantum state can be written as $\rho_I(x,y) = \psi_I(x)\overline{\psi_I(y)}$, the density matrix is given by $\rho(x,y,t) = \psi(x,t)\overline{\psi(y,t)}$, where ψ solves the Schrödinger equation (10.8). The particle density equals $n(x,t) = 2\rho(x,x,t) = 2|\psi(x,t)|^2$ and the particle current density

$$J = -\frac{\hbar q}{m} \text{Im}(\overline{\psi}\nabla_x \psi).$$

We refer to such a situation as a *single state* as the single wave function ψ completely describes the quantum state.

For self-consistent modeling, the Poisson equation for the electric potential is added to the Schrödinger equations (10.8). Let V be the sum of an external potential V_{ex}, modeling, for instance, semiconductor heterostructures, and the self-consistent potential V_{sc}, which is given by

$$\varepsilon_s \Delta V_{sc} = q(n - C(x)), \quad x \in \mathbb{R}^3. \tag{10.10}$$

The electron density n is computed according to (10.9). The system of equations, consisting of the Schrödinger equations (10.8), the Poisson equation (10.10) with (10.9), is referred to as the *mixed-state Schrödinger-Poisson system*.

10.2 Transparent Boundary Conditions

In this section, we consider a quantum system consisting of an active region of an electronic device, which is connected to the exterior medium through access zones that can be assumed to be at equilibrium and are modeled by waveguides (which confine waves in a certain direction). The access zones allow for the injection of charge carriers into the active region. Instead of solving the Schrödinger–Poisson system in the whole domain, consisting of the access zones and the active region, we wish to solve the problem only in the active region in order to reduce the computational cost. Then transparent boundary conditions at the interface between the access and active zones have to be prescribed in order to model the continuous electron injection.

Such a situation is referred to as an *open quantum system*. Open quantum systems are characterized by the fact that elements of the system interact with an environment. Here, this notion refers to the influence of the boundaries. We refer to [3] for a detailed discussion of boundary conditions for open quantum systems. The same notion is employed to describe an electron ensemble whose motion is influenced by external sources, like a phonon heat bath, modeled by collision terms in the evolution equation (see Sect. 11.3). If the elements of a quantum system do not interact with the environment, this situation is termed a *closed quantum system*.

Consider the stationary Schrödinger equation

$$-\frac{\hbar^2}{2m^*}\Delta\psi - qV(x)\psi = E\psi \quad \text{in } \Omega,$$

where V is a given potential, E the energy, and $\Omega \subset \mathbb{R}^3$ a bounded domain. We assume that the effective mass m^* is a constant. In heterostructures, however, it might be space dependent or even induce nonlocal effects. For convenience, we scale the Schrödinger equation. Choosing the reference length $\lambda = \text{diam}(\Omega)$, the reference potential $k_B T_L/q$, and the reference energy $k_B T_L$ and introducing the scaled variables

$$x = \lambda x_s, \quad V = \frac{k_B T_L}{q}V_s, \quad E = k_B T_L E_s,$$

the Schrödinger equation becomes, after omitting the index s,

$$-\frac{\varepsilon^2}{2}\Delta\psi - V(x)\psi = E\psi \quad \text{in } \Omega,$$

where $\varepsilon = \hbar/\sqrt{m^* k_B T_L \lambda^2}$ is the scaled Planck constant.

One-dimensional stationary problem. The one-dimensional Schrödinger equation reads as

$$-\frac{\varepsilon^2}{2}\psi'' - V(x)\psi = E\psi, \quad x \in \mathbb{R}. \tag{10.11}$$

We assume that the active region is modeled by the interval $(0,1)$, and the access zones are the intervals $(-\infty,0)$ and $(1,\infty)$. The access zones are waveguides in which the potential is constant. Then, the Schrödinger equation can be solved in these intervals, and it is possible to reduce the Schrödinger problem on the whole line to a Schrödinger problem on the interval $(0,1)$. Since we do not know a priori the wave function at the boundary $x=0$ and $x=1$, this constitutes an open quantum system. Boundary conditions at $x=0$ and $x=1$ can be derived by specifying the injection conditions. This derivation was first performed by Lent and Kirkner [4], called the *quantum transmitting boundary method*. The one-dimensional situation was analyzed by Ben Abdallah et al. [5]. Following [5], we assume that electron waves with positive crystal momentum $p>0$ are injected at $x=0$. They exit the interval at $x=1$ or they are reflected by the potential at $x=0$ (see Fig. 10.1). In a similar way, electrons with $p<0$ are injected at $x=1$ and either transmitted or reflected at $x=1$.

Since the access zones model waveguides, the potential is constant in these intervals,

$$V(x)=V(0) \quad \text{for } x<0, \quad V(x)=V(1) \quad \text{for } x>1. \tag{10.12}$$

Therefore, the Schrödinger equation (10.11) can be solved explicitly and the solutions are plane waves in $(-\infty,0)$ and $(1,\infty)$ (see Sect. 1.2). This motivates the following ansatz. First, let the crystal momentum p be positive. Then we define

$$\psi_p(x) = \begin{cases} e^{ipx/\varepsilon} + r(p)e^{-ipx/\varepsilon} & \text{for } x<0, \\ t(p)e^{ip_+(p)(x-1)/\varepsilon} & \text{for } x>1, \end{cases} \tag{10.13}$$

where $p_+(p)$ has to be determined. This ansatz means that a wave with amplitude 1 is coming from $-\infty$ (since we assumed that $p>0$) and is either transmitted to $+\infty$ with amplitude $t(p)$ or reflected by the potential and travels back to $-\infty$ with amplitude $r(p)$. The reflection-transmission coefficients $r(p)$ and $t(p)$ can be deduced from the Schrödinger equation. Inserting the above ansatz into (10.11) yields

$$(E+V(0))\psi_p = -\frac{\varepsilon^2}{2}\psi_p'' = \frac{p^2}{2}\psi_p \quad \text{for } x<0,$$

and thus the energy is given by $E=p^2/2-V(0)$. Furthermore, we obtain

Fig. 10.1 Electrons with $p>0$ are injected at $x=0$ and either reflected at $x=0$ or transmitted at $x=1$

$$(E+V(1))\psi_p = -\frac{\varepsilon^2}{2}\psi_p'' = \frac{p_+(p)^2}{2}\psi_p \quad \text{for } x > 1,$$

which gives an expression for $p_+(p)$:

$$p_+(p) = \sqrt{2(E+V(1))} = \sqrt{p^2 + 2(V(1) - V(0))}.$$

We take the positive root since the wave travels to $+\infty$ and hence $p_+(p) > 0$ is required.

If the momentum p is negative, we make an analogous ansatz:

$$\psi_p(x) = \begin{cases} t(p)e^{-ip_-(p)x/\varepsilon} & \text{for } x < 0, \\ e^{-ip(x-1)/\varepsilon} + r(p)e^{ip(x-1)/\varepsilon} & \text{for } x > 1, \end{cases} \tag{10.14}$$

where $p_-(p)$ has to be determined. This ansatz models a wave coming from $+\infty$ and being either transmitted to $-\infty$ or reflected at $x = 1$ and traveling back to $+\infty$. Inserting this ansatz into (10.11) gives, after a similar computation as above,

$$E = \frac{p^2}{2} - V(1), \quad p_-(p) = \sqrt{p^2 - 2(V(1) - V(0))}.$$

The boundary conditions at $x = 0$ and $x = 1$ can be determined from the continuity of ψ_p in $\mathbb{R}$. Indeed, for $p > 0$ and $x \to 0$, $x < 0$, we conclude from (10.13) that

$$\varepsilon\psi_p'(0) = ip(1 - r(p)), \quad ip\psi_p(0) = ip(1 + r(p)).$$

Eliminating $r(p)$ leads to the boundary condition

$$\varepsilon\psi_p'(0) + ip\psi_p(0) = 2ip.$$

For $x \to 1, x > 1$, we infer that

$$\varepsilon\psi_p'(1) = ip_+(p)t(p) = ip_+(p)\psi_p(1).$$

For $p < 0$ we obtain

$$\varepsilon\psi_p'(1) - ip\psi_p(1) = -ip(1 - r(p)) - ip(1 + r(p)) = -2ip,$$
$$\varepsilon\psi_p'(0) = -ip_-(p)t(p) = -ip_-(p)\psi_p(0).$$

This leads to the following result.

Proposition 10.3 (Lent–Kirkner boundary conditions). *Let V be a given potential satisfying (10.12). Then the solution (ψ_p, E_p) of the eigenvalue problem*

$$-\frac{\varepsilon^2}{2}\psi_p'' - V(x)\psi_p = E_p\psi_p, \quad x \in \mathbb{R},$$

can be written equivalently, on the interval $(0, 1)$, as the solution of

$$-\frac{\varepsilon^2}{2}\psi_p'' - V(x)\psi_p = E_p\psi_p, \quad x \in (0,1),$$

with the boundary conditions

$$\varepsilon\psi_p'(0) + \mathrm{i}p\psi_p(0) = 2\mathrm{i}p, \quad \varepsilon\psi_p'(1) = \mathrm{i}p_+(p)\psi_p(1) \qquad \text{for } p > 0, \quad (10.15)$$

$$-\varepsilon\psi_p'(1) + \mathrm{i}p\psi_p(1) = 2\mathrm{i}p, \quad \varepsilon\psi_p'(0) = -\mathrm{i}p_-(p)\psi_p(0) \quad \text{for } p < 0, \quad (10.16)$$

and outside of the interval $(0,1)$, it equals (10.13), $E_p = p^2/2 - V(0)$ if $p > 0$ and (10.14), $E_p = p^2/2 - V(1)$ if $p < 0$, where

$$p_\pm(p) = \sqrt{p^2 \pm 2(V(1) - V(0))}.$$

The reflection and transmission amplitudes $r(p)$ and $t(p)$, respectively, are determined by

$$r(p) = \frac{1}{2}\left(\psi_p(0) + \mathrm{i}\frac{\varepsilon}{p}\psi_p'(0)\right), \qquad t(p) = \psi_p(1) \qquad \text{for } p > 0, \qquad (10.17)$$

$$r(p) = \frac{1}{2}\left(\psi_p(1) - \mathrm{i}\frac{\varepsilon}{p}\psi_p'(1)\right), \qquad t(p) = \psi_p(0) \qquad \text{for } p < 0. \qquad (10.18)$$

Equations (10.15) and (10.16) are called the *Lent–Kirkner boundary conditions* [4]. Formulas (10.17) and (10.18) follow immediately from the definition of ψ_p in (10.13). It is shown in [6] that the above model allows for an interpretation in terms of a family of dissipative operators, the so-called *quantum transmitting boundary operator family*, leading to the quantum transmitting Schrödinger–Poisson system analyzed in [6] and related to a dissipative Schrödinger–Poisson system.

From the solution ψ_p of the Schrödinger equation, the electron and current densities can be computed. In physical variables, they read as follows:

$$n(x) = \int_{\mathbb{R}} f(p)|\psi_p(x)|^2 \, \mathrm{d}p,$$

$$J(x) = \frac{q\hbar}{m^*}\int_{\mathbb{R}} f(p)\mathrm{Im}(\overline{\psi_p(x)}\nabla\psi_p(x)) \, \mathrm{d}p,$$

where $f(p)$ describes the statistics of the electrons. For instance, in a quantum well, in which electrons are confined in one direction, the statistics is

$$f(p) = \frac{m^*k_B T}{\pi\hbar^2}\ln\left(1 + \mathrm{e}^{(-p^2/2m^* + E_F)/k_B T}\right),$$

where E_F is the Fermi energy (cf. Lemma 1.17).

Multi-dimensional stationary problem. The model of Lent and Kirkner [4] was generalized to arbitrary space dimension by Ben Abdallah [7]. The quantum device is supposed to occupy the domain $\Omega \subset \mathbb{R}^d$ with the active region Ω_0 and access

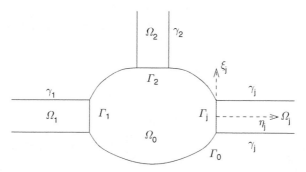

Fig. 10.2 Quantum domain Ω with the active region Ω_0 and access zones Ω_j. In each access zone, a local coordinate system $(\xi_j, \eta_j) \in \mathbb{R}^{d-1} \times \mathbb{R}$ is introduced

zones Ω_j, $j = 1, \ldots, N$. The boundary of the active region consists of a part Γ_0 and N flat surfaces Γ_j of dimension $d - 1$. The access zones Ω_j are assumed to be semi-infinite cylinders with basis Γ_j and lateral boundary γ_j (see Fig. 10.2). The boundary of Ω is given by the union $\Gamma_0 \cup \gamma_1 \cup \cdots \cup \gamma_N$. For instance, the electron beam is injected at the waveguide Ω_1 and splits into several beams exiting the device by the leads Ω_j, $j \geq 2$.

For given potential $V(x)$, we wish to solve the eigenvalue problem

$$-\frac{\varepsilon^2}{2}\Delta\psi - V(x)\psi = E\psi \quad \text{in } \Omega, \quad \psi = 0 \quad \text{on } \partial\Omega.$$

The Dirichlet boundary condition on $\partial\Omega$ means that the domain Ω is like a quantum well. We require that ψ is only bounded and may be nonintegrable to represent a scattering state. The main assumption is that the given potential only depends on the transversal direction in each access zone, i.e.,

$$V = V_j(\xi_j) \quad \text{in } \Omega_j, \ j = 1, \ldots, N, \tag{10.19}$$

where $(\xi_j, \eta_j) \in \mathbb{R}^{d-1} \times \mathbb{R}$ are local coordinates. Thus, the zone Ω_j behaves like a waveguide, and the Schrödinger equation can be solved explicitly in Ω_j. Indeed, let (ψ_m^j, E_m^j) be the eigenfunction–eigenvalue pairs of the transversal Schrödinger problem

$$-\frac{\varepsilon^2}{2}\Delta_{\xi_j}\psi - V_j(\xi_j)\psi = E\psi \quad \text{in } \Gamma_j, \quad \psi = 0 \quad \text{on } \partial\Gamma_j, \tag{10.20}$$

where $m \in \mathbb{N}$ and $j = 1, \ldots, N$. The functions $(\psi_m^j)_j$ form a complete orthonormal set of $L^2(\Gamma_j)$, and for each j, the energies E_m^j are nondecreasing with respect to m and tend to infinity as $m \to \infty$. In order to solve the Schrödinger equation in the waveguide Ω_j, we insert the ansatz $\psi(\xi_j, \eta_j) = \psi_m^j(\xi_j)\lambda(\eta_j)$ in that equation, which gives

$$\left(-\frac{\varepsilon^2}{2}\Delta_{\xi_j}\psi_m^j - V_j(\xi_j)\psi_m^j\right)\lambda - \frac{\varepsilon^2}{2}\psi_m^j\lambda'' = E\psi_m^j\lambda.$$

By (10.20), the expression in the brackets equals $E_m^j\psi_m^j$. Hence,

$$\left(-\frac{\varepsilon^2}{2}\lambda'' - (E - E_m^j)\lambda\right)\psi_m^j = 0$$

and $-(\varepsilon^2/2)\lambda'' = (E - E_m^j)\lambda$. Thus, any solution ψ of the Schrödinger equation in the waveguide Ω_j can be written as the infinite sum

$$\psi(\xi_j, \eta_j) = \sum_{m=1}^{\infty} \psi_m^j(\xi_j)\lambda_m^j(\eta_j),$$

where λ_m^j solves the differential equation

$$-\frac{\varepsilon^2}{2}\frac{d^2\lambda_m^j}{d\eta_j^2} = (E - E_m^j)\lambda_m^j, \quad \eta_j > 0.$$

This equation can be solved explicitly. Setting

$$p_m^j(E) = \sqrt{2|E - E_m^j|} \quad \text{and} \quad N^j(E) = \sup\{m \geq 1 : E > E_m^j\}, \qquad (10.21)$$

we make the ansatz

$$\lambda_m^j(\eta_j) = \begin{cases} a_m^j \exp\left(-ip_m^j(E)\eta_j/\varepsilon\right) + b_m^j \exp\left(ip_m^j(E)\eta_j/\varepsilon\right) & \text{for } m \leq N^j(E), \\ b_m^j \exp\left(-p_m^j(E)\eta_j/\varepsilon\right) & \text{for } m > N^j(E). \end{cases}$$
$$(10.22)$$

If $m > N^j(E)$, the energy E is smaller than E_m^j corresponding to a bound-state behavior, whereas in the opposite case the state behaves like a free particle. Since we have assumed that ψ is bounded, the positive exponential term can be neglected. The coefficients a_m^j are supposed to be known since they describe the incoming waves. The reflection-transmission coefficients b_m^j are unknown and will be deduced from the equations. Elimination of b_m^j then leads to the boundary conditions for ψ on Γ_j. To this end, we observe that the wave function and its derivative at $\eta_j = 0$ or Γ_j can be written as

$$\psi|_{\Gamma_j} = \sum_{m=1}^{N^j}(a_m^j + b_m^j)\psi_m^j(\xi_j) + \sum_{m=N^j+1}^{\infty} b_m^j \psi_m^j(\xi_j), \qquad (10.23)$$

$$\varepsilon\frac{\partial\psi}{\partial\eta_j}\Big|_{\Gamma_j} = \sum_{m=1}^{N^j} ip_m^j(E)(-a_m^j + b_m^j)\psi_m^j(\xi_j) - \sum_{m=N^j+1}^{\infty} p_m^j(E)b_m^j \psi_m^j(\xi_j). \qquad (10.24)$$

On the other hand, since $(\psi_m^j)_m$ is a complete orthonormal set of $L^2(\Gamma_j)$, we can develop $\psi|_{\Gamma_j}$ according to

$$\psi|_{\Gamma_j} = \sum_{m=1}^{\infty} c_m^j(\psi)\psi_m^j(\xi_j).$$

The coefficients $c_m^j(\psi)$ are given by

$$c_m^j(\psi) = \int_{\Gamma_j} \psi\psi_m^j \, d\xi_j = \begin{cases} a_m^j + b_m^j & \text{for } m \leq N^j(E), \\ b_m^j & \text{for } m > N^j(E). \end{cases}$$

Thus, the coefficients b_m^j in (10.24) can be eliminated, which gives transparent boundary conditions for ψ on Γ_j. We have shown the following result.

Proposition 10.4 (Transparent boundary conditions). *Let Ω, Ω_0, and Γ_j be defined as above and let the potential V only depend on the transversal direction, i.e., V satisfies condition (10.19). Then the solution ψ of the eigenvalue problem*

$$-\frac{\varepsilon^2}{2}\Delta\psi - V(x)\psi = E\psi \quad \text{in } \Omega, \quad \psi = 0 \quad \text{on } \partial\Omega,$$

is a solution of

$$-\frac{\varepsilon^2}{2}\Delta\psi - V(x)\psi = E\psi \quad \text{in } \Omega_0, \quad \psi|_{\Gamma_0} = 0,$$

$$\varepsilon\frac{\partial\psi}{\partial\eta_j}\Big|_{\Gamma_j} = \sum_{m=1}^{N^j(E)} ip_m^j(E)(-2a_m^j + c_m^j(\psi))\psi_m^j(\xi_j)$$

$$- \sum_{m=N^j(E)+1}^{\infty} p_m^j(E)c_m^j(\psi)\psi_m^j(\xi_j), \quad j = 1,\ldots,N,$$

where ψ_m^j are the transversal wave functions solving (10.20), a_m^j are the coefficients of the incoming waves, according to (10.22), $p_m^j(E)$ and $N^j(E)$ are defined in (10.21), and

$$c_m^j(\psi) = \int_{\Gamma_j} \psi\psi_m^j \, d\xi_j.$$

Multi-dimensional transient problem. We assume a geometry of the quantum device as in Fig. 10.2. For a given potential $V(x,t)$, consider the open time-dependent Schrödinger problem

$$i\varepsilon\partial_t\psi = -\frac{\varepsilon^2}{2}\Delta\psi - V(x,t)\psi \quad \text{in } \Omega,\, t \in \mathbb{R}, \quad \psi(\cdot,0) = \psi_I \quad \text{in } \Omega.$$

The aim is to give an equivalent formulation of the Schrödinger problem on the active region Ω_0 only, with transparent boundary conditions. This task is performed in

[8–10] for homogeneous transparent boundary conditions. Nier has analyzed a more general version of this problem using the density-matrix formulation and scattering theory techniques [11]. Here, we present the results of [12] for inhomogeneous transparent boundary conditions in several space dimensions.

We assume that the potential V is the sum of a given stationary potential V^0 only depending on the transversal direction and a time-dependent potential V_j in each lead Ω_j, i.e., for any $j = 1, \ldots, N$, we write $V(x,t) = V^0(x) + V_j(t)$ for $x \in \Omega_j$, where $V^0(x) = V_j^0(\xi_j)$ in Ω_j. Then, let ψ_m^0 be a solution of the stationary Schrödinger equation in the waveguides,

$$-\frac{\varepsilon^2}{2}\Delta\psi_m^0 - V^0(x)\psi_m^0 = E(m)\psi_m^0 \quad \text{in } \Omega_j, \ j = 1, \ldots, N.$$

The functions ψ_m^0 are the initial data of our problem. We define the phase factor

$$\theta_m^j(t) = \exp\left(-\frac{i}{\varepsilon}\int_0^t (E(m) - V_j(s))\,ds\right)$$

and the plane wave functions

$$\psi_m^{pw}(x,t) = \psi_m^0 \sum_{j=1}^N \theta_m^j(t)\mu_j(x),$$

where $(\mu_1, \ldots, \mu_N)$ is a partition of unity of Ω, i.e., $0 \leq \mu_j \leq 1$, $\sum_j \mu_j = 1$ in Ω, $\mu_j = 1$ in Ω_j, and $\mu_j = 0$ in Ω_ℓ for all $\ell \neq j$. We notice that the plane wave functions solve the equations

$$i\varepsilon\partial_t\psi_m^{pw} = -\frac{\varepsilon^2}{2}\Delta\psi_m^{pw} - V(x,t)\psi_m^{pw}, \quad \psi_m^{pw}(\cdot,0) = \psi_m^0 \quad \text{in } \Omega \setminus \Omega_0.$$

Furthermore, we introduce the functions

$$\chi_m^j(\xi_j,t) = \psi_m^j(\xi_j)\exp\left(-\frac{i}{\varepsilon}\int_0^t (E_m^j - V_j(s))\,ds\right),$$

which form a basis of $L^2(\Gamma_j)$. Here, ψ_m^j are the transversal eigenmodes of the waveguide Ω_j, i.e., they are the solutions of the problem

$$-\frac{\varepsilon^2}{2}\Delta\psi_m^j - V_j^0(\xi_j)\psi_m^j = E_m^j\psi_m^j \quad \text{in } \Gamma_j, \quad \psi_m^j = 0 \quad \text{on } \partial\Gamma_j.$$

It is shown by Ben Abdallah et al. [12] that the Schrödinger problem

$$i\varepsilon\partial_t\psi_m = -\frac{\varepsilon^2}{2}\Delta\psi_m - V(x,t)\psi_m \quad \text{in } \Omega, \ t \in \mathbb{R}, \quad \psi_m(\cdot,0) = \psi_m^0 \quad \text{in } \Omega,$$

can be formulated as a boundary-value problem on the active region Ω_0 with a Dirichlet boundary condition on Γ_0 and

$$\frac{\partial}{\partial \eta_j}(\psi_m - \psi_m^{\mathrm{pw}}) = -\mathrm{e}^{-i\pi/4} \sum_{\ell=1}^{\infty} \chi_\ell^j(\cdot,t) \sqrt{\frac{\partial}{\partial t}} \int_{\Gamma_j} (\psi_m - \psi_m^{\mathrm{pw}})(\xi_j) \overline{\chi_m^j(\xi_j,t)} \, \mathrm{d}\xi_j.$$

Here, the fractional derivative $\sqrt{\partial/\partial t}$ is defined by

$$\sqrt{\frac{\partial}{\partial t}} f = \frac{1}{\sqrt{\pi}} \frac{\mathrm{d}}{\mathrm{d}t} \int_0^t \frac{f(s)}{\sqrt{t-s}} \, \mathrm{d}s.$$

This derivative can be also written as a time convolution of the boundary data with the kernel $t^{-3/2}$. The above boundary condition is derived by solving explicitly a Laplace-transformed Schrödinger problem for the Dirichlet–Neumann operator and applying the inverse Laplace transform. The transparent boundary conditions are, in contrast to the boundary conditions for the stationary problem, nonlocal in time and of memory type, thus requiring the storage of all the past history at the boundary in a numerical simulation. We refer to the review [13] for numerical approximations and efficient implementations of such boundary conditions.

Finally, we remark that the electric potential may be also given self-consistently. Then the potential V is the sum of the given potential $V^0(x)$, the waveguide potential $V_j(t)$, and the self-consistent potential V_{sc}, defined by

$$\lambda_D^2 \Delta V_{\mathrm{sc}} = n - C(x) \quad \text{in } \Omega_0, \quad V_{\mathrm{sc}} = 0 \quad \text{on } \Gamma_0,$$

where the electron density is given by

$$n = \sum_{m=1}^{\infty} \int_{\mathbb{R}} f_m(p) |\psi_m|^2 \, \mathrm{d}p,$$

and $f_m(p)$ describes the statistics of the electrons.

References

1. A. Arnold. Mathematical properties of quantum evolution equations. In: G. Allaire, A. Arnold, P. Degond, and T. Hou (eds.), *Quantum Transport – Modelling, Analysis and Asymptotics*, Lecture Notes Math. 1946, 45–110. Springer, Berlin, 2008.
2. R. Dautray and J.-L. Lions. *Mathematical Analysis and Numerical Methods for Science and Technology*. Springer, Berlin, 1985.
3. W. Frensley. Boundary conditions for open quantum systems driven far from equilibrium. *Rev. Modern Phys.* 62 (1990), 745–791.
4. C. Lent and D. Kirkner. The quantum transmitting boundary method. *J. Appl. Phys.* 67 (1990), 6353–6359.
5. N. Ben Abdallah, P. Degond, and P. Markowich. On a one-dimensional Schrödinger-Poisson scattering model. *Z. Angew. Math. Phys.* 48 (1997), 135–155.
6. M. Baro, H.-C. Kaiser, H. Neidhardt, and J. Rehberg. A quantum transmitting Schrödinger-Poisson system. *Rev. Math. Phys.* 16 (2004), 281–330.
7. N. Ben Abdallah. On a multidimensional Schrödinger-Poisson scattering model for semiconductors. *J. Math. Phys.* 41 (2000), 4241–4261.

8. X. Antoine and C. Besse. Construction, structure and asymptotic approximations of a micro-differential transparent boundary condition for the linear Schrödinger equation. *J. Math. Pure Appl.* 80 (2001), 701–738.
9. A. Arnold. Numerical absorbing boundary conditions for quantum evolution equations. *VLSI Design* 6 (1998), 313–319.
10. V. Baskakov and A. Popov. Implementation of transparent boundaries for numerical solution of the Schrödinger equation. *Wave Motion* 14 (1991), 123–128.
11. F. Nier. The dynamics of some quantum open systems with short-range nonlinearities. *Nonlinearity* 11 (1998), 1127–1172.
12. N. Ben Abdallah, F. Méhats, and O. Pinaud. On an open transient Schrödinger-Poisson system. *Math. Models Meth. Appl. Sci.* 15 (2005), 667–688.
13. X. Antoine, A. Arnold, C. Besse, M. Ehrhardt, and A. Schädle. A review of transparent and artificial boundary conditions techniques for linear and nonlinear Schrödinger equations. *Commun. Comput. Phys.* 4 (2008), 729–796.

Chapter 11
The Wigner Equation

The quantum mechanical motion of an electron ensemble can be described by the Schrödinger or the density-matrix formulation (see Sect. 10.1). There is an alternative description based on the quantum-kinetic Wigner formalism, which we present and discuss in this chapter. There are two main reasons for using this framework in applications (mostly for transient problems). First, the Wigner picture allows, in contrast to Schrödinger models, for a modeling of scattering phenomena in the form of a quantum Boltzmann equation. Second, the quantum-kinetic framework makes it easier to formulate boundary conditions at the device contacts, which may be inspired from classical kinetic considerations [1]. In this chapter, following [2], we formulate the quantum Liouville equation, the quantum Vlasov equation, and quantum Boltzmann models and discuss their relations to the classical kinetic equations introduced in Chaps. 3 and 4.

11.1 The Quantum Liouville Equation

The quantum Liouville equation is the quantum analogue of the Liouville equation presented in Sect. 3.1. It will be derived from the electron-ensemble Liouville–von Neumann equation for the density matrix ρ,

$$i\hbar \partial_t \rho(r,s,t) = (H_r - H_s)\rho(r,s,t), \quad \rho(r,s,0) = \rho_I(r,s), \quad r,s \in \mathbb{R}^{3M}, \quad (11.1)$$

for an ensemble consisting of M electrons with mass m in a vacuum (see Sect. 10.1 for the introduction of this equation). We define the *Fourier transform* of a function $f : \mathbb{R}^{3M} \to \mathbb{C}$ as

$$(\mathscr{F}(f))(p) = \int_{\mathbb{R}^{3M}} f(y)e^{-iy\cdot p/\hbar}\, dy,$$

Jüngel, A.: *The Wigner Equation.* Lect. Notes Phys. **773**, 231–247 (2009)
DOI 10.1007/978-3-540-89526-8_11   © Springer-Verlag Berlin Heidelberg 2009

and its inverse,

$$(\mathscr{F}^{-1}(g))(y) = \frac{1}{(2\pi\hbar)^{3M}} \int_{\mathbb{R}^{3M}} g(p) e^{iy \cdot p/\hbar} dp,$$

for functions $g : \mathbb{R}^{3M} \to \mathbb{C}$.

For the kinetic formulation of the Liouville–von Neumann equation, we need the so-called *Wigner function* introduced by Wigner in 1932 [3]:

$$w(x,p,t) = \int_{\mathbb{R}^{3M}} \rho\left(x+\frac{y}{2}, x-\frac{y}{2}, t\right) e^{-iy \cdot p/\hbar} dy. \tag{11.2}$$

Setting

$$u(x,y,t) = \rho\left(x+\frac{y}{2}, x-\frac{y}{2}, t\right), \tag{11.3}$$

the Wigner function can be written as the Fourier transform of u, $w = \mathscr{F}(u)$. Furthermore, $u = \mathscr{F}^{-1}(w)$. Since y has the dimension of a length, $p/\hbar$ in the Fourier transform has the dimension of an inverse length and thus, p has the dimension of a momentum. We interpret p as the crystal momentum $\hbar k$. We notice that the transformation $\rho \mapsto w$ is called the *Wigner–Weyl transform*.

The evolution equation for the Wigner function is obtained by transforming the Liouville–von Neumann equation to the (x, y) variables and applying Fourier transformation. The result is expressed in the following proposition.

Proposition 11.1 (Many-particle quantum Liouville equation). *Let ρ be a solution of the Liouville–von Neumann equation* (11.1). *Then the Wigner function* (11.2) *is formally a solution of*

$$\partial_t w + \frac{p}{m} \cdot \nabla_x w + q\theta[V]w = 0, \quad t > 0, \quad w(x,p,0) = w_I(x,p) \tag{11.4}$$

for x, $p \in \mathbb{R}^{3M}$, where the initial datum is given by

$$w_I(x,p) = \int_{\mathbb{R}^{3M}} \rho_I\left(x+\frac{y}{2}, x-\frac{y}{2}\right) e^{-iy \cdot p/\hbar} dy,$$

and $\theta[V]$ is a pseudo-differential operator, defined by

$$(\theta[V]w)(x,p,t) = \frac{1}{(2\pi\hbar)^{3M}} \int_{\mathbb{R}^{3M} \times \mathbb{R}^{3M}} (\delta V)(x,y,t) w(x,p',t) e^{iy \cdot (p-p')/\hbar} dp' dy, \tag{11.5}$$

where

$$\delta V(x,y,t) = \frac{i}{\hbar}\left(V\left(x+\frac{y}{2}, t\right) - V\left(x-\frac{y}{2}, t\right)\right).$$

Equation (11.4) is called the *many-particle Wigner equation* or *many-particle quantum Liouville equation*. The local term $(p/m) \cdot \nabla_x w$ is the quantum analogue of the classical transport term of the Liouville equation (see (3.7)). The nonlocal term $q\theta[V]w$ models the influence of the electric potential. The nonlocality has the effect that the electron ensemble "feels" an upcoming potential barrier.

Before we prove the proposition, we discuss the pseudo-differential operator. It can be written, by slight abuse of notation, as

$$(\theta[V]w)(x,p,t) = \int_{\mathbb{R}^{3M} \times \mathbb{R}^{3M}} (\delta V)(x,y,t)u(x,-y,t)e^{iy \cdot p/\hbar} dy$$
$$= (2\pi\hbar)^{3M} \mathscr{F}^{-1}\left((\delta V)(x,y,t)u(x,-y,t)\right).$$

Therefore, it acts in the Fourier space essentially as a multiplication operator. The multiplicator δV is called the *symbol* of the operator. The symbol δV is a discrete directional derivative, since in the formal limit "$\hbar \to 0$", we find

$$\delta V(x,\hbar y,t) \to i\nabla_x V(x,t) \cdot y.$$

We refer to [4] for a mathematical theory of pseudo-differential operators. In particular, the Wigner equation (11.4) is a linear pseudo-differential equation.

Proof (of Proposition 11.1). First, we derive the evolution equation for u, defined in (11.3), and then take the inverse Fourier transform. We compute, for $r = x+y/2$ and $s = x-y/2$,

$$\text{div}_y(\nabla_x u)(x,y,t) = \text{div}_y(\nabla_r \rho + \nabla_s \rho)\left(x+\frac{y}{2},x-\frac{y}{2},t\right)$$
$$= \frac{1}{2}(\Delta_r \rho - \Delta_s \rho)\left(x+\frac{y}{2},x-\frac{y}{2},t\right).$$

Then the transformed Liouville–von Neumann equation for u becomes

$$\partial_t u(x,y,t) = \partial_t \rho(r,s,t) = -\frac{i}{\hbar}\left(-\frac{\hbar^2}{2m}(\Delta_r - \Delta_s) - qV(r,t) + qV(s,t)\right)\rho(r,s,t)$$
$$= \frac{i\hbar}{m}\text{div}_y(\nabla_x u)(x,y,t) + q\delta V(x,y,t)u(x,y,t)$$

or

$$\partial_t u - \frac{i\hbar}{m}\text{div}_y(\nabla_x u) - q(\delta V)u = 0, \quad x,y \in \mathbb{R}^{3M}, \ t > 0.$$

The Fourier transform gives

$$\partial_t \mathscr{F}(u) - \frac{i\hbar}{m}\mathscr{F}(\text{div}_y \nabla_x u) - q\mathscr{F}((\delta V)u) = 0. \tag{11.6}$$

The second term on the left-hand side can be written, by integrating by parts, as

$$\mathscr{F}(\text{div}_y \nabla_x u)(x,p,t) = \int_{\mathbb{R}^{3M}} \text{div}_y(\nabla_x u)(x,y,t)e^{-iy \cdot p/\hbar} dy$$
$$= \frac{i}{\hbar}\int_{\mathbb{R}^{3M}} p \cdot \nabla_x u(x,y,t)e^{-iy \cdot p/\hbar} dy = \frac{i}{\hbar}p \cdot \nabla_x \mathscr{F}(u)(x,p,t)$$
$$= \frac{i}{\hbar}p \cdot \nabla_x w(x,v,t).$$

The third term on the left-hand side of (11.6) becomes, by (11.5),

$$
\begin{aligned}
\mathscr{F}((\delta V)u)(x,p,t) &= \int_{\mathbb{R}^{3M}} (\delta V)(x,y,t)u(x,y,t)e^{-iy\cdot p/\hbar}\,\mathrm{d}y \\
&= (2\pi\hbar)^{-3M} \int_{\mathbb{R}^{3M}} (\delta V)(x,y,t)w(x,p',t)e^{iy\cdot(p'-p)/\hbar}\,\mathrm{d}p'\,\mathrm{d}y \\
&= (2\pi\hbar)^{-3M} \int_{\mathbb{R}^{3M}} (\delta V)(x,-y,t)w(x,p',t)e^{iy\cdot(p-p')/\hbar}\,\mathrm{d}p'\,\mathrm{d}y \\
&= -(\theta[V]w)(x,p,t).
\end{aligned}
$$

Therefore, (11.6) equals the Wigner equation (11.4). □

Lemma 11.2. *The ensemble particle density n and the ensemble current density J, defined in (10.3) and (10.4), respectively, can be expressed in terms of the Wigner function as*

$$
n(x,t) = \frac{2}{(2\pi\hbar)^{3M}} \int_{\mathbb{R}^{3M}} w(x,p,t)\,\mathrm{d}p, \quad J(x,t) = -\frac{2}{(2\pi\hbar)^{3M}}\frac{q}{m} \int_{\mathbb{R}^{3M}} w(x,p,t)p\,\mathrm{d}p.
$$

The above integrals are called the zeroth and first moments of the Wigner function, respectively, in analogy to the classical situation (see Sect. 2.1).

Proof. The first identity follows from

$$
n(x,t) = 2\rho(x,x,t) = 2u(x,0,t) = 2(2\pi\hbar)^{-3M} \int_{\mathbb{R}^{3M}} w(x,p,t)\,\mathrm{d}p.
$$

For the proof of the second identity, we compute

$$
\begin{aligned}
J(x,t) &= \frac{i\hbar q}{m}(\nabla_r - \nabla_s)\rho(x,x,t) = \frac{2i\hbar q}{m}\nabla_y u(x,0,t) \\
&= \frac{2i\hbar q}{m}\frac{1}{(2\pi\hbar)^{3M}} \int_{\mathbb{R}^{3M}} w(x,p,t)\nabla_y e^{iy\cdot p/\hbar}\Big|_{y=0}\,\mathrm{d}p \\
&= -\frac{q}{m}\frac{2}{(2\pi\hbar)^{3M}} \int_{\mathbb{R}^{3M}} w(x,p,t)p\,\mathrm{d}p,
\end{aligned}
$$

finishing the proof. □

We discuss three questions related to the quantum Liouville equation:

- How can we formalize the classical limit "$\hbar \to 0$" and which is the limit equation?
- Are the solutions of the quantum Liouville equation nonnegative if this property holds true initially?
- How does the quantum Liouville equation change when taking into account the semiconductor crystal?

The classical limit. The limit "$\hbar \to 0$" can be formalized in an appropriate scaling. We choose the reference length λ, the reference time τ, the reference momentum

$m\lambda/\tau$, and the reference voltage $k_B T_L/q$. We assume that the reference wave energy $\hbar/\tau$ is much smaller than the thermal and kinetic energies, i.e.,

$$\frac{\hbar/\tau}{k_B T_L} = \varepsilon \quad \text{and} \quad \frac{\hbar/\tau}{m(\lambda/\tau)^2} = \varepsilon \quad \text{with } \varepsilon \ll 1$$

(this fixes λ for given τ and vice versa). Thus, introducing the scaling

$$x = \lambda x_s, \quad t = \tau t_s, \quad p = \frac{m\lambda}{\tau} p_s, \quad V = \frac{k_B T_L}{q} V_s,$$

we obtain, after omitting the index s, the scaled Wigner equation

$$\partial_t w + p \cdot \nabla_x w + \theta[V]w = 0, \tag{11.7}$$

where $\theta[V]w$ is given by

$$(\theta[V]w)(x,p,t) = \frac{1}{(2\pi)^{3M}} \int_{\mathbb{R}^{3M} \times \mathbb{R}^{3M}} (\delta V)(x,\eta,t)w(x,p',t)e^{i\eta \cdot (p-p')} \, dp' \, d\eta,$$

with the symbol

$$\delta V(x,\eta,t) = \frac{i}{\varepsilon} \left(V\left(x + \frac{\varepsilon}{2}\eta,t\right) - V\left(x - \frac{\varepsilon}{2}\eta,t\right) \right).$$

The classical limit $\varepsilon \to 0$ in the symbol δV yields $\delta V(x,\eta,t) \to i\nabla_x V(x,t) \cdot \eta$ and hence, by integrating by parts,

$$\begin{aligned}
(\theta[V]w)(x,p,t) &\to \frac{i}{(2\pi)^{3M}} \int_{\mathbb{R}^{3M} \times \mathbb{R}^{3M}} \nabla_x V(x,t) \cdot \eta w(x,p',t)e^{i\eta \cdot (p-p')} \, dp' \, d\eta \\
&= -\frac{1}{(2\pi)^{3M}} \nabla_x V(x,t) \cdot \int_{\mathbb{R}^{3M} \times \mathbb{R}^{3M}} \nabla_p e^{-i\eta \cdot p'} w(x,p',t) \, dp' e^{i\eta \cdot p} \, d\eta \\
&= \frac{1}{(2\pi)^{3M}} \nabla_x V(x,t) \cdot \int_{\mathbb{R}^{3M} \times \mathbb{R}^{3M}} \nabla_p w(x,p',t)e^{i\eta \cdot (p-p')} \, dp' \, d\eta \\
&= \nabla_x V(x,t) \cdot \nabla_p w(x,p,t).
\end{aligned}$$

This limit was made rigorous by Lions and Paul [5]. For a quadratic potential, the operator $\theta[V]$ takes exactly the form of its classical counterpart,

$$\theta\left[\frac{\lambda}{2}|x|^2\right]w = \lambda x \cdot \nabla_p w,$$

such that in this situation, the Wigner equation equals formally the classical Liouville equation (see Sect. 3.1)

$$\partial_t w + p \cdot \nabla_x w + \nabla_x V \cdot \nabla_p w = 0.$$

In the general case, this equation follows from (11.7) in the formal limit $\varepsilon \to 0$. This was made rigorous by Markowich and Ringhofer [6, 7] for smooth potentials. The limit was also performed in [2, Sect. 1.4] by an asymptotic expansion of δV and u in powers of ε. For references on the mathematical analysis of the Wigner equation (or the Wigner–Poisson system), we refer to the review of Arnold [8].

Nonnegativity of the Wigner function. The solution of the classical Liouville equation stays nonnegative for all times if the initial distribution function is nonnegative. Unfortunately, this property does generally *not* hold for the solution of the quantum Liouville equation. In the case of a pure quantum state it is possible to characterize those states for which the Wigner function is nonnegative. It was shown by Hudson [9] that

$$w(x,p,t) = \int_{\mathbb{R}^{3M}} \psi\left(x + \frac{y}{2}, t\right) \overline{\psi}\left(x - \frac{y}{2}, t\right) e^{-iy \cdot p/\hbar} dy$$

is nonnegative if and only if either $\psi = 0$ or

$$\psi(x,t) = \exp(-x^\top A(t)x - a(t) \cdot x - b(t)), \quad x \in \mathbb{R}^{3M}, \ t > 0,$$

where $A(t) \in \mathbb{C}^{3M \times 3M}$ is a matrix with symmetric positive definite real part and $a(t) \in \mathbb{C}^{3M}$, $b(t) \in \mathbb{C}$. Inserting this ansatz into the Schrödinger equation shows that the potential has to be quadratic in x, i.e.,

$$V(x,t) = x^\top \widetilde{A}(t)x + \widetilde{a}(t) \cdot x + \widetilde{b}(t)$$

for some $\widetilde{A}(t) \in \mathbb{C}^{3M \times 3M}$, $\widetilde{a}(t) \in \mathbb{C}^{3M}$, $\widetilde{b}(t) \in \mathbb{C}$, in order to obtain a nonnegative Wigner solution.

The case of mixed quantum states, i.e., for arbitrary initial data $w_I \in L^2(\mathbb{R}^{3M} \times \mathbb{R}^{3M})$, is more involved. In fact, a necessary condition for the nonnegativity of w seems not to be known.

By an appropriate averaging of the Wigner function over sufficiently large phase-space regions, which is realized by the so-called *Husimi transformation*, the oscillations of the Wigner functions are smoothed out, leading to a nonnegative function, and thus avoiding negative values of the Wigner function (see, for instance, [5, 10]).

The semi-classical quantum Liouville equation. The quantum Liouville equation (11.4) models the motion of electrons in a vacuum under the influence of an electric field. We discuss now the case of a single electron moving in a crystal. Let us consider a single electron in a fixed energy band $E(k)$ with $k \in B$, where B is the Brillouin zone (see Sect. 1.1). In this situation, the (scaled) semi-classical Hamiltonian $H(x,k,t) = |k|^2/2 - V(x,t)$ has to be replaced by $H(x,k,t) = E(k) - V(x,t)$. Let ψ be the Schrödinger wave function corresponding to this energy band Hamiltonian and define the single-state density matrix $\rho(r,s,t) = \psi(r,t)\overline{\psi(s,t)}$, where r and s are elements of the Bravais lattice L (see Sect. 1.1). Next, we define similar as above the function

$$u(x,y,t) = \rho\left(x+\frac{y}{2}, x-\frac{y}{2}, t\right)$$

for $x \in \frac{1}{2}L$ and for all y which can be represented as a difference of two points in L. We introduce artificial grid points such that $(x,y) \in \frac{1}{2}L \times L$ and set $u(x,y,t) = 0$ on all artificial grid points. The Wigner function is then the Fourier transform of u,

$$w(x,k,t) = \sum_{y \in L} u(x,y,t)e^{-iy\cdot k}, \quad x \in \frac{1}{2}L, \ k \in B.$$

It is shown by Arnold et al. [11] (also see [2, Sect. 1.4]) that the Wigner function satisfies the dimensionless equation

$$\partial_t w + \frac{i}{\alpha}\left(\beta E\left(k+\frac{\alpha}{2i}\nabla_x\right) - \beta E\left(k-\frac{\alpha}{2i}\nabla_x\right)\right.$$
$$\left. + \gamma V\left(x+\frac{\alpha}{2i}\nabla_k\right) - \gamma V\left(x-\frac{\alpha}{2i}\nabla_k\right)\right)w = 0,$$

where α, β, and γ are dimensionless parameters. Typically, β and γ are of order one, whereas α, which is defined as the ratio of a characteristic wave vector and a typical device length, is much smaller than one. We have employed the notations

$$E\left(k\pm\frac{\alpha}{2i}\nabla_x\right)w(x,k,t) = \frac{1}{2\text{meas}(B)}\int_{2B}\sum_{x'\in L/2} E\left(k\pm\frac{\alpha y}{2}\right)w(x',k,t)e^{i\alpha y\cdot(x-x')}\,dy,$$

$$V\left(x\pm\frac{\alpha}{2i}\nabla_k\right)w(x,k,t) = \frac{1}{\text{meas}(B)}\int_B\sum_{y\in L} V\left(x\pm\frac{\alpha y}{2},t\right)w(x,k',t)e^{i\alpha y\cdot(k-k')}\,dk'.$$

In order to obtain a numerically more treatable equation, one might perform *partially* the limit $\alpha \to 0$. More precisely, we let $\alpha \to 0$ in the lattice $L = \alpha L_0$, where L_0 has a lattice spacing of order one. Then the lattice becomes finer and the discretely defined Wigner function is expected to converge formally to a continuous function defined on $\mathbb{R}^3 \times B$. The limit $\alpha \to 0$ in the band operator

$$\frac{i\beta}{\alpha}\left(E\left(k+\frac{\alpha}{2i}\nabla_x\right) - E\left(k-\frac{\alpha}{2i}\nabla_x\right)\right)$$

formally gives $\beta\nabla_k E(k)\cdot\nabla_x$. The same limit in the potential operator

$$\theta[V] = \frac{i\gamma}{\alpha}\left(V\left(x+\frac{\alpha}{2i}\nabla_k\right) - V\left(x-\frac{\alpha}{2i}\nabla_k\right)\right)$$

would lead to $\gamma\nabla_x V\cdot\nabla_k$. As the potential operator for $\alpha > 0$ accounts for quantum effects, we wish to retain this term and perform the limit $\alpha \to 0$ only in the lattice and the band operator. This leads to the *semi-classical quantum Liouville equation for single states*:

$$\partial_t w + \beta\nabla_k E(k)\cdot\nabla_x w + \theta[V]w = 0, \quad x \in \mathbb{R}^3, \ k \in B, \ t > 0.$$

The complete limit $\alpha \to 0$, leading to the semi-classical Liouville equation, was mathematically analyzed by Steinrück [12]. The Wigner equation in a crystal, taking into account several energy bands, was discussed by Ringhofer in [13].

11.2 The Quantum Vlasov Equation

The quantum Liouville equation has the same disadvantage as its classical analogue, namely that the equation needs to be solved in a very high dimensional electron-ensemble phase space which makes its numerical solution almost unfeasible. In this section we derive the quantum analogue of the classical Vlasov equation, the quantum Vlasov equation, which acts on the six-dimensional phase space. We proceed similarly as in [2, Sect. 1.5].

Consider an ensemble of M electrons with mass m moving in a vacuum and influenced by a (real-valued) potential $V(x,t)$. The motion of the particle ensemble is described by the density matrix as a solution of the Liouville–von Neumann equation (11.1). We impose the following assumptions:

1. The potential can be decomposed into a sum of external potentials acting on one particle and of two-particle interaction potentials:

$$V(x_1,\ldots,x_M,t) = \sum_{j=1}^{M} V_{\text{ext}}(x_j,t) + \frac{1}{2}\sum_{j,\ell=1}^{M} V_{\text{int}}(x_j,x_\ell), \qquad (11.8)$$

where the interaction potential V_{int} is symmetric, i.e., $V_{\text{int}}(x_j,x_\ell) = V_{\text{int}}(x_\ell,x_j)$ for all $j,\ell = 1,\ldots,M$.

2. The limit $V_0 = \lim_{M\to\infty} MV_{\text{int}}$ exists, i.e., the interaction potential is of order $1/M$.

3. The electrons of the ensemble are initially indistinguishable in the sense of

$$\rho(r_1,\ldots,r_M,s_1,\ldots,s_M,0) = \rho(r_{\pi(1)},\ldots,r_{\pi(M)},s_{\pi(1)},\ldots,s_{\pi(M)},0) \qquad (11.9)$$

for all permutations π of $\{1,\ldots,M\}$ and all $r_j, s_j \in \mathbb{R}^3$.

4. The initial subensemble density matrices

$$\rho_I^{(a)}(r^{(a)},s^{(a)}) = \int_{\mathbb{R}^{3(M-a)}} \rho_I(r^{(a)},u_{a+1},\ldots,u_M,s^{(a)},u_{a+1},\ldots,u_M)\,du_{a+1}\cdots du_M,$$

where $r^{(a)} = (r_1,\ldots,r_a)$ and $s^{(a)} = (s_1,\ldots,s_a)$, can be factorized for all $1 \le a \le M-1$:

$$\rho_I^{(a)}(r^{(a)},s^{(a)}) = \prod_{j=1}^{a} R_I(r_j,s_j),$$

where R_I is a given function.

We discuss these hypotheses. The factor $\frac{1}{2}$ in (11.8) is necessary since each electron–electron pair in the sum of two-particle interactions is counted twice. The symmetry of the interaction potentials implies that

$$V(x_1,\ldots,x_M,t) = V(x_{\pi(1)},\ldots,x_{\pi(M)},t) \quad \text{for all } t \geq 0$$

and for all permutations π. It can be shown that this property and (11.9) have the consequence that

$$\rho(r_1,\ldots,r_M,s_1,\ldots,s_M,t) = \rho(r_{\pi(1)},\ldots,r_{\pi(M)},s_{\pi(1)},\ldots,s_{\pi(M)},t) \qquad (11.10)$$

holds for all $t > 0$. Physically, this means that the electrons are indistinguishable for all times.

We wish to derive an evolution equation for the subensemble density matrix

$$\rho^{(a)}(r^{(a)},s^{(a)},t) = \int_{\mathbb{R}^{3(M-a)}} \rho(r^{(a)},u_{a+1},\ldots,u_M,s^{(a)},u_{a+1},\ldots,u_M,t)\,du_{a+1}\cdots du_M.$$

Clearly, in view of the indistinguishable property (11.10), the subensemble density matrices satisfy

$$\rho^{(a)}(r_1,\ldots,r_a,s_1,\ldots,s_a,t) = \rho^{(a)}(r_{\pi(1)},\ldots,r_{\pi(a)},s_{\pi(1)},\ldots,s_{\pi(a)},t) \qquad (11.11)$$

for all permutations π of $\{1,\ldots,a\}$ and all r_j, $s_j \in \mathbb{R}^3$, $t \geq 0$.

We recall that the evolution of the complete electron ensemble is governed by the Liouville–von Neumann equation (11.1), rewritten as

$$i\hbar\partial_t\rho = -\frac{\hbar^2}{2m}\sum_{j=1}^{M}(\Delta_{r_j} - \Delta_{s_j})\rho - q\sum_{j=1}^{M}(V_{\text{ext}}(r_j,t) - V_{\text{ext}}(s_j,t))\rho$$

$$-\frac{q}{2}\sum_{j,\ell=1}^{M}(V_{\text{int}}(r_j,r_\ell) - V_{\text{int}}(s_j,s_\ell))\rho. \qquad (11.12)$$

We set $u_j = s_j = r_j$ for $j = a+1,\ldots,M$ in the above equation, integrate over $(u_{a+1},\ldots,u_M) \in \mathbb{R}^{3(M-a)}$, and use the property (11.11) to obtain, after an analogous computation as in Sect. 3.2, the quantum equivalent of the BBGKY hierarchy,

$$i\hbar\partial_t\rho^{(a)} = -\frac{\hbar^2}{2m}\sum_{j=1}^{a}(\Delta_{r_j} - \Delta_{s_j})\rho^{(a)} - q\sum_{j=1}^{a}(V_{\text{ext}}(r_j,t) - V_{\text{ext}}(s_j,t))\rho^{(a)}$$

$$- q(M-a)\sum_{j=1}^{a}\int_{\mathbb{R}^3}(V_{\text{int}}(r_j,u_*) - V_{\text{int}}(s_j,u_*))\rho_*^{(a+1)}\,du_*$$

for all $1 \leq a \leq M-1$, where we have set

$$\rho_*^{(a+1)} = \rho^{(a+1)}(r^{(a)},u_*,s^{(a)},u_*,t).$$

As in the classical case, the quantum BBGKY hierarchy does not simplify the quantum Liouville equation. A simplification is obtained in the limit $M \to \infty$. Since, by assumption, MV_{int} converges to V_0 as $M \to \infty$, and assuming that the density matrices and their derivatives converge pointwise to some limit functions, the quantum BBGKY hierarchy becomes in the limit $M \to \infty$

$$i\hbar \partial_t \rho^{(a)} = -\frac{\hbar^2}{2m} \sum_{j=1}^{a} (\Delta_{r_j} - \Delta_{s_j}) \rho^{(a)} - q \sum_{j=1}^{a} (V_{\text{ext}}(r_j,t) - V_{\text{ext}}(s_j,t)) \rho^{(a)}$$

$$- q \sum_{j=1}^{a} \int_{\mathbb{R}^3} (V_0(r_j,u_*) - V_0(s_j,u_*)) \rho_*^{(a+1)} \, du_*. \tag{11.13}$$

We claim that a one-particle density matrix contains all the dynamics of the many-particle problem given by (11.13).

Theorem 11.3 (Quantum Vlasov equation). *Let the hypotheses on page 238 hold and let W be a solution of the* quantum Vlasov equation

$$\partial_t W + \frac{p}{m} \cdot \nabla_x W + q\theta[V_{\text{eff}}]W = 0, \quad x,p \in \mathbb{R}^3, \ t > 0, \tag{11.14}$$

$$W(x,p,0) = W_I(x,p), \quad x,p \in \mathbb{R}^3,$$

where the pseudo-differential operator $\theta[V_{\text{eff}}]$ is defined as in (11.5) with $M = 1$, the effective potential V_{eff} is given by

$$V_{\text{eff}}(x,t) = V_{\text{ext}}(x,t) + \int_{\mathbb{R}^3} n(z,t)V_0(x,z) \, dz, \tag{11.15}$$

the electron density is

$$n(x,t) = \frac{2}{(2\pi\hbar)^3} \int_{\mathbb{R}^3} W(x,p,t) \, dp, \tag{11.16}$$

and the initial datum equals

$$W_I(x,p) = \int_{\mathbb{R}^3} R_I \left(x + \frac{y}{2}, x - \frac{y}{2} \right) e^{-iy \cdot p/\hbar} \, dy, \quad x,p \in \mathbb{R}^3.$$

We define the single-state density matrix R as the inverse Fourier transform of the Wigner function,

$$W(x,p,t) = \int_{\mathbb{R}^3} R \left(x + \frac{y}{2}, x - \frac{y}{2}, t \right) e^{-iy \cdot p/\hbar} \, dy, \quad x,p \in \mathbb{R}^3, \ t > 0.$$

Then the density matrix

$$\rho^{(a)}(r^{(a)}, s^{(a)}, t) = \prod_{j=1}^{a} R(r_j, s_j, t) \tag{11.17}$$

is a solution of the limit BBGKY Liouville–von Neumann equation (11.13) with initial datum $\rho^{(a)}(\cdot, \cdot, 0) = \rho_I^{(a)}$ in $\mathbb{R}^{3a} \times \mathbb{R}^{3a}$.

The expression (11.17) is also called a *Hartree ansatz*. As the effective potential depends on the function W through (11.16), the quantum Vlasov equation is a *nonlinear* pseudo-differential equation.

Proof. Similar to the classical case, it can be seen that (11.13) is satisfied by the ansatz (11.17) if R solves the equation

$$i\hbar\partial_t R = -\frac{\hbar^2}{2m}(\Delta_r - \Delta_s)R - q\left(V_{\text{eff}}(r,t) - V_{\text{eff}}(s,t)\right)R, \quad r,s \in \mathbb{R}^3,\ t > 0, \quad (11.18)$$

where the effective potential V_{eff} is defined in (11.15) and (11.16). The kinetic formulation of (11.18) is derived as in Sect. 11.1. $\square$

In contrast to the classical Vlasov equation, the quantum Vlasov equation does not preserve the nonnegativity of the solution; see the discussion in Sect. 11.1. However, if the initial single-particle density matrix $R(r,s,0)$ is positive semi-definite, the electron density n, defined in (11.16), remains nonnegative for all times.

Remark 11.4 (Weak-coupling and low-density limits). A related limit in the M-particle Schrödinger equation or the corresponding BBGKY hierarchy was considered in [14–16] (also see the references in these works). The scaling, however, is different to the scaling used above. More precisely, first the hyperbolic space–time scaling $x \mapsto \varepsilon x$, $t \mapsto \varepsilon t$ was introduced. For the so-called *weak-coupling limit*, the potential is assumed to be scaled by $V_{\text{int}} \mapsto \sqrt{\varepsilon} V_{\text{int}}$, and the number of particles scales like $M = \varepsilon^{-3}$. The weak-coupling limit $\varepsilon \to 0$ is characterized by the fact that the potential interaction is weak, namely of order $\sqrt{\varepsilon}$, and the particle density is of order one. Then the number of collisions per time unit is ε^{-1}. Another scaling is the so-called *low-density limit*. In this case, the potential remains unscaled but $M = \varepsilon^{-2}$. Then the particle density is of order ε, and the particles collide once per time unit. In the classical context, the latter limit corresponds to the *Boltzmann-Grad limit* [17]. In both cases, by applying a kinetic approach [14], the M-particle Schrödinger equation reduces in the limit $\varepsilon \to 0$ to the classical Boltzmann equation in which the cross-section of the collision operator is the only quantum factor. $\square$

Similar to the quantum Liouville equation, the solution of the quantum Vlasov equation converges (at least formally) as "$\hbar \to 0$" to a solution of the classical Vlasov equation

$$\partial_t W + \frac{p}{m} \cdot \nabla_x W + q\nabla_x V_{\text{eff}} \cdot \nabla_p W = 0.$$

The limit "$\hbar \to 0$" has to be understood in the sense explained in Sect. 11.1.

As in Sect. 3.2, a usual choice for the two-particle interaction potential is the Coulomb potential

$$V_0(x,y) = -\frac{q}{4\pi\varepsilon_s}\frac{1}{|x-y|}, \quad x,y \in \mathbb{R}^3,\ x \neq y,$$

where ε_s denotes the permittivity of the semiconductor material. In Sect. 3.2 it is shown that the effective potential

$$V_{\text{eff}}(x,t) = V_{\text{ext}}(x,t) - \frac{q}{4\pi\varepsilon_s} \int_{\mathbb{R}^3} \frac{n(z,t)}{|z-x|} \, dz \qquad (11.19)$$

solves the Poisson equation

$$\varepsilon_s \Delta V_{\text{eff}} = q(n - C(x)),$$

where $C(x) = -(\varepsilon_s/q)V_{\text{ext}}(x)$ is the doping concentration if V_{ext} is generated by ions of charge $+q$ in the semiconductor. The initial-value problem (11.14) and (11.19) (together with (11.16)) is called the *quantum Vlasov–Poisson system*.

11.3 Wigner–Boltzmann Equations

In the previous sections, we have considered only ballistic and hence reversible quantum transport of electrons. However, if the characteristic device length is much larger than the mean free path of the electrons, scattering phenomena between electrons and phonons or among the electrons have to be taken into account. Inspired from classical kinetic theory, collisions may be modeled by an appropriate collision operator, which is added to the right-hand side of the quantum Liouville or Vlasov equation. This gives the *Wigner–Boltzmann equation*

$$\partial_t w + v(p) \cdot \nabla_x w + q\theta[V_{\text{eff}}]w = Q(w), \quad x, p \in \mathbb{R}^3, \ t > 0,$$

which we write here for the effective single-state potential V_{eff} derived in Sect. 11.2. The velocity may depend on the pseudo-wave vector, $v(p) = v(\hbar k)$. In the parabolic band case, $v(p) = \hbar k/m$. This model is an example of an open quantum system since the electron ensemble interacts with some environment, such as an external phonon bath, through the collision operator $Q(w)$.

There are many approaches including scattering in quantum models. One method is to take into account the phase-breaking time of the electrons in the system by adding an imaginary term to the Hamiltonian, which can be translated as a particular collision term in the Wigner equation [18, 19]. This approach has the disadvantages that it is reasonable for a system which is close to equilibrium but maybe not out of equilibrium, and that the imaginary term is constant throughout the device and therefore fails to model the inhomogeneous particle density in the out-of-equilibrium system. Dissipation may be also included by the so-called Büttiker probes [20, 21]. Compared to the phase-breaking term, this approach is current conserving, but one has to take into account the space inhomogeneity of the density, and a fitting parameter has to be used. In device simulations, often recursive techniques are employed, for the Schrödinger wave function [22], for the Green's function [23], or including a self-energy term [24].

In the following, we present some spatially local collision operators which are employed in numerical simulations of the Wigner equation.

Wigner–Fokker–Planck model. The first model is the *quantum Fokker–Planck collision operator*

$$Q(w) = D_{pp}\Delta_p w + 2\gamma \mathrm{div}_p(pw) + D_{qq}\Delta_x w + 2D_{pq}\mathrm{div}_x(\nabla_p w), \qquad (11.20)$$

where $\gamma > 0$ is a friction parameter and the nonnegative coefficients D_{pp}, D_{pq}, and D_{qq} constitute the phase-space diffusion matrix of the system. The first term models classical diffusion and the last two terms quantum diffusion. The corresponding *Wigner–Fokker–Planck equation* governs the dynamical evolution of an electron ensemble interacting dissipatively with an idealized heat bath consisting of an ensemble of harmonic oscillators and modeling the semiconductor lattice.

Without quantum diffusion, the collision operator

$$Q(w) = D_{pp}\Delta_p w + 2\gamma \mathrm{div}_p(pw) \qquad (11.21)$$

was derived by Caldeira and Leggett [25] and Diósi [26]. The Wigner equation with the Caldeira–Leggett operator is also known under the name of *quantum Brownian motion* or *quantum Langevin equation* and it received a large interest in the context of interaction between light and matter (see, for instance, [27]).

The Caldeira–Leggett scattering term does not satisfy the so-called *Lindblad condition*

$$D_{pp}D_{qq} - D_{pq}^2 \geq \frac{\gamma^2}{4},$$

which is a generic condition for quantum systems to preserve complete positivity of the density matrix along the evolution. Such a property has to be satisfied for a true quantum evolution. Thus the Wigner–Caldeira–Leggett equation is quantum mechanically not correct. The reason for this shortcoming comes from the inconsistency that the equation contains the temperature T, through its coefficients, but the $1/T \to 0$ limit was performed in [25] along the derivation of the model. In [28], the approach was improved by deriving the Fokker–Planck operator (11.20) with finite temperature.

Another scattering operator related to (11.20) was derived in [29] from quantized one-dimensional linearly damped harmonic oscillators:

$$Q(w) = \hbar\gamma\Omega \frac{\partial^2 w}{\partial x^2} + 2\gamma \frac{\partial}{\partial p}(pw) + \Omega^2 x \frac{\partial w}{\partial p},$$

where Ω denotes the frequency of the harmonic oscillator. Notice that the term $\Omega^2 x(\partial w/\partial p)$ is not contained in (11.20) but may be interpreted as a drift term in the Wigner equation coming from a quadratic potential.

The existence of solutions of the transient Wigner–Fokker–Planck equation was proved by Arnold, López, and co-workers [30–33]. The stationary case was treated in [34].

Wigner–BGK model. The second model is the relaxation-time approximation or *Bhatnagar–Gross–Krook (BGK) model* [13, 35]

$$Q(w) = \frac{1}{\tau}\left(\frac{n}{n_0}w_0 - w\right),$$

where the particle density and the equilibrium density are given by, respectively,

$$n(x,t) = \frac{2}{(2\pi\hbar)^3}\int_{\mathbb{R}^3} w(x,p,t)\,\mathrm{d}p, \quad n_0(x,t) = \frac{2}{(2\pi\hbar)^3}\int_{\mathbb{R}^3} w_0(x,p,t)\,\mathrm{d}p,$$

and w_0 is the Wigner function of the quantum mechanical thermal equilibrium, defined, for instance in the mixed state, by the thermal equilibrium density matrix

$$\rho_{eq}(r,s) = \sum_j f(E_j)\psi_j(r)\overline{\psi_j(s)}$$

(by means of the Wigner–Weyl transformation), where (ψ_j, E_j) are the eigenfunction–eigenvalue pairs of the quantum Hamiltonian and $f(E_j)$ is the Fermi–Dirac or Maxwell–Boltzmann distribution. The above collision operator expresses the tendency of the system to relax, in the absence of external forces, to the quantum thermal equilibrium since the solution of the Wigner–BGK model with vanishing transport and electric field,

$$\partial_t w = \frac{1}{\tau}\left(\frac{n}{n_0}w_0 - w\right),$$

tends to the equilibrium Wigner function nw_0/n_0. We remark that the relaxation term can be represented in Lindblatt form, such that positivity preservation is guaranteed for this model [36].

The relaxation-time collision operator in the Wigner equation is employed in numerical simulations of resonant tunneling diodes [1, 35]. In fact, virtually all Wigner function-based device simulations were carried out for one-dimensional tunneling diodes. The earliest approaches in the mid-1980s were based on finite-difference schemes [1, 35, 37]. Later, spectral collocation methods were developed as an efficient alternative for the discretization of the nonlocal operator $\theta[V]$ [38]. This approach was combined with an operator-splitting technique for the transport term $v \cdot \nabla_x$ and the pseudo-differential operator $\theta[V]$ by Arnold and Ringhofer [39]. An analysis of a discrete-momentum Wigner model was performed in [40, 41]. With the finite-difference method, scattering was restricted to the relaxation-time approximation and the one-dimensional momentum space. On the other hand, the Monte Carlo method, which was implemented from 2002 on [42, 43], allows for the inclusion of more detailed scattering processes. For more recent results, see [44, 45] and the review [46]. The Monte Carlo method has the potential to make multi-dimensional simulations feasible.

Other collision models. A third model is given by a semi-classical Boltzmann scattering term of the form

$$Q(w) = \int_{\mathbb{R}^3}\left(s(p,p')w(x,p',t) - s(p',p)w(x,p,t)\right)\mathrm{d}p',$$

where $s(p, p')$ is the scattering rate, describing the electron–phonon interactions, for instance. Such Boltzmann-type operators give good numerical results [44], but they are quantum mechanically not correct in the sense of positivity preservation of the density matrix [8].

A fourth scattering model is derived by Degond and Ringhofer [47] under the assumptions that the quantum collision operator conserves a set of moments $\kappa(p)$ (like conservation of mass, momentum, and energy),

$$\int_{\mathbb{R}^3} Q(w)\kappa(p)\,dp = 0 \quad \text{for all functions } w \text{ and all } x \in \mathbb{R}^3,$$

and that it dissipates the quantum entropy, i.e.,

$$\text{Tr}\left(\log(\widehat{\rho})W^{-1}[Q(W(\widehat{\rho}))]\right) \leq 0 \quad \text{for all density-matrix operators } \widehat{\rho},$$

where Tr is the trace of an operator, $W(\widehat{\rho})$ denotes the Wigner transform of $\widehat{\rho}$ and W^{-1} its inverse, and log is the logarithm of a self-adjoint operator defined in the standard way by its spectral decomposition. Then the scattering operator can be written as follows:

$$Q(w) = \int_{\mathbb{R}^3}\int_{\mathbb{R}^3}\int_{\mathbb{R}^3} \sigma(x, p, p_1, p', p_1')(g'g_1' - gg_1)\,dp'\,dp_1'\,dp_1,$$

where $g = \exp(W[\log(W^{-1}(w))])$ and σ denotes the scattering cross-section. The nonlocal nature of the quantum collisions is reflected by the spatial nonlocality due to the operator logarithm although the conservation properties are (spatially) local. In the classical limit, the quantum collision operator reduces to the usual Boltzmann operator, since we expect that $\exp(W[\log(W^{-1}(w))])$ tends (at least formally) to the function w.

The kinetic quantum models considered in this chapter and the semi-classical kinetic models studied in Chaps. 3 and 4 are summarized in Fig. 11.1.

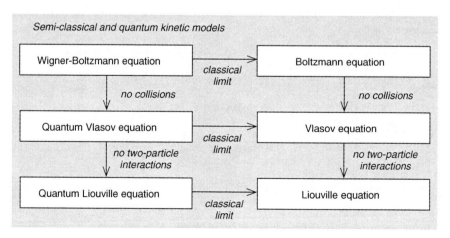

Fig. 11.1 Semi-classical and quantum-kinetic models and their relations

References

1. W. Frensley. Wigner-function model of a resonant-tunneling semiconductor device. *Phys. Rev. B* 36 (1987), 1570–1580.
2. P. Markowich, C. Ringhofer, and C. Schmeiser. *Semiconductor Equations*. Springer, Vienna, 1990.
3. E. Wigner. On the quantum correction for the thermodynamic equilibrium. *Phys. Rev.* 40 (1932), 749–759.
4. M. Taylor. *Pseudodifferential Operators*. Princeton University Press, Princeton, 1981.
5. P.-L. Lions and T. Paul. Sur les mesures de Wigner. *Rev. Mat. Iberoamer.* 9 (1993), 553–618.
6. P. Markowich. On the equivalence of the Schrödinger and the quantum Liouville equations. *Math. Meth. Appl. Sci.* 11 (1989), 459–469.
7. P. Markowich and C. Ringhofer. An analysis of the quantum Liouville equation. *Z. Angew. Math. Mech.* 69 (1989), 121–127.
8. A. Arnold. Mathematical properties of quantum evolution equations. In: G. Allaire, A. Arnold, P. Degond, and T. Hou (eds.), *Quantum Transport – Modelling, Analysis and Asymptotics*, Lecture Notes Math. 1946, 45–110. Springer, Berlin, 2008.
9. R. Hudson. When is the Wigner quasiprobability nonnegative? *Reports Math. Phys.* 6 (1974), 249–252.
10. P. Markowich, N. Mauser, and F. Poupaud. A Wigner-function approach to (semi) classical limits: electrons in a periodic potential. *J. Math. Phys.* 35 (1994), 1066–1094.
11. A. Arnold, P. Degond, P. Markowich, and H. Steinrück. The Wigner-Poisson problem in a crystal. *Appl. Math. Letters* 2 (1989), 187–191.
12. H. Steinrück. The Wigner-Poisson problem in a crystal: existence, uniqueness, semiclassical limit in the one-dimensional case. *Z. Angew. Math. Mech.* 72 (1992), 93–102.
13. C. Ringhofer. Computational methods for semiclassical and quantum transport in semiconductor devices. *Acta Numerica* (1997), 485–521.
14. D. Benedetto, F. Castella, R. Esposito, and M. Pulvirenti. A short review on the derivation of the nonlinear quantum Boltzmann equations. *Commun. Math. Sci.* 5 (2007), 55–71.
15. D. Benedetto, F. Castella, R. Esposito, and M. Pulvirenti. From the *N*-body Schrödinger equation to the quantum Boltzmann equation: a term-by-term convergence result in the weak coupling regime. *Commun. Math. Phys.* 277 (2008), 1–44.
16. L. Erdös, M. Salmhofer, and H.-T. Yau. On the quantum Boltzmann equation. *J. Stat. Phys.* 116 (2004), 367–380.
17. C. Cercignani, R. Illner, and M. Pulvirenti. *The Mathematical Theory of Dilute Gases*. Springer, New York, 1994.
18. R. Akis, J. Bird, and D. Ferry. The effects of inelastic scattering in open quantum dots: reduction of conductance fluctuations and disruption of wave-function 'scarring'. *J. Phys.: Condens. Matter* 8 (1996), L667-L674.
19. G. Neofotistos, R. Lake, and S. Datta. Inelastic-scattering effects on single-barrier tunneling. *Phys. Rev. B* 43 (1991), 2442–2445.
20. M. Büttiker. Four-terminal phase-coherent conductance. *Phys. Rev. Letters* 57 (1986), 1761–1764.
21. R. Venugopal, M. Paulsson, S. Goasquen, S. Datta, and M. Lundstrom. A simple quantum mechanical treatment of scattering in nanoscale transistors. *J. Appl. Phys.* 93 (2003), 5613–5625.
22. M. Gilbert and D. Ferry. Efficient quantum three-dimensional modeling of fully depleted ballistic silicon-on-insulator metal-oxide-semiconductor field-effect-transistors. *J. Appl. Phys.* 95 (2004), 7954–7960.
23. D. Fisher and P. Lee. Relation between conductivity and transmission matrix. *Phys. Rev. B* 23 (1981), 6851–6854.
24. M. Gilbert, R. Akis, and D. Ferry. Phonon-assisted ballistic to diffusive crossover in silicon nanowire transistors. *J. Appl. Phys.* 98 (2005), 094303.

25. A. Caldeira and A. Leggett. Path integral approach to quantum Brownian motion. *Physica A* 121 (1983), 587–616.
26. L. Diósi. On high-temperature Markovian equation for quantum Brownian motion. *Europhys. Letters* 22 (1993), 1–3.
27. C. Cohern-Tannoudji, J. Dupont-Roc, and G. Grynberg. *Processus d'interaction entre photons et atomes.* Savoirs actuels, Intereditions/Editions du CNRS, 1988.
28. F. Castella, L. Erdös, F. Frommlet, and P. Markowich. Fokker–Planck equations as scaling limits of reversible quantum systems. *J. Stat. Phys.* 100 (2000), 543–601.
29. H. Dekker. Quantization of the linearly damped harmonic oscillator. *Phys. Rev. A* 16 (1977), 2126–2134.
30. A. Arnold, E. Dhamo, and C. Manzini. The Wigner-Poisson-Fokker–Planck system: global-in-time solutions and dispersive effects. *Ann. Inst. H. Poincaré, Anal. non linéaire* 24 (2007), 645–676.
31. A. Arnold, J.-L. López, P. Markowich, and J. Soler. An analysis of quantum Fokker–Planck models: a Wigner function approach. *Rev. Mat. Iberoamer.* 20 (2004), 771–814.
32. A. Arnold and C. Sparber. Conservative quantum dynamical semigroups for mean-field quantum diffusion models. *Commun. Math. Phys.* 251 (2004), 179–207.
33. J. A. Cañizo, J.-L. López, and J. Nieto. Global L^1 theory and regularity for the 3D nonlinear Wigner-Poisson-Fokker-Planck system. *J. Diff. Eqs.* 198 (2004), 356–373.
34. I. Gamba, M. Gualdani, and C. Sparber. A note on the time-decay of solutions for the linearized Wigner-Poisson system. To appear in *Kinetic and Related Models*, 2009.
35. N. Kluksdahl, A. Kriman, D. Ferry, and C. Ringhofer. Self-consistent study of the resonant tunneling diode. *Phys. Rev. B* 39 (1989), 7720–7735.
36. A. Arnold. The relaxation-time von Neumann-Poisson equation. In: O. Mahrenholtz and R. Mennicken (eds.), *Proceedings of ICIAM 95*, Hamburg. *Z. Angew. Math. Mech.* 76 Supp. 2 (1996), 293–296.
37. U. Ravaioli, M. Osman, W. Pötz, N. Kluksdahl, and D. Ferry. Investigation of ballistic transport through resonant-tunnelling quantum wells using Wigner function approach. *Physica B* 134 (1985), 36–40.
38. C. Ringhofer. A spectral method for the numerical simulation of quantum tunneling phenomena. *SIAM J. Numer. Anal.* 27 (1990), 32–50.
39. A. Arnold and C. Ringhofer. Operator splitting methods applied to spectral discretizations of quantum transport equations. *SIAM J. Numer. Anal.* 32 (1995), 1876–1894.
40. A. Arnold, H. Lange, and P. Zweifel. A discrete-velocity, stationary Wigner equation. *J. Math. Phys.* 41 (2000), 7167–7180.
41. T. Goudon. Analysis of a semi-discrete version of the Wigner equation. *SIAM J. Numer. Anal.* 40 (2002), 2007–2025.
42. M. Nedjalkov, R. Kosik, H. Kosina, and S. Selberherr. Wigner transport through tunneling structures – scattering interpretation of the potential operator. In: *International Conference on Simulation of Semiconductor Processes and Devices SISPAD* (2002), 187–190.
43. L. Shifren and D. Ferry. A Wigner function based ensemble Monte Carlo approach for accurate incorporation of quantum effects in device simulation. *J. Comp. Electr.* 1 (2002), 55–58.
44. H. Kosina and M. Nedjalkov. Wigner function-based device modeling. In: M. Rieth and W. Schommers (eds), *Handbook of Theoretical and Computational Nanotechnology* 10, 731–763. American Scientific Publishers, Los Angeles, 2006.
45. M. Nedjalkov, D. Vasileska, D. Ferry, C. Jacoboni, C. Ringhofer, I. Dimov, and V. Palanovski. Wigner transport models of the electron–phonon kinetics in quantum wires. *Phys. Rev. B* 74 (2006), 035311.
46. V. Sverdlov, E. Ungersboeck, H. Kosina, and S. Selberherr. Current transport models for nanoscale semiconductor devices. *Materials Sci. Engin. R* 58 (2008), 228–270.
47. P. Degond and C. Ringhofer. Binary quantum collision operators conserving mass momentum and energy. *C. R. Acad. Sci. Paris, Sér. I* 336 (2003), 785–790.

Part V
Macroscopic Quantum Models

In the following chapters, we derive macroscopic quantum models from a Wigner–Boltzmann equation. In analogy to the classical situation, we define quantum equilibrium states and employ moment methods and Chapman–Enskog expansion techniques. As a result, we obtain quantum analogues of the semi-classical model hierarchy, consisting of (quantum) drift-diffusion, energy-transport, and hydrodynamic models.

Chapter 12
Quantum Drift-Diffusion Equations

The main aim of this and the following chapters is to derive macroscopic quantum models from the parabolic band Wigner–Boltzmann equation (see Sect. 11.3), written in the crystal momentum $p = \hbar k$,

$$\partial_t w + \frac{p}{m^*} \cdot \nabla_x w + q\theta[V]w = Q(w), \quad x, p \in \mathbb{R}^3, \, t > 0, \quad w(\cdot,\cdot,0) = w_I, \quad (12.1)$$

where V is the effective electric potential of the electron ensemble. Here, $\theta[V]$ denotes the pseudo-differential operator defined by

$$(\theta[V]w)(x,p,t) = \frac{1}{(2\pi)^3} \int_{\mathbb{R}^3 \times \mathbb{R}^3} (\delta V)(x,\eta,t) w(x,p',t) e^{i(p-p')\cdot\eta} \, dp' \, d\eta,$$

$$(\delta V)(x,\eta,t) = \frac{i}{\hbar} \left(V\left(x + \frac{\hbar\eta}{2},t\right) - V\left(x - \frac{\hbar\eta}{2},t\right) \right),$$

and $Q(w)$ is a collision operator. Semi-classical macroscopic models are derived in Chaps. 5–9 by using a moment method and by specifying a closure condition. The idea is to mimic this procedure in the quantum case.

Employing the same scaling as in Sect. 11.1, the dimensionless Wigner–Boltzmann equation becomes

$$\partial_t w + p \cdot \nabla_x w + \theta[V]w = Q(w), \quad x, p \in \mathbb{R}^3, \, t > 0, \quad (12.2)$$

with the scaled pseudo-differential operator

$$(\theta[V]w)(x,p,t) = \frac{1}{(2\pi)^3} \int_{\mathbb{R}^3 \times \mathbb{R}^3} (\delta V)(x,\eta,t) w(x,p',t) e^{i(p-p')\cdot\eta} \, dp' \, d\eta, \quad (12.3)$$

$$(\delta V)(x,\eta,t) = \frac{i}{\varepsilon} \left(V\left(x + \frac{\varepsilon}{2}\eta,t\right) - V\left(x - \frac{\varepsilon}{2}\eta,t\right) \right),$$

and the scaled Planck constant

Jüngel, A.: *Quantum Drift-Diffusion Equations.* Lect. Notes Phys. **773**, 251–274 (2009)
DOI 10.1007/978-3-540-89526-8_12   © Springer-Verlag Berlin Heidelberg 2009

$$\varepsilon = \frac{\hbar/\tau}{m^*(\lambda/\tau)^2},$$

expressing the ratio of the wave and kinetic energy. We write the scaled Wigner equation as in Chap. 11 in terms of the dimensionless crystal momentum p instead of the scaled pseudo-wave vector k as in Chaps. 4–9. The latter is related to the former by the formula $p = \varepsilon k$.

12.1 Quantum Maxwellians

Inspired from the classical situation, the quantum Maxwellian is defined by that Wigner function which maximizes the quantum entropy subject to the constraint that its moments are given. This idea is due to Degond and Ringhofer [1]. In order to define the quantum Maxwellian, we use the Wigner transform $W(\hat\rho)$ of an operator $\hat\rho$ on $L^2(\mathbb{R}^3)$ with integral kernel ρ (satisfying certain regularity assumptions). Then we can write

$$(\hat\rho\phi)(x) = \int_{\mathbb{R}^3} \rho(x,y)\phi(y)\,dy, \quad \phi \in L^2(\mathbb{R}^3).$$

The *Wigner transform* of $\hat\rho$ is defined as the Fourier transform of the transformed function ρ:

$$W(\hat\rho)(x,p) = \int_{\mathbb{R}^3} \rho\left(x+\frac{\varepsilon}{2}\eta, x-\frac{\varepsilon}{2}\eta\right) e^{-i\eta\cdot p}\,d\eta.$$

Its inverse W^{-1}, also called Weyl quantization, is defined as an operator on $L^2(\mathbb{R}^3)$,

$$(W^{-1}(f)\phi)(x) = \frac{1}{(2\pi\varepsilon)^3} \int_{\mathbb{R}^3\times\mathbb{R}^3} f\left(\frac{x+y}{2}\right) \phi(y) e^{ip\cdot(x-y)/\varepsilon}\,dp\,dy, \quad \phi \in L^2(\mathbb{R}^3).$$

With these definitions, we are able to introduce as in [1] the *quantum exponential* Exp and the *quantum logarithm* Log formally by

$$\operatorname{Exp} f = W(\exp W^{-1}(f)), \quad \operatorname{Log} f = W(\log W^{-1}(f)),$$

where exp and log are the operator exponential and logarithm, respectively, defined by their corresponding spectral decomposition. The quantum exponential and logarithm have the following formal properties.

Lemma 12.1 (Properties of Exp and Log). *The quantum logarithm is the inverse of the quantum exponential. Furthermore,* Exp *and* Log *are formally (Fréchet) differentiable and*

$$\frac{d}{dw}\operatorname{Log} w = \frac{1}{w}, \quad \frac{d}{dw}\operatorname{Exp} w = \operatorname{Exp} w.$$

Proof. The first assertion follows from the formal computation

$$\operatorname{Log}(\operatorname{Exp} w) = W\left(\exp W^{-1}\left(W(\log W^{-1}(w))\right)\right) = W\left(\exp(\log W^{-1}(w))\right) = w.$$

Since the Wigner transform and its inverse are linear operations, the second assertion follows from the properties of the operator exponential and logarithm (see [1,
Lemma 3.3] or [2, Theorem 1] for a more precise argument).   □

Let a quantum mechanical state be described by the Wigner function w solving
the Wigner–Boltzmann equation (12.2). The scaled *quantum entropy* or *von Neumann entropy* of the quantum mechanical state is given by

$$S(w) = -\frac{2}{(2\pi\varepsilon)^3} \int_{\mathbb{R}^3 \times \mathbb{R}^3} w(x,p,\cdot) \left((\text{Log}\, w)(x,p,\cdot) - 1 + \frac{|p|^2}{2} - V(x,\cdot) \right) dx\,dp,$$

(12.4)

where V is the electric potential. Whereas the classical entropy is a function on
the configuration space, the above quantum entropy at given time is a real number,
underlining the nonlocal nature of quantum mechanics. We set

$$\langle g(p) \rangle = \frac{2}{(2\pi\varepsilon)^3} \int_{\mathbb{R}^3} w(x,p,t) g(p)\,dp$$

for functions $g = g(p)$. The above notation is consistent with the notation from Part
III, $\langle f(k) \rangle = \int f(k)\,dk/4\pi^3$, since in scaled variables, $p = \varepsilon k$ holds. Let some weight
functions $\kappa(p) = (\kappa_0(p),\ldots,\kappa_N(p))$ be given.

Lemma 12.2. *Let $w = w(x,p,t)$ be given and let the moments $m = (m_0,\ldots,m_N)$ of
w be defined by*

$$m_j(x,t) = \langle w(x,p,t)\kappa_j(p) \rangle, \quad j = 0,\ldots,N.$$

The formal solution of the constrained maximization problem

$$S(M[w]) = \max \left\{ S(f) : \langle f(x,p,t)\kappa(p) \rangle = m(x,t) \text{ for all } x \in \mathbb{R}^3, t > 0 \right\}, \quad (12.5)$$

if it exists, is given by

$$M[w] = \text{Exp}\left(\tilde{\lambda} \cdot \kappa + V - \frac{|p|^2}{2} \right),$$

where $\tilde{\lambda} = (\tilde{\lambda}_0,\ldots,\tilde{\lambda}_N)$ are some Lagrange multipliers.

We call $M[w]$ the *quantum Maxwellian* of w. If we assume that $w_0(p) = 1$ and
$w_2(p) = \frac{1}{2}|p|^2$, setting $\lambda_0 = \tilde{\lambda}_0 + V$, $\lambda_2 = \tilde{\lambda}_2 - 1$, and $\lambda_j = \tilde{\lambda}_j$ otherwise, we can
write the quantum Maxwellian more compactly as

$$M[w] = \text{Exp}(\lambda \cdot \kappa(p)).$$

Proof. We define for $\tilde{\lambda} = (\tilde{\lambda}_0,\ldots,\tilde{\lambda}_N)$ and $m = (m_0,\ldots,m_N)$ the Lagrange functional

$$F(w,\tilde{\lambda}) = S(w) + \int_{\mathbb{R}^3} \tilde{\lambda}(x) \cdot (m - \langle w(x,p,t)\kappa(p) \rangle)\,dx.$$

Using Lemma 12.1, the derivative of the quantum entropy is given by

$$\left(\frac{d}{dw}S(w)\right)(u) = -\frac{2}{(2\pi\varepsilon)^3}\int_{\mathbb{R}^3\times\mathbb{R}^3}\left(\operatorname{Log}w + \frac{|p|^2}{2} - V\right)u(x,p)\,dx\,dp.$$

Therefore, the necessary condition for an extremal point reads as

$$0 = \left(\frac{d}{dw}F(w^*,\tilde{\lambda}^*)\right)(u)$$

$$= \int_{\mathbb{R}^3\times\mathbb{R}^3}\left(\operatorname{Log}w + \frac{|p|^2}{2} - V - \tilde{\lambda}^*(x)\cdot\kappa(p)\right)u(x,p)\,dx\,dp$$

for all functions $u(x,p)$. This implies that

$$\operatorname{Log}w^* + \frac{|p|^2}{2} - V(x,t) - \tilde{\lambda}^*(x)\cdot\kappa(p) = 0$$

and finally,

$$w^* = \operatorname{Exp}\left(\tilde{\lambda}^*\cdot\kappa + V - \frac{|p|^2}{2}\right),$$

finishing the proof.   □

Example 12.3. We give some examples of quantum Maxwellians which are used in the following sections. We define, for given w, the local particle density n, mean velocity u, and energy density ne by

$$\begin{pmatrix}n\\nu\\ne\end{pmatrix}(x,t) = \frac{2}{(2\pi\varepsilon)^3}\int_{\mathbb{R}^3}w(x,p,t)\begin{pmatrix}1\\p\\\frac{1}{2}|p|^2\end{pmatrix}dp.$$

If only the electron density is prescribed, we obtain the quantum Maxwellian

$$M[w](x,p,t) = \operatorname{Exp}\left(A(x,t) - \frac{|p|^2}{2}\right),$$

where the Lagrange multiplier A is uniquely determined by the zeroth moment of w. This Maxwellian will be employed for the derivation of the quantum drift-diffusion model. In the case of prescribed particle density, velocity, and energy density, we obtain the quantum Maxwellian

$$M[w] = \operatorname{Exp}\left(A(x,t) - \frac{|p - v(x,t)|^2}{2T(x,t)}\right),$$

where A, v, and T are determined by the moments of w. This Maxwellian is taken as the thermal equilibrium state in the quantum hydrodynamic equations. Finally, prescribing zeroth- and second-order moments, the quantum Maxwellian reads as follows:

$$M[w] = \mathrm{Exp}\left(A(x,t) - \frac{|p|^2}{2T(x,t)}\right),$$

which we use for the derivation of the quantum energy-transport equations. $\square$

The quantum Maxwellian is a nonlocal operator. It can be made more explicit when expanding it in terms of the scaled Planck constant ε, which appears in the definition of the Wigner transform. The expansion is done by means of the following lemma, which is adopted from [1].

Lemma 12.4 (Expansion of the quantum exponential). *Let $f(x,p)$ be a smooth function. Then the quantum exponential $\mathrm{Exp}(f)$ can be expanded as*

$$\mathrm{Exp}(f) = \exp(f) - \frac{\varepsilon^2}{8}\exp(f)R_2 + \mathscr{O}\left(\varepsilon^4\right),$$

where

$$R_2 = \sum_{j,\ell=1}^{3}\left(\frac{\partial^2 f}{\partial x_j x_\ell}\frac{\partial^2 f}{\partial p_j \partial p_\ell} - \frac{\partial^2 f}{\partial x_j \partial p_\ell}\frac{\partial^2 f}{\partial p_j \partial x_\ell} + \frac{1}{3}\frac{\partial^2 f}{\partial x_j \partial x_\ell}\frac{\partial f}{\partial p_j}\frac{\partial f}{\partial p_\ell}\right.$$
$$\left. - \frac{2}{3}\frac{\partial^2 f}{\partial x_j \partial p_\ell}\frac{\partial f}{\partial p_j}\frac{\partial f}{\partial x_\ell} + \frac{1}{3}\frac{\partial^2 f}{\partial p_j \partial p_\ell}\frac{\partial f}{\partial x_j}\frac{\partial f}{\partial x_\ell}\right).$$

Depending on the specific structure of the function f, the above expansion can be made more explicit.

Proposition 12.5. *The correction term R_2 of Lemma 12.4 can be written for $f(x,p)$ $= A(x) - |p - v(x)|^2/2T(x)$ as follows:*

$$R_2 = T^{-1}\left(X^0 + \sum_{j=1}^{3}X_j^1 s_j + \sum_{j,\ell=1}^{3}X_{j\ell}^2 s_j s_\ell + \sum_{j,\ell,m=1}^{3}X_{j\ell m}^3 s_j s_\ell s_m\right.$$
$$\left. + Y^0|s|^2 + \sum_{j=1}^{3}Y_j^1|s|^2 s_j + \sum_{j,\ell=1}^{3}Y_{j\ell}^2|s|^2 s_j s_\ell + Z^0|s|^4\right),$$

where $s = (p - v)/\sqrt{T}$ and the coefficients are defined by

$$X^0 = -\Delta A - \frac{1}{3}|\nabla A|^2 + \frac{1}{2T}\mathrm{Tr}(R^\top R),$$

$$X_j^1 = \frac{2}{\sqrt{T}}\sum_{m=1}^{3}\frac{\partial}{\partial x_m}\left(\frac{1}{3}A - \log T\right)R_{mj} - \frac{\Delta w_j}{\sqrt{T}},$$

$$X_{j\ell}^2 = \frac{1}{3}\frac{\partial^2 A}{\partial x_j \partial x_\ell} + \frac{2}{3}\frac{\partial(\log A)}{\partial x_j}\frac{\partial A}{\partial x_\ell} - \frac{\partial(\log T)}{\partial x_j}\frac{\partial(\log T)}{\partial x_\ell} - \frac{1}{3T}(R^\top R)_{j\ell},$$

$$X_{j\ell m}^3 = \frac{1}{3\sqrt{T}}\frac{\partial^2 w_m}{\partial x_j \partial x_\ell},$$

$$Y^0 = \nabla\left(\frac{1}{2}\log T - \frac{1}{3}A\right)\cdot\nabla(\log T) - \frac{1}{2}\Delta(\log T),$$

$$Y_j^1 = \frac{1}{3\sqrt{T}}\sum_{m=1}^{3}\frac{\partial(\log T)}{\partial x_m}R_{mj},$$

$$Y_{j\ell}^2 = \frac{1}{6}\left(\frac{\partial^2(\log T)}{\partial x_j\partial x_\ell} + \frac{\partial(\log T)}{\partial x_j}\frac{\partial(\log T)}{\partial x_\ell}\right),$$

$$Z^0 = -\frac{1}{12}|\nabla(\log T)|^2,$$

$R = (R_{j\ell})$ and $R_{j\ell} = \partial f_j/\partial x_\ell - \partial f_\ell/\partial x_j$.

The proposition follows after computing the derivatives of f with respect to x_j and p_ℓ and employing Lemma 12.4 (see [3] for a complete proof). If $v = 0$ or $T = 1$, the expansion of the quantum Maxwellian simplifies.

Corollary 12.6. *The following formal expansions hold:*

$$\text{Exp}\left(A - \frac{|p|^2}{2T}\right) = \exp\left(A - \frac{|p|^2}{2T}\right)\left[1 + \frac{\varepsilon^2}{8T}\left(\Delta A + \frac{1}{3}|\nabla A|^2 - \frac{1}{3}p^\top(\nabla\otimes\nabla)Ap\right.\right.$$

$$+ \frac{|p|^2}{2}\Delta\beta + T(p\cdot\nabla\beta)^2 + \frac{|p|^2}{3T}p^\top(\nabla\otimes\nabla)\beta p$$

$$+ \frac{2}{3}(p\cdot\nabla\beta)(p\cdot\nabla A) - \frac{|p|^2}{3}(p\cdot\nabla\beta)^2 - \frac{|p|^2}{3}\nabla A\cdot\nabla\beta$$

$$\left.\left.+ \frac{|p|^4}{3}|\nabla\beta|^2\right)\right] + \mathscr{O}\left(\varepsilon^4\right),$$

$$\text{Exp}\left(A - \frac{|p|^2}{2}\right) = \exp\left(A - \frac{|p|^2}{2}\right)\left[1 + \frac{\varepsilon^2}{8}\left(\Delta A + \frac{1}{3}|\nabla A|^2 - \frac{1}{3}p^\top(\nabla\otimes\nabla)Ap\right)\right]$$

$$+ \mathscr{O}\left(\varepsilon^4\right),$$

where $\beta = 1/T$, *and* $(\nabla\otimes\nabla)A$ *and* $(\nabla\otimes\nabla)\beta$ *denote the Hessians of A and β, respectively.*

Remark 12.7. For later reference, we mention another equilibrium state, first derived by Wigner in 1932 [4]. Maximizing the quantum entropy (12.4) with the temperature constraint $T = 1$ gives the expression

$$M_0 = \text{Exp}\left(V - \frac{1}{2}|p|^2\right).$$

By Corollary 12.6, its expansion becomes

$$M_0 = e^{V - |p|^2/2}\left[1 + \frac{\varepsilon^2}{8}\left(\Delta V + \frac{1}{3}|\nabla V|^2 - \frac{1}{3}p^\top(\nabla\otimes\nabla)Vp\right)\right] + \mathscr{O}\left(\varepsilon^4\right).$$

This corresponds to formula (25) in Wigner's paper [4]. In the classical limit $\varepsilon \to 0$, the expression reduces to the classical equilibrium state $e^{V-|p|^2/2}$. $\square$

12.2 Derivation from the Wigner–Boltzmann Equation

The quantum drift-diffusion model is derived from the parabolic band Wigner–Boltzmann equation in the diffusion scaling with a relaxation-time scattering operator. We follow here the derivation of Degond et al. in [5]. More precisely, we employ in the Wigner–Boltzmann equation (12.2) the collision operator

$$Q(w) = M[w] - w. \tag{12.6}$$

The quantum Maxwellian $M[w]$ is defined by the constrained maximization problem (12.5) with $N = 0$ and $\kappa_0(p) = 1$. Thus, the quantum Maxwellian is defined by

$$M[w] = \mathrm{Exp}\left(A - \frac{1}{2}|p|^2\right),$$

where the Lagrange multiplier A is determined through

$$\int_{\mathbb{R}^3} M[w]\,\mathrm{d}p = \int_{\mathbb{R}^3} w\,\mathrm{d}p.$$

We consider a diffusion scaling, i.e., we replace the time t and $Q(w)$ by t/α and $Q(w)/\alpha$, respectively. Then (12.2) becomes

$$\alpha^2 \partial_t w_\alpha + \alpha(p \cdot \nabla_x w_\alpha + \theta[V_\alpha]w_\alpha) = M[w_\alpha] - w_\alpha, \quad x, p \in \mathbb{R}^3, t > 0, \tag{12.7}$$

with initial datum $w(\cdot,\cdot,0) = w_I$ in $\mathbb{R}^3 \times \mathbb{R}^3$. The potential operator $\theta[V_\alpha]$ is defined in (12.3). The electric potential V_α is a solution of the Poisson equation

$$\lambda_D^2 \Delta V_\alpha = \langle w_\alpha \rangle - C(x), \tag{12.8}$$

where $\langle g(p) \rangle = 2(2\pi\varepsilon)^{-3} \int g(p)\,\mathrm{d}p$. We wish to perform the limit $\alpha \to 0$. As in the semi-classical case, this limit relies on the following properties of the collision operator which are immediate consequences of its definition (12.6).

Lemma 12.8. *The collision operator* (12.6) *satisfies the following:*

    (i) Collisional invariant: $\langle Q(w) \rangle = 0$,

    (ii) Null space: $Q(w) = 0$ *if and only if* $w = M[w]$.

The quantum drift-diffusion model is derived in the formal limit $\alpha \to 0$ from the moment equation

$$\alpha \partial_t \langle w_\alpha \rangle + \mathrm{div}_x \langle p w_\alpha \rangle + \langle \theta[V_\alpha]w_\alpha \rangle = \alpha^{-1} \langle Q(w_\alpha) \rangle = 0, \tag{12.9}$$

obtained from (12.7) by integration over $p \in \mathbb{R}^3$. First, we prove some properties of the moments of $\theta[V]$.

Lemma 12.9 (Moments of the potential operator). *Let the operator $\theta[V]$ be defined by (12.3). Then, for all functions $w = w(x, p, t)$,*

$$\langle \theta[V]w \rangle = 0, \quad \langle p\theta[V]w \rangle = -n\nabla_x V, \tag{12.10}$$

where $n = \langle w \rangle$.

Proof. We notice that an integration of the Fourier inversion formula

$$\phi(x, p, t) = \frac{1}{(2\pi)^3} \int_{\mathbb{R}^3} (\mathscr{F}(\phi))(x, \eta, t) e^{ip \cdot \eta} \, d\eta$$

over $\mathbb{R}^3$ with respect to p gives

$$(\mathscr{F}(\phi))(x, 0, t) = \int_{\mathbb{R}^3} \phi(x, p, t) \, dp = \frac{1}{(2\pi)^3} \int_{\mathbb{R}^3 \times \mathbb{R}^3} (\mathscr{F}(\phi))(x, \eta, t) e^{ip \cdot \eta} \, d\eta \, dp. \tag{12.11}$$

With this expression we compute

$$\langle \theta[V]w \rangle = \frac{2}{(2\pi\varepsilon)^3} \frac{1}{(2\pi)^3} \int_{\mathbb{R}^3 \times \mathbb{R}^3} (\delta V)(x, \eta, t)(\mathscr{F}(w))(x, \eta, t) e^{ip \cdot \eta} \, d\eta \, dp$$

$$= \frac{2}{(2\pi\varepsilon)^3} (\delta V)(x, 0, t)(\mathscr{F}(w))(x, 0, t)$$

$$= \frac{2}{(2\pi\varepsilon)^3} \frac{i}{\varepsilon} \left[V\left(x + \frac{\varepsilon}{2}\eta, t\right) - V\left(x - \frac{\varepsilon}{2}\eta, t\right) \right]_{\eta=0} (\mathscr{F}(w))(x, 0, t) = 0.$$

Furthermore, since $pe^{ip \cdot \eta} = -i\nabla_\eta e^{ip \cdot \eta}$, integration by parts yields

$$\langle p\theta[V]w \rangle = \frac{2}{(2\pi\varepsilon)^3} \frac{i}{(2\pi)^3}$$

$$\times \int_{\mathbb{R}^3 \times \mathbb{R}^3} ((\nabla_\eta \delta V)\mathscr{F}(w) + (\delta V)\nabla_\eta \mathscr{F}(w)) (x, \eta, t) e^{ip \cdot \eta} \, d\eta \, dp.$$

Then the expression (12.11) implies that

$$\langle p\theta[V]w \rangle = 2i(2\pi\varepsilon)^{-3} ((\nabla_\eta \delta V)\mathscr{F}(w) + (\delta V)\nabla_y \mathscr{F}(w)) (x, 0, t).$$

Employing $(\delta V)(x, 0, t) = 0$ and

$$(\nabla_\eta \delta V)(x, 0, t) = \frac{i}{2} \left[\nabla_x V\left(x + \frac{\varepsilon}{2}\eta, t\right) + \nabla_x V\left(x - \frac{\varepsilon}{2}\eta, t\right) \right]_{\eta=0} = i\nabla_x V(x, t)$$

finally yields

$$\langle p\theta[V]w\rangle = -2(2\pi\eta)^{-3}\nabla_x V(x,t)\,(\mathscr{F}(w))\,(x,0,t) = -\nabla_x Vn,$$

finishing the proof. □

Theorem 12.10 (Nonlocal quantum drift-diffusion equations). *Let (w_α,V_α) be a solution of the Wigner–Boltzmann–Poisson system (12.7)–(12.8) with initial datum $w_\alpha(\cdot,\cdot,0) = w_I$. Then, formally, $w_\alpha \to w$ and $V_\alpha \to V$ as $\alpha \to 0$, where $w(x,p,t) = \mathrm{Exp}(A(x,t) - |p|^2/2)$ and (A,V) is a solution of the* quantum drift-diffusion equations

$$\partial_t n - \mathrm{div}\, J_n = 0, \quad J_n = \mathrm{div}\, P - n\nabla V, \quad \lambda_D^2 \Delta V = n - C(x), \quad t > 0, \qquad (12.12)$$

$$n(\cdot,0) = n_I \quad in\ \mathbb{R}^3,$$

the particle density and quantum stress tensor are, respectively,

$$n = \frac{2}{(2\pi\varepsilon)^3}\int_{\mathbb{R}^3} \mathrm{Exp}\left(A - \frac{|p|^2}{2}\right) dp, \quad P = \frac{2}{(2\pi\varepsilon)^3}\int_{\mathbb{R}^3} p\otimes p\,\mathrm{Exp}\left(A - \frac{|p|^2}{2}\right) dp,$$

$$(12.13)$$

where the matrix $p\otimes p$ consists of the elements $(p\otimes p)_{j\ell} = p_j p_\ell$, and the initial datum is given by $n_I = \langle w_I \rangle$.

The quantum stress tensor is a nonlocal operator involving the Lagrange multiplier A which relates to the particle density n through (12.13). In the isothermal semi-classical case, P is given by the matrix coefficients $n\delta_{j\ell}$ yielding the drift-diffusion equations of Chap. 5.

Proof. The derivation is performed in three steps. The first step is the formal limit $\alpha \to 0$ in (12.7) giving $Q(w) = 0$ and hence, by Lemma 12.8 (ii), $w = M[w]$. This gives $w = \mathrm{Exp}(A - |p|^2/2)$ for some Lagrange multiplier A. In the second step, we employ the Chapman–Enskog method. We introduce the first-order correction

$$w_\alpha = M[w_\alpha] + \alpha g_\alpha,$$

which in fact is a definition of g_α. We denote by g the limit of g_α as $\alpha \to 0$. The simple form of the collision operator allows us to derive an explicit expression for g. Indeed, the definition of g_α implies that

$$g_\alpha = -\alpha^{-1}Q(w_\alpha) = -\alpha\partial_t w_\alpha - (p\cdot\nabla_x w_\alpha + \theta[V_\alpha]w_\alpha)$$

converges to

$$g = -(p\cdot\nabla_x w + \theta[V]w), \qquad (12.14)$$

where $V = \lim_{\alpha\to 0} V_\alpha$.

In the third step, we turn to the moment equation (12.9). By Lemma 12.9, the integral over $\theta[V_\alpha]w_\alpha$ vanishes, $\langle\theta[V_\alpha]w_\alpha\rangle = 0$. It can be verified that the function $p \mapsto pM[w_\alpha](p)$ is odd. Hence, its integral over $\mathbb{R}^3$ vanishes and

$$\langle pw_\alpha\rangle = \langle p(M[w_\alpha] + \alpha g_\alpha)\rangle = \alpha\langle pg_\alpha\rangle.$$

Thus, the moment equation (12.9) becomes, after division by α,

$$\partial_t \langle w_\alpha \rangle + \text{div}_x \langle p g_\alpha \rangle = 0,$$

and the limit $\alpha \to 0$ gives, inserting (12.14),

$$\partial_t \langle w \rangle - \text{div}_x \langle p(p \cdot \nabla_x w + \theta[V]w) \rangle = 0.$$

By Lemma 12.9, we have $\langle p\theta[V]w \rangle = -n\nabla_x V$, and hence,

$$\partial_t \langle w \rangle - \text{div}_x \langle p \otimes p\nabla_x w \rangle + n\nabla_x V = 0,$$

which equals (12.12). $\quad\square$

The quantum drift-diffusion model (12.12) is nonlocal due to the relation between A and n or P. A local model is obtained in the $\mathcal{O}(\varepsilon^4)$ expansion of the quantum Maxwellian.

Theorem 12.11 (Local quantum drift-diffusion equations). *Let (n, J_n, V) be a solution of the nonlocal quantum drift-diffusion equations (12.12). Then, formally, $J_n = J_0 + \mathcal{O}(\varepsilon^4)$ and (n, J_0, V) solves the (local) quantum drift-diffusion equations*

$$\partial_t n - \text{div} J_0 = 0, \quad J_0 = \nabla n - n\nabla V - \frac{\varepsilon^2}{6} n\nabla\left(\frac{\Delta\sqrt{n}}{\sqrt{n}}\right), \tag{12.15}$$

$$\lambda_D^2 \Delta V = n - C(x), \quad n(\cdot, 0) = n_I \quad \text{in } \mathbb{R}^3, \, t > 0. \tag{12.16}$$

For the proof of this theorem, we need the following simple lemma.

Lemma 12.12. *The following integral identities hold:*

$$\int_{\mathbb{R}^3} e^{-|p|^2/2}\, dp = (2\pi)^{3/2},$$

$$\int_{\mathbb{R}^3} p_j p_\ell e^{-|p|^2/2}\, dp = (2\pi)^{3/2}\delta_{j\ell},$$

$$\int_{\mathbb{R}^3} p_j p_\ell p_m p_n e^{-|p|^2/2}\, dp = (2\pi)^{3/2}(\delta_{j\ell}\delta_{mn} + \delta_{jm}\delta_{\ell n} + \delta_{jn}\delta_{\ell m}),$$

where $\delta_{j\ell}$ denotes the Kronecker delta.

Proof (of Theorem 12.11). We need to expand the electron density and the stress tensor in powers of ε^2. By Corollary 12.6, the $\mathcal{O}(\varepsilon^4)$ expansion of the quantum Maxwellian is given by

$$\text{Exp}\left(A - \frac{|p|^2}{2}\right) = \exp\left(A - \frac{|p|^2}{2}\right)\left[1 + \frac{\varepsilon^2}{8}\left(\Delta A + \frac{1}{3}|\nabla A|^2 - \frac{1}{3}p^\top(\nabla \otimes \nabla)Ap\right)\right]$$
$$+ \mathcal{O}(\varepsilon^4).$$

Thus, using Lemma 12.12, the electron density $n = \langle \mathrm{Exp}(A - |p|^2/2) \rangle$ can be expanded as follows:

$$
\begin{aligned}
n &= e^A \left(1 + \frac{\varepsilon^2}{8} \left(\Delta A + \frac{1}{3} |\nabla A|^2 \right) \right) \left\langle e^{-|p|^2/2} \right\rangle \\
&\quad - \frac{\varepsilon^2}{24} e^A \sum_{j,\ell=1}^{3} \frac{\partial^2 A}{\partial x_j \partial x_\ell} \left\langle p_j p_\ell e^{-|p|^2/2} \right\rangle + \mathscr{O}(\varepsilon^4) \\
&= 2 (2\pi\varepsilon^2)^{-3/2} e^A \left(1 + \frac{\varepsilon^2}{8} \left(\Delta A + \frac{1}{3}|\nabla A|^2 \right) - \frac{\varepsilon^2}{24} \Delta A \right) + \mathscr{O}(\varepsilon^4) \\
&= 2 (2\pi\varepsilon^2)^{-3/2} e^A \left(1 + \frac{\varepsilon^2}{12} \left(\Delta A + \frac{1}{2}|\nabla A|^2 \right) \right) + \mathscr{O}(\varepsilon^4).
\end{aligned} \tag{12.17}
$$

Next, we develop the quantum stress tensor P in powers of ε^2. By its definition (12.13) and by Lemma 12.12,

$$
\begin{aligned}
P_{j\ell} &= e^A \left(1 + \frac{\varepsilon^2}{8} \left(\Delta A + \frac{1}{3}|\nabla A|^2 \right) \right) \left\langle p_j p_\ell e^{-|p|^2/2} \right\rangle \\
&\quad - \frac{\varepsilon^2}{24} \sum_{m,n=1}^{3} \frac{\partial^2 A}{\partial x_m \partial x_n} \left\langle p_j p_\ell p_m p_n e^{-|p|^2/2} \right\rangle + \mathscr{O}(\varepsilon^4) \\
&= 2 (2\pi\varepsilon^2)^{-3/2} e^A \left(1 + \frac{\varepsilon^2}{8} \left(\Delta A + \frac{1}{3}|\nabla A|^2 \right) \right) \delta_{j\ell} \\
&\quad - \frac{\varepsilon^2}{12} (2\pi\varepsilon^2)^{-3/2} e^A \left(\delta_{j\ell} \Delta A + 2 \frac{\partial^2 A}{\partial x_j \partial x_\ell} \right) + \mathscr{O}(\varepsilon^4) \\
&= 2 (2\pi\varepsilon^2)^{-3/2} e^A \left(1 + \frac{\varepsilon^2}{12} \left(\Delta A + \frac{1}{2}|\nabla A|^2 \right) \right) \delta_{j\ell} \\
&\quad - \frac{\varepsilon^2}{6} (2\pi\varepsilon^2)^{-3/2} e^A \frac{\partial^2 A}{\partial x_j \partial x_\ell} + \mathscr{O}(\varepsilon^4).
\end{aligned}
$$

The $\mathscr{O}(\varepsilon^4)$ expansion (12.17) leads to

$$
P_{j\ell} = n \delta_{j\ell} - \frac{\varepsilon^2}{12} n \frac{\partial^2 A}{\partial x_j \partial x_\ell} + \mathscr{O}(\varepsilon^4).
$$

Differentiating the $\mathscr{O}(\varepsilon^2)$ expansion of n with respect to x, we arrive at $\nabla n = n\nabla A + \mathscr{O}(\varepsilon^2)$. Hence,

$$
\begin{aligned}
(\mathrm{div}\, P)_j &= \sum_{\ell=1}^{3} \frac{\partial P_{j\ell}}{\partial x_\ell} = \frac{\partial n}{\partial x_j} - \frac{\varepsilon^2}{12} \sum_{\ell=1}^{3} \left(\frac{\partial n}{\partial x_\ell} \frac{\partial^2 A}{\partial x_j \partial x_\ell} + n \frac{\partial^3 A}{\partial x_j \partial x_\ell^2} \right) + \mathscr{O}(\varepsilon^4) \\
&= \frac{\partial n}{\partial x_j} - \frac{\varepsilon^2}{12} \sum_{\ell=1}^{3} n \frac{\partial}{\partial x_j} \left(\frac{1}{2} \left(\frac{\partial A}{\partial x_\ell} \right)^2 + \frac{\partial^2 A}{\partial x_\ell^2} \right) + \mathscr{O}(\varepsilon^4).
\end{aligned}
$$

In vector form, this reads as

$$\operatorname{div} P = \nabla n - \frac{\varepsilon^2}{12} n \nabla \left(\Delta A + \frac{1}{2} |\nabla A|^2 \right) + \mathcal{O}\left(\varepsilon^4\right).$$

It remains to express A in terms of n. We already noticed that $\nabla A = \nabla n / n + \mathcal{O}\left(\varepsilon^2\right)$ from which we conclude that

$$\begin{aligned}
\Delta A + \frac{1}{2} |\nabla A|^2 &= \frac{\Delta n}{n} - \frac{|\nabla n|^2}{n^2} + \frac{1}{2} \left| \frac{\nabla n}{n} \right|^2 + \mathcal{O}\left(\varepsilon^2\right) \\
&= \frac{\Delta n}{n} - \frac{|\nabla n|^2}{2n^2} + \mathcal{O} = 2 \frac{\Delta \sqrt{n}}{\sqrt{n}} + \mathcal{O}\left(\varepsilon^2\right).
\end{aligned} \tag{12.18}$$

Therefore,

$$J_n = \operatorname{div} P - n\nabla V = \nabla n - n\nabla V - \frac{\varepsilon^2}{6} n \nabla \left(\frac{\Delta \sqrt{n}}{\sqrt{n}} \right) + \mathcal{O}\left(\varepsilon^4\right),$$

and the conclusion follows. □

The quantum correction $\Delta \sqrt{n} / \sqrt{n}$ can be interpreted as a quantum potential, the so-called *Bohm potential*, which is well known in quantum mechanics. We will show in Sect. 14.1 that the Bohm potential also arises from the fluid dynamical formulation of the single-state Schrödinger equation. We notice that (12.15) and (12.16) reduce to the classical drift-diffusion model in the classical limit $\varepsilon \to 0$.

Quantum drift-diffusion models were first used by electro-engineers to simulate strong inversion layers near the oxide of MOS transistors [6]. The nonlocality of quantum mechanics is approximated by the fact that the equations of state do depend not only on the particle density but also on its gradient. Therefore, Eqs. (12.15) and (12.16) are also called the *density-gradient model*. It was employed to model field emissions from metals and steady-state tunneling in metal–insulator–metal structures [7] and to simulate ultrasmall semiconductor devices [8–10].

The quantum drift-diffusion model allows for hybrid models. This means that the classical drift-diffusion may be employed in device regions which can be modeled semi-classically and the quantum drift-diffusion model is taken in domains in which quantum mechanical effects seem to be important. The only difference between both models is the Bohm potential term. The quantum drift-diffusion equations can be also coupled to Schrödinger models, see [11, 12].

In bounded domains $\Omega \subset \mathbb{R}^3$, boundary conditions from classical models can be employed in the quantum drift-diffusion model, thus avoiding the problem of artificial reflection of wave functions from Schrödinger models. For instance, we can impose mixed Dirichlet–Neumann boundary conditions

$$n = n_D, \quad V = V_D \quad \text{on } \Gamma_D, \quad J_0 \cdot \eta = \nabla V \cdot \eta = 0 \quad \text{on } \Gamma_N,$$

where η denotes the exterior unit normal vector on $\partial \Omega = \Gamma_D \cup \Gamma_N$, Γ_D the union of contacts, and Γ_N the insulating boundary segments. As the quantum drift-diffusion equations contain fourth-order derivatives, an additional boundary condition for the particle density is needed. In [13], it is proposed that the normal component of the "quantum current density" $n\nabla(\Delta\sqrt{n}/\sqrt{n}) \cdot \eta$ vanishes on the insulating boundary and that no quantum effects occur at the contacts:

$$\nabla\left(\frac{\Delta\sqrt{n}}{\sqrt{n}}\right) \cdot \eta = 0 \quad \text{on } \Gamma_N, \quad \Delta\sqrt{n} = 0 \quad \text{on } \Gamma_D.$$

Another choice is to prescribe the so-called *quantum quasi-Fermi potential* $F = \log n - V - (\varepsilon^2/6)\Delta\sqrt{n}/\sqrt{n}$ on the boundary:

$$\nabla F \cdot \eta = 0 \quad \text{on } \Gamma_N, \quad F = F_D \quad \text{on } \Gamma_D,$$

where the boundary potential F_D is the sum of the (constant) equilibrium Fermi potential and the external potential [14].

We should also mention some shortcomings of the quantum drift-diffusion model. The quantum correction is only of low order such that no quantum interference phenomena are included in the model. The simulation results for tunneling diodes are not very satisfactory, since negative resistance effects can be only observed at very low lattice temperature or appropriately fitted effective mass or mobility [10, 15–18]. This may be due to the fact that the diffusive effects of the model are too strong for such devices whose functionality is based on quantum interference. Furthermore, the simulation of single-electron devices seems to be out of reach.

Mathematically, the quantum drift-diffusion model (12.15) is a parabolic fourth-order equation. Without classical diffusion and electric field, the model can be written as

$$\partial_t n + \text{div}\left(n\nabla\left(\frac{\Delta\sqrt{n}}{\sqrt{n}}\right)\right) = 0. \tag{12.19}$$

Its one-dimensional version appears in the modeling of interface fluctuations of certain spin systems [19] and has attracted the interest of mathematicians since the equation possesses some remarkable mathematical properties (see [20] for a review).

The first existence result for positive local-in-time solutions of the one-dimensional version of (12.19) was shown by Bleher et al. [21]. The result was improved to global-in-time nonnegative solutions in [22]. Existence of solutions of the multidimensional equation was shown recently [23, 24]. The stationary quantum drift-diffusion model was analyzed in [14] and the time-dependent bipolar equations in [25]. The classical limit was studied, for instance, in [26]. In the papers [13, 27–29], the equations were numerically discretized by finite-element methods and numerical convergence results were given. Ultrasmall double-gate MOSFET structures computed from the quantum drift-diffusion model were simulated in several space dimensions [30] and using high-resolution schemes [31]. An optimal control problem for the stationary quantum drift-diffusion model was presented in [32].

Finally, we write the quantum drift-diffusion equations in unscaled form:

$$\partial_t n - \frac{1}{q}\operatorname{div} J_n = 0, \quad J_n = \frac{q\tau}{m^*}\left(k_B T_L \nabla n - qn\nabla V - \frac{\hbar^2}{6m^*}n\nabla\left(\frac{\Delta\sqrt{n}}{\sqrt{n}}\right)\right),$$

where τ is the relaxation time originating from the relaxation-time collision operator.

12.3 Quantum Fluid Entropy

The quantum-kinetic entropy or quantum free energy is defined in (12.4) by

$$S(w) = -\frac{2}{(2\pi\varepsilon)^3}\int_{\mathbb{R}^3\times\mathbb{R}^3} w\left(\operatorname{Log} w - 1 + \frac{|p|^2}{2} - V\right)\,dx\,dp.$$

The quantum fluid entropy according to the quantum drift-diffusion model is obtained by inserting $w_0 = \operatorname{Exp}(A - |p|^2/2)$, which is equivalent to $\operatorname{Log} w_0 = A - |p|^2/2$:

$$S(w_0) = -\frac{2}{(2\pi\varepsilon)^3}\int_{\mathbb{R}^3\times\mathbb{R}^3} w_0(A - 1 - V)\,dx\,dp = -\int_{\mathbb{R}^3} n(A - 1 - V)\,dx,$$

since $\langle w_0\rangle = n$. It is shown by Degond and Ringhofer [1, Prop. 4.3] that, if n is a solution of the nonlocal quantum drift-diffusion equation (12.12), for given potential V, then

$$\frac{d}{dt}S(w_0) \geq \int_{\mathbb{R}^3} n\partial_t V\,dx.$$

In particular, if V is time independent, the macroscopic quantum entropy is nondecreasing in time.

We claim that also the ε^2 expansion of the above quantum fluid entropy is nondecreasing in time along the solutions of the quantum drift-diffusion model (12.15). First, we expand the entropy in powers of ε^2.

Lemma 12.13. *The $\mathcal{O}\left(\varepsilon^4\right)$ expansion of the quantum fluid entropy*

$$S(w_0) = -\int_{\mathbb{R}^3} n(A - 1 - V)\,dx$$

reads as follows (up to integrals over n):

$$S = -\int_{\mathbb{R}^3}\left(n(\log n - 1) + \frac{\varepsilon^2}{6}|\nabla\sqrt{n}|^2 - nV\right)\,dx + \mathcal{O}\left(\varepsilon^4\right).$$

The term $n(\log n - 1)$ corresponds to the thermodynamic entropy, the second term to the quantum energy, and the last term to the electric energy.

Proof. We recall that by (12.17), the particle density can be expanded as

$$n = 2(2\pi\varepsilon^2)^{-3/2}e^A\left(1 + \frac{\varepsilon^2}{12}\left(\Delta A + \frac{1}{2}|\nabla A|^2\right)\right) + \mathscr{O}\left(\varepsilon^4\right).$$

The logarithm of this expression becomes, since $\log(1 + \varepsilon^2 x) = \varepsilon^2 x + \mathscr{O}\left(\varepsilon^4\right)$,

$$\log n = \log\left(2(2\pi\varepsilon^2)^{-3/2}\right) + A + \frac{\varepsilon^2}{12}\left(\Delta A + \frac{1}{2}|\nabla A|^2\right) + \mathscr{O}\left(\varepsilon^4\right).$$

Employing (12.18), the ε^2 terms can be expressed in terms of $\sqrt{n}$:

$$\log n = \log\left(2(2\pi\varepsilon^2)^{-3/2}\right) + A + \frac{\varepsilon^2}{6}\frac{\Delta\sqrt{n}}{\sqrt{n}} + \mathscr{O}\left(\varepsilon^4\right).$$

Inserting this expression for A in the definition of the quantum fluid entropy and integrating by parts, we arrive at

$$S(w_0) = -\int_{\mathbb{R}^3}\left(n(\log n - 1) - nV + \frac{\varepsilon^2}{6}|\nabla\sqrt{n}|^2\right.$$
$$\left. -\log\left(2\left(2\pi\varepsilon^2\right)^{-3/2}\right)n\right)dx + \mathscr{O}\left(\varepsilon^4\right).$$

Since mass conservation holds, the integral $\int\log(2(2\pi\varepsilon^2)^{-3/2})n\,dx$ is constant and can be removed from the definition of the entropy. □

The $\mathscr{O}\left(\varepsilon^4\right)$ approximation of the quantum fluid entropy is indeed nondecreasing in time if the electric potential is time independent, as shown in the following proposition.

Proposition 12.14. *Let V be a given function and let n be a solution of the local quantum drift-diffusion equations (12.15). Furthermore, let the quantum fluid entropy be given by*

$$S_0 = -\int_{\mathbb{R}^3}\left(n(\log n - 1) + \frac{\varepsilon^2}{6}|\nabla\sqrt{n}|^2 - nV\right)dx.$$

Then

$$\frac{dS_0}{dt} - \int_{\mathbb{R}^3}n\left|\nabla\left(\log n - V - \frac{\varepsilon^2}{6}\frac{\Delta\sqrt{n}}{\sqrt{n}}\right)\right|^2 dx = \int_{\mathbb{R}^3}n\partial_t V\,dx. \qquad (12.20)$$

Proof. The particle density n satisfies the equation

$$\partial_t n = \mathrm{div}\left(n\nabla\left(\log n - V - \frac{\varepsilon^2}{6}\frac{\Delta\sqrt{n}}{\sqrt{n}}\right)\right).$$

Then, integrating by parts in the quantum term,

$$\frac{dS_0}{dt} = -\int_{\mathbb{R}^3} \left(\log n \partial_t n + \frac{\varepsilon^2}{3} \nabla \sqrt{n} \cdot \partial_t \nabla \sqrt{n} - V \partial_t n - n \partial_t V \right) dx$$

$$= -\int_{\mathbb{R}^3} \left(\partial_t n \left(\log n - \frac{\varepsilon^2}{6} \frac{\Delta \sqrt{n}}{\sqrt{n}} - V \right) - n \partial_t V \right) dx$$

$$= \int_{\mathbb{R}^3} n \left| \nabla \left(\log n - V - \frac{\varepsilon^2}{6} \frac{\Delta \sqrt{n}}{\sqrt{n}} \right) \right|^2 dx + \int_{\mathbb{R}^3} n \partial_t V \, dx,$$

showing the claim. □

The second integral on the left-hand side of (12.20) is called the *entropy production*. In the self-consistent case $\lambda_D^2 \Delta V = n - C(x)$, we have to modify the entropy:

$$S_1 = -\int_{\mathbb{R}^3} \left(n(\log n - 1) + \frac{\varepsilon^2}{6} |\nabla \sqrt{n}|^2 - \frac{1}{2}(n - C)V \right) dx$$

$$= -\int_{\mathbb{R}^3} \left(n(\log n - 1) + \frac{\varepsilon^2}{6} |\nabla \sqrt{n}|^2 + \frac{\lambda_D^2}{2} |\nabla V|^2 \right) dx.$$

Then the proof of the above proposition shows that

$$\frac{dS_1}{dt} - \int_{\mathbb{R}^3} n \left| \nabla \left(\log n - V - \frac{\varepsilon^2}{6} \frac{\Delta \sqrt{n}}{\sqrt{n}} \right) \right|^2 dx$$

$$= \int_{\mathbb{R}^3} \left(V \partial_t n - \frac{\lambda_D^2}{2} \partial_t |\nabla V|^2 + \lambda_D^2 \partial_t |\nabla V|^2 \right) dx$$

$$= \int_{\mathbb{R}^3} \left(V \partial_t n + \lambda_D^2 \Delta V \partial_t V - \lambda_D^2 \partial_t (V \Delta V) \right) dx$$

$$= \int_{\mathbb{R}^3} \left(\partial_t ((n - C)V) - \lambda_D^2 \partial_t (V \Delta V) \right) dx = 0.$$

Thus, the quantum fluid entropy is also nondecreasing in the self-consistent case.

12.4 The Entropic Quantum Drift-Diffusion Model

The nonlocal quantum drift-diffusion model (12.12) of Sect. 12.2 can be written in a different way which facilitates a numerical approximation. The derivation needs some operator calculus and therefore, we start from the collisional Liouville–von Neumann equation

$$i\varepsilon \partial_t \widehat{\rho} = [H, \widehat{\rho}] + i\varepsilon Q(\widehat{\rho}), \tag{12.21}$$

where $[H, \widehat{\rho}]$ is the commutator between the Hamiltonian $H = -(\varepsilon^2/2)\Delta - V(x,t)$ and the density-matrix operator $\widehat{\rho}$ (see Sect. 10.1). We follow the arguments of Degond et al. [33, Appendix]. The entropic quantum drift-diffusion model can be also derived directly from a development of the quantum stress tensor in (12.12), but this requires some lengthy operator computations (see [34, Sect. 3.3]).

The collision operator $Q(\widehat{\rho})$ in (12.21) is assumed to be of relaxation-time type. To make this precise, we need to define the quantum Maxwellian operator. We recall that the trace $\mathrm{Tr}(\widehat{\rho}_1\widehat{\rho}_2)$ of two self-adjoint operators $\widehat{\rho}_1$ and $\widehat{\rho}_2$ is defined, through the Parseval equality, by

$$\mathrm{Tr}(\widehat{\rho}_1\widehat{\rho}_2) = \frac{2}{(2\pi\varepsilon)^3} \int_{\mathbb{R}^3} W(\widehat{\rho}_1)W(\widehat{\rho}_2)\,\mathrm{d}p. \qquad (12.22)$$

In view of (12.4), the quantum-kinetic entropy or free energy is given by

$$S = -\frac{2}{(2\pi\varepsilon)^3} \int_{\mathbb{R}^3} w\left(\mathrm{Log}\,w - 1 + \frac{|p|^2}{2} - V\right) \mathrm{d}p,$$

where $w = W(\widehat{\rho})$ is the Wigner function and $\mathrm{Log}\,w = W(\log W^{-1}(w)) = W(\log\widehat{\rho})$ the quantum logarithm introduced in Sect. 12.1. In terms of the density-matrix operator, the quantum entropy can be written, using (12.22), as

$$S(\widehat{\rho}) = -\mathrm{Tr}\left(\widehat{\rho}(\log\widehat{\rho} - 1 + H)\right).$$

Let $\widehat{\rho}$ be a given density-matrix operator with the associated electron density n, defined by $\mathrm{Tr}(\widehat{\rho}\phi) = \int n\phi\,\mathrm{d}x$ for all functions ϕ. The notation $\mathrm{Tr}(\widehat{\rho}\phi)$ has to be understood as the multiplication of the operator $\widehat{\rho}$ with the multiplication operator $(\widehat{\phi(\psi)})(x) = \phi(x)\psi(x)$. For simplicity, we do not distinguish between the operator $\widehat{\phi}$ and the function ϕ. The quantum Maxwellian is the formal solution (if it exists) of the constrained maximization problem

$$S(\widehat{\rho}^*) = \max\left\{S(\widehat{\rho}) :\ \mathrm{Tr}(\widehat{\rho}\phi) = \int_{\mathbb{R}^3} n\phi\,\mathrm{d}x \text{ for all } \phi\right\}.$$

The constraint expresses a moment condition in operator formulation, corresponding to the constraint in the Wigner function formulation of (12.5). In order to solve the extremal problem formally, we differentiate the Lagrangian

$$F(\widehat{\rho},\lambda) = S(\widehat{\rho}) - \left(\mathrm{Tr}(\widehat{\rho}W^{-1}(\lambda)) - \int_{\mathbb{R}^3} n\lambda\,\mathrm{d}x\right).$$

It is shown in [1, 2] that the derivative of an operator $G(\widehat{\rho}) = \mathrm{Tr}(g(\widehat{\rho}))$ is given by $DG(\widehat{\rho})(\widehat{\sigma}) = \mathrm{Tr}(Dg(\widehat{\rho})\widehat{\sigma})$ for all (self-adjoint) operators $\widehat{\rho}$ and $\widehat{\sigma}$. The necessary condition for an extremal point reads as

$$0 = \frac{\mathrm{d}F}{\mathrm{d}\widehat{\rho}}(\widehat{\rho}^*,\lambda^*)(\widehat{\sigma}) = -\mathrm{Tr}\left((\log\widehat{\rho}^* + H)\widehat{\sigma}\right) - \mathrm{Tr}\left(\widehat{\sigma}W^{-1}(\lambda^*)\right)$$

for all $\widehat{\sigma}$. Hence, $\log\widehat{\rho}^* + H + W^{-1}(\lambda^*) = 0$ and

$$\widehat{\rho}^* = \exp(-H - W^{-1}(\lambda^*)).$$

We define $A = -W^{-1}(\lambda^*) + V$ and $H(A) = H + W^{-1}(\lambda^*) = H - A + V$. Then the *quantum Maxwellian* of $\widehat{\rho}$ is defined by

$$M[\widehat{\rho}] = \widehat{\rho}^* = \exp(-H(A)) = \exp(-H + A - V) = \exp\left(\frac{\varepsilon^2}{2}\Delta + A\right).$$

This corresponds to the quantum Maxwellian of Sect. 12.1, formulated as an operator. Indeed, observing that the inverse Wigner transform of $-|p|^2/2 + A$ equals $(\varepsilon^2/2)\Delta + A$,

$$W^{-1}\left(-\frac{|p|^2}{2} + A\right) = \frac{\varepsilon^2}{2}\Delta + A,$$

we infer that

$$\begin{aligned}
\mathrm{Exp}\left(A - \frac{|p|^2}{2}\right) &= W\left(\exp W^{-1}\left(A - \frac{|p|^2}{2}\right)\right) \\
&= W\left(\exp\left(\frac{\varepsilon^2}{2}\Delta + A\right)\right) = W(M[\widehat{\rho}]). \quad (12.23)
\end{aligned}$$

We can now specify the collision operator:

$$Q(\widehat{\rho}) = M[\widehat{\rho}] - \widehat{\rho}.$$

We assume a diffusion scaling for the collisional Liouville–von Neumann equation,

$$i\varepsilon\alpha\partial_t\widehat{\rho}_\alpha = [H, \widehat{\rho}_\alpha] + \frac{i\varepsilon}{\alpha}(M[\widehat{\rho}_\alpha] - \widehat{\rho}_\alpha), \quad t > 0, \quad \widehat{\rho}_\alpha(0) = \widehat{\rho}_I \quad \text{in } \mathbb{R}^3, \quad (12.24)$$

and wish to perform the limit $\alpha \to 0$. The main result is as follows.

Theorem 12.15 (Entropic quantum drift-diffusion equations). *Let $\widehat{\rho}_\alpha$ be a solution of the collisional Liouville–von Neumann equation* (12.24) *with the Hamiltonian $H = -(\varepsilon^2/2)\Delta - V$, where V is a given potential. Then, in the formal limit $\alpha \to 0$, $\widehat{\rho}_\alpha$ converges to $\widehat{\rho} = \exp(-H + A - V)$, and A is a solution of the* entropic quantum drift-diffusion equations

$$\partial_t n - \mathrm{div} J_n = 0, \quad J_n = n\nabla(A - V), \quad t > 0, \quad n(\cdot, 0) = n_I \quad \text{in } \mathbb{R}^3, \quad (12.25)$$

where

$$\mathrm{Tr}(\widehat{\rho}\phi) = \int_{\mathbb{R}^3} n\phi \, dx, \quad \mathrm{Tr}(\widehat{\rho}_I\phi) = \int_{\mathbb{R}^3} n_I\phi \, dx \quad \text{for all } \phi.$$

Using (12.22), the moment reconstruction problem

$$\mathrm{Tr}(\exp(-H(A))\phi) = \mathrm{Tr}(M[\widehat{\rho}]\phi) = \int_{\mathbb{R}^3} n\phi \, dx \quad \text{for all } \phi$$

can be formulated as

$$\frac{2}{(2\pi\varepsilon)^3} \int_{\mathbb{R}^3} \mathrm{Exp}\left(A - \frac{1}{2}|p|^2\right) \mathrm{d}p = n, \tag{12.26}$$

since, by (12.23), $W(\exp(-H(A))) = \mathrm{Exp}(A - |p|^2/2)$.

Proof. Let $\widehat{\rho}_\alpha$ be a solution of (12.24) and let A_α be defined by the relation $M[\widehat{\rho}_\alpha] = \exp(-H(A_\alpha))$. As in the proof of Theorem 12.10, we introduce the Chapman–Enskog expansion

$$\widehat{\rho}_\alpha = M[\widehat{\rho}_\alpha] + \alpha R_\alpha,$$

thus defining the operator R_α. Since, by the definition of $M[\widehat{\rho}_\alpha]$,

$$\mathrm{Tr}(Q(\widehat{\rho}_\alpha)\phi) = \mathrm{Tr}(M[\widehat{\rho}_\alpha]\phi) - \mathrm{Tr}(\widehat{\rho}_\alpha\phi) = \int_{\mathbb{R}^3} n\phi \, \mathrm{d}x - \int_{\mathbb{R}^3} n\phi \, \mathrm{d}x = 0,$$

the moment equation in operator formulation is

$$i\varepsilon\alpha\partial_t \mathrm{Tr}(\widehat{\rho}_\alpha\phi) = \mathrm{Tr}([H,\widehat{\rho}_\alpha]\phi) = \mathrm{Tr}([H,M[\widehat{\rho}_\alpha]]\phi) + \alpha\mathrm{Tr}([H,R_\alpha]\phi). \tag{12.27}$$

This equation can be simplified. Indeed, using $H(A_\alpha) = H - A_\alpha + V$, we obtain

$$\begin{aligned}
[H,M[\widehat{\rho}_\alpha]] &= [H(A_\alpha), \exp(-H(A_\alpha))] + [A_\alpha - V, \exp(-H(A_\alpha))] \\
&= [A_\alpha - V, \exp(-H(A_\alpha))], \tag{12.28}
\end{aligned}$$

since the commutator of an operator and its exponential vanishes. We claim that the trace of the right-hand side of this equation also vanishes. For this, we employ the cyclicity formulas $\mathrm{Tr}([a,b]c) = \mathrm{Tr}(a[b,c]) = \mathrm{Tr}([c,a]b)$. Then

$$\mathrm{Tr}\left([A_\alpha - V, \exp(-H(A_\alpha))]\phi\right) = \mathrm{Tr}\left([\phi, A_\alpha - V]\exp(-H(A_\alpha))\right) = 0.$$

Here, we have used the fact that the commutator of two multiplication operators is zero. Therefore, the first term on the right-hand side of the moment equation (12.27) vanishes, and we obtain

$$i\varepsilon\partial_t \mathrm{Tr}(\widehat{\rho}_\alpha\phi) = \mathrm{Tr}([H,R_\alpha]\phi).$$

The formal limit $\alpha \to 0$ gives

$$i\varepsilon\partial_t \mathrm{Tr}(\widehat{\rho}\phi) = \mathrm{Tr}([H,R]\phi), \tag{12.29}$$

where $\widehat{\rho} = \lim_{\alpha\to 0} \widehat{\rho}_\alpha$ and $R = \lim_{\alpha\to 0} R_\alpha$.

In the following, we determine the limit operator R. The limit $\alpha \to 0$ in the Liouville–von Neumann equation (12.24) shows that $\widehat{\rho} = M[\widehat{\rho}] = \exp(-H(A))$, where $A = \lim_{\alpha\to 0} A_\alpha$. Furthermore, inserting the Chapman–Enskog expansion in the Liouville–von Neumann equation gives

$$i\varepsilon\alpha\partial_t (M[\widehat{\rho}_\alpha] + \alpha R_\alpha) = [H, M[\widehat{\rho}_\alpha] + \alpha R_\alpha] - i\varepsilon R_\alpha,$$

and thus, the limit $\alpha \to 0$ leads to

$$R = -\frac{i}{\varepsilon}[H, M[\widehat{\rho}]].$$

We employ this expression in the limit equation (12.29):

$$i\varepsilon\partial_t \text{Tr}(\widehat{\rho}\phi) = -\frac{i}{\varepsilon}\text{Tr}\left([H, [H, M[\widehat{\rho}]]]\phi\right).$$

It remains to compute the double commutator. By (12.28), $[H, M[\widehat{\rho}]] = [A - V, \exp(-H(A))] = [A - V, \widehat{\rho}]$. Hence,

$$i\varepsilon\partial_t \text{Tr}(\widehat{\rho}\phi) = -\frac{i}{\varepsilon}\text{Tr}([H, [A - V, \widehat{\rho}]]\phi) \qquad (12.30)$$

$$= -\frac{i}{\varepsilon}\text{Tr}\left(\left[-\frac{\varepsilon^2}{2}\Delta, [A - V, \widehat{\rho}]\right]\phi\right) + \frac{i}{\varepsilon}\text{Tr}([V, [A - V, \widehat{\rho}]]\phi).$$

The cyclicity formulas imply that the second term on the right-hand side vanishes:

$$\text{Tr}([V, [A - V, \widehat{\rho}]]\phi) = \text{Tr}([\phi, V][A - V, \widehat{\rho}]) = 0.$$

We claim that the first term can be written as

$$\text{Tr}\left(\left[-\frac{\varepsilon^2}{2}\Delta, [A - V, \widehat{\rho}]\right]\phi\right) = \varepsilon^2\text{Tr}(\widehat{\rho}\nabla\phi \cdot \nabla(A - V)). \qquad (12.31)$$

In order to show this equation, we employ the cyclicity formulas, arriving at

$$\text{Tr}\left(\left[-\frac{\varepsilon^2}{2}\Delta, [A - V, \widehat{\rho}]\right]\phi\right) = \text{Tr}\left(\left[\left[\frac{\varepsilon^2}{2}\Delta, \phi\right], A - V\right]\widehat{\rho}\right).$$

Then, employing the identity

$$[[\Delta, \phi], A - V]\psi = 2\nabla\phi \cdot \nabla(A - V)\psi \quad \text{for smooth functions } \psi,$$

which follows after a direct calculation, we obtain (12.31). Equation (12.30) now becomes

$$i\varepsilon\partial_t \text{Tr}(\widehat{\rho}\phi) = -i\varepsilon\text{Tr}(\widehat{\rho}\nabla\phi \cdot \nabla(A - V)).$$

This can be written as

$$\partial_t \int_{\mathbb{R}^3} n\phi \, dx = -\int_{\mathbb{R}^3} n\nabla\phi \cdot \nabla(A - V) \, dx,$$

which is the weak formulation of (12.25). $\quad\square$

The moment reconstruction problem (12.26) possesses a simpler expression if we suppose that the modified Hamiltonian $H(A)$ has a discrete spectrum with eigenvalues $\lambda_j(A)$ and a complete set of orthonormal eigenfunctions $\psi_j(A)$, $j \in \mathbb{N}$. Then the particle density is the weighted sum over all $|\psi_j(A)|^2$:

$$n(x) = \sum_{j=1}^{\infty} e^{-\lambda_j(A)} |\psi_j(A)|^2, \qquad (12.32)$$

since, for all functions ϕ,

$$\int_{\mathbb{R}^3} n\phi \, dx = \mathrm{Tr}\left(\exp(-H(A))\phi\right) = \sum_{j \in \mathbb{N}} \left(\phi \exp(-H(A))\psi_j(A), \psi_j(A)\right)_{L^2}$$

$$= \sum_{j \in \mathbb{N}} \int_{\mathbb{R}^3} \phi e^{-\lambda_j} |\psi_j(A)|^2 \, dx = \int_{\mathbb{R}^3} \left(\sum_{j \in \mathbb{N}} e^{-\lambda_j} |\psi_j(A)|^2\right) \phi \, dx.$$

Then the entropic quantum drift-diffusion model is the system of Eqs. (12.25) and (12.32), where $(\psi_j(A), \lambda_j(A))$ are the eigenfunction–eigenvalue pairs of $H(A)$.

Remark 12.16 (Links to other quantum drift-diffusion models). The equilibrium state of the entropic quantum drift-diffusion model is characterized by $J_n = 0$ or $A = V$. The moment reconstruction problem then becomes

$$n = \sum_{j=1}^{\infty} e^{-\lambda_j(V)} |\psi_j(V)|^2,$$

where $(\psi_j(V), \lambda_j(V))$ are the eigenfunction–eigenvalue pairs of the original Hamiltonian $H = -(\varepsilon^2/2)\Delta - V$. If, additionally, the density n is related to V through the Poisson equation, this leads to the Schrödinger–Poisson problem characterizing equilibrium states.

Close to equilibrium, one may approximate A by V in the moment reconstruction problem, which gives the system

$$\partial_t n - \mathrm{div}\left(n\nabla(A - V)\right) = 0, \quad n = \sum_{j=1}^{\infty} e^{A-V-\lambda_j(V)} |\psi_j(V)|^2,$$

where the eigenvalue problem is associated with the original Hamiltonian. This system is known as the *Schrödinger–Poisson drift-diffusion model* investigated in [12, 17, 35].

As a side product of the proof of Theorem 12.15, it follows that $\mathrm{div} P = n\nabla A$, where P is the quantum stress tensor of the nonlocal quantum drift-diffusion model. In this sense, the entropic and the nonlocal quantum models coincide. We infer that the local quantum drift-diffusion model derived in Sect. 12.2 is the $\mathcal{O}(\varepsilon^4)$ approximation of the entropic quantum drift-diffusion model. $\square$

When the entropic model is considered in a bounded domain, appropriate boundary conditions need to be specified. In [33], homogeneous boundary conditions for the wave functions ψ_j are prescribed,

$$\nabla \psi_j \cdot \eta = 0 \quad \text{on } \partial\Omega, \, j \in \mathbb{N}.$$

For the potentials, two types of boundary conditions were suggested. First, the number of particles in the domain is enforced to be constant by employing Neumann boundary conditions:

$$\nabla(A - V) \cdot \eta = 0 \quad \text{on } \partial\Omega.$$

The electric potential is the sum of the self-consistent potential V_{sc}, solving the Poisson equation, and an external potential V_{ext}, modeling heterostructures. If the equilibrium state is considered, no bias is applied to the device, which is expressed by the boundary conditions

$$V_{sc} = 0 \quad \text{on } \partial\Omega.$$

Second, in order to allow for a current flow, Dirichlet conditions are assumed for the particle density,

$$n = \sum_{j=1}^{\infty} e^{-\lambda_j(A)} |\psi_j(A)|^2 = n_D \quad \text{on } \partial\Omega,$$

and for the self-consistent potential,

$$V_{sc} = V_D \quad \text{on } \partial\Omega.$$

In [33], the one-dimensional entropic quantum model was discretized with a finite-difference scheme and current–voltage characteristics of a resonant tunneling diode were numerically computed. Similar to the quantum drift-diffusion model (12.15), the current–voltage curves exhibit negative differential resistance effects only for small temperature or appropriately chosen effective mass. The numerical scheme is shown to preserve the physical properties of the continuous model, such as charge conservation, positivity of the electron density, and dissipation of the quantum fluid entropy [36].

References

1. P. Degond and C. Ringhofer. Quantum moment hydrodynamics and the entropy principle. *J. Stat. Phys.* 112 (2003), 587–628.
2. P. Degond and C. Ringhofer. A note on quantum moment hydrodynamics and the entropy principle. *C. R. Acad. Sci. Paris, Sér. I* 335 (2002), 967–972.
3. A. Jüngel, D. Matthes, and J.-P. Milišić. Derivation of new quantum hydrodynamic equations using entropy minimization. *SIAM J. Appl. Math.* 67 (2006), 46–68.
4. E. Wigner. On the quantum correction for the thermodynamic equilibrium. *Phys. Rev.* 40 (1932), 749–759.
5. P. Degond, F. Méhats, and C.Ringhofer. Quantum energy-transport and drift-diffusion models. *J. Stat. Phys.* 118 (2005), 625–665.
6. M. Ancona. Diffusion-drift modeling of strong inversion layers. *COMPEL* 6 (1987), 11–18.
7. M. Ancona. Density-gradient analysis of field emission from metals. *Phys. Rev. B* 46 (1992), 4874–4883.
8. A. Asenov, S. Kaya, J. Davies, and S. Saini. Oxide thickness variation induced threshold voltage fluctuations in decanano MOSFET: a 3D density gradient simulation study. *Superlatt. Microstruct.* 28 (2000), 507–515.

9. H. Tsuchiya and T. Miyoshi. Quantum transport modeling of ultrasmall semiconductor devices. *IEICE Trans. Electr.* E82-C (1999), 880–888.
10. A. Wettstein, A. Schenk, and W. Fichtner. Quantum device-simulation with the density-gradient model on unstructured grids. *IEEE Trans. Electr. Devices* 48 (2001), 279–284.
11. A. El Ayyadi and A. Jüngel. Semiconductor simulations using a coupled quantum drift-diffusion Schrödinger-Poisson model. *SIAM J. Appl. Math.* 66 (2005), 554–572.
12. C. de Falco, E. Gatti, A. Lacaita, and R. Sacco. Quantum-corrected drift-diffusion model for transport in semiconductor devices. *J. Comput. Phys.* 204 (2005), 533–561.
13. A. Jüngel and R. Pinnau. A positivity-preserving numerical scheme for a nonlinear fourth order parabolic equation. *SIAM J. Numer. Anal.* 39 (2001), 385–406.
14. N. Ben Abdallah and A. Unterreiter. On the stationary quantum drift diffusion model. *Z. Angew. Math. Phys.* 49 (1998), 251–275.
15. M. Ancona, Z. Yu, W.-C. Lee, R. Dutton, and P. Voorde. Density-gradient simulations of quantum effects in ultra-thin-oxide MOS structures. *SISPAD '97* (1997), 97–100.
16. A. Jüngel and J.-P. Milišić. Macroscopic quantum models with and without collisions. *Bulletin Inst. Math. Acad. Sinica (New Series)* 2 (2007), 251–279.
17. S. Micheletti, R. Sacco, and P. Simioni. Numerical simulation of resonant tunneling diodes with a quantum drift-diffusion model. In: W. Schilders, E. ten Maten, and S. Houben (eds.), *Scientific Computing in Electrical Engineering*, 313–321. Springer, 2004.
18. J. Watling, A. Brown, and A. Asenov. Can the density gradient approach describe the source-drain tunnelling in decanano double-gate MOSFETs? *J. Comput. Electr.* 1 (2002), 289–293.
19. B. Derrida, J. Lebowitz, E. Speer, and H. Spohn. Fluctuations of a stationary nonequilibrium interface. *Phys. Rev. Letters* 67 (1991), 165–168.
20. A. Jüngel and D. Matthes. A review on results for the Derrida-Lebowitz-Speer-Spohn equation. To appear in *Proceedings of Equadiff 2007*, 2009.
21. P. Bleher, J. Lebowitz, and E. Speer. Existence and positivity of solutions of a fourth-order nonlinear PDE describing interface fluctuations. *Commun. Pure Appl. Math.* 47 (1994), 923–942.
22. A. Jüngel and R. Pinnau. Global non-negative solutions of a nonlinear fourth-order parabolic equation for quantum systems. *SIAM J. Math. Anal.* 32 (2000), 760–777.
23. U. Gianazza, G. Savaré, and G. Toscani. The Wasserstein gradient flow of the Fisher information and the quantum drift-diffusion equation. To appear in *Arch. Rat. Mech. Anal.*, 2009.
24. A. Jüngel and D. Matthes. The Derrida-Lebowitz-Speer-Spohn equation: existence, non-uniqueness, and decay rates of the solutions. *SIAM J. Math. Anal.* 39 (2008), 1996–2015.
25. A. Jüngel and I. Violet. The quasineutral limit in the quantum drift-diffusion equations. *Asympt. Anal.* 53 (2007), 139–157.
26. L. Chen and Q. Ju. Existence of weak solution and semiclassical limit for quantum drift-diffusion model. *Z. Angew. Math. Phys.* 58 (2007), 1–15.
27. R. Pinnau. Uniform convergence of the exponentially fitted scheme for the quantum drift-diffusion model. *SIAM J. Numer. Anal.* 42 (2004), 1648–1668.
28. R. Pinnau and J. M. Ruiz. Convergent finite element discretizations of the density gradient equation for quantum semiconductors. To appear in *Proceedings of SIMAI 2006*, 2009.
29. R. Pinnau and A. Unterreiter. The stationary current-voltage characteristics of the quantum drift-diffusion model. *SIAM J. Numer. Anal.* 37 (1999), 211–245.
30. S. Odanaka. Multidimensional discretization of the stationary quantum drift-diffusion model for ultrasmall MOSFET structures. *IEEE Trans. Comp. Aided Design Integr. Circuits Sys.* 23 (2004), 837–842.
31. S. Odanaka. A high-resolution method for quantum confinement transport simulations in MOSFETs. *EEE Trans. Comp. Aided Design Integr. Circuits Sys.* 26 (2007), 80–85.
32. A. Unterreiter and S. Volkwein. Optimal control of the stationary quantum drift-diffusion model. *Commun. Math. Sci.* 5 (2007), 85–111.
33. P. Degond, S. Gallego, and F. Méhats. An entropic quantum drift-diffusion model for electron transport in resonant tunneling diodes. *J. Comput. Phys.* 221 (2007), 226–249.

34. P. Degond, S. Gallego, F. Méhats, and C. Ringhofer. Quantum hydrodynamic and diffusion models derived from the entropy principle. Lecture notes for a summer school in Cetraro, Italy, 2006.
35. A. Pirovano, A. Lacaita, and A. Spinelli. Two-dimensional quantum effects in nanoscale MOSFETs. *IEEE Trans. Electron Devices* 47 (2002), 25–31.
36. S. Gallego and F. Méhats. Entropic discretization of a quantum drift-diffusion model. *SIAM J. Numer. Anal.* 43 (2005), 1828–1849.

Chapter 13
Quantum Diffusive Higher-Order Moment Equations

The strategy of the previous chapter to derive quantum drift-diffusion equations can be generalized. Similar to the semi-classical diffusive limit, we impose several moment constraints, leading to a system of diffusive equations. In this chapter, we explain the general strategy and consider in more detail a specific quantum model, consisting of the balance equations of mass and energy, the so-called quantum energy-transport equations.

13.1 Derivation from the Wigner–Boltzmann Equation

We consider the Wigner–Boltzmann equation in the diffusion scaling:

$$\alpha^2 \partial_t w + \alpha \left(p \cdot \nabla_x w + \theta[V]w \right) = Q(w), \quad x, p \in \mathbb{R}^3, \ t > 0, \tag{13.1}$$

$$w(x, p, 0) = w_I(x, p), \quad x, p \in \mathbb{R}^3. \tag{13.2}$$

This scaling is also used in Chap. 8 to derive semi-classical higher-order moment equations. The electric potential V can be given or be a solution of the Poisson equation. Let *even* weight functions $\kappa(p) = (\kappa_0(p), \ldots, \kappa_N(p))$ be given, for instance $\kappa(p) = (1, \frac{1}{2}|p|^2, \ldots)$. Given the Wigner function w, we introduce the quantum Maxwellian $M[w] = \mathrm{Exp}(\lambda \cdot \kappa(p))$, where $\lambda = (\lambda_0, \ldots, \lambda_N)$ are some Lagrange multipliers. The quantum Maxwellian is the formal solution of the constrained maximization problem (12.5). In particular, the moments of w and $M[w]$ coincide, $\langle w \kappa_j \rangle = \langle M[w] \kappa_j \rangle$ for all $j = 0, \ldots, N$, where we recall that $\langle g(p) \rangle = 2(2\pi\varepsilon)^{-3} \int_{\mathbb{R}^3} g(p)\, dp$.

We assume that the collision operator can be decomposed into two parts:

$$Q(w) = Q_0(w) + \alpha^2 Q_1(w), \quad \text{where} \quad Q_0(w) = M[w] - w \tag{13.3}$$

Jüngel, A.: *Quantum Diffusive Higher-Order Moment Equations*. Lect. Notes Phys. **773**, 275–282 (2009)
DOI 10.1007/978-3-540-89526-8_13

is a relaxation-time or BGK operator. It satisfies the following properties: Its kernel is given by the Maxwellians, i.e., $Q_0(w) = 0$ if and only if $w = M[w]$, and all moments of $Q_0(w)$ vanish, i.e., $\langle Q_0(w)\kappa(p)\rangle = 0$. The operator Q_1 remains unspecified.

Theorem 13.1 (Quantum diffusive moment equations). *Let V be a given potential and w_α be a solution of the Wigner–Boltzmann equation (13.1) and (13.2). Then, in the formal limit $\alpha \to 0$, w_α converges to $M[w]$ which solves*

$$\partial_t m - \operatorname{div}\left(\operatorname{div}\langle p \otimes p\kappa M[w]\rangle + \langle p\kappa\theta[V]M[w]\rangle\right)$$
$$- \langle \kappa\theta[V](p\cdot\nabla_x M[w] + \theta[V]M[w])\rangle = \langle \kappa Q_1(M[w])\rangle, \quad x\in\mathbb{R}^3,\ t>0,\ (13.4)$$

where $m = \langle\kappa M[w]\rangle$ and $m(\cdot,0) = \langle\kappa w_I\rangle$.

Proof. Multiplying (13.1) by $\kappa(p)/\alpha^2$, integrating over the momentum space, and using the conservation property of Q_0, we obtain the moment equations

$$\partial_t\langle\kappa w_\alpha\rangle + \alpha^{-1}\left(\operatorname{div}_x\langle\kappa p w_\alpha\rangle + \langle\kappa\theta[V]w_\alpha\rangle\right) = \langle\kappa Q_1(w_\alpha)\rangle.$$

The derivation of the diffusive model is performed in three steps. In the first step, we let $\alpha \to 0$ in the Wigner–Boltzmann equation (13.1), leading to $w = M[w]$, where $w = \lim_{\alpha\to 0} w_\alpha$. For the second step, we insert the Chapman–Enskog expansion

$$w_\alpha = M[w_\alpha] + \alpha g_\alpha$$

in the above moment equations. Observing that $\langle\kappa p M[w_\alpha]\rangle$ and $\langle\kappa\theta[V]M[w_\alpha]\rangle$ vanish (since $\kappa(p)$ and $M[w_\alpha]$ are even in p), we obtain in the limit $\alpha \to 0$

$$\partial_t\langle\kappa M[w]\rangle + \operatorname{div}_x\langle p\kappa g\rangle + \langle\kappa\theta[V]g\rangle = \langle\kappa Q_1(M[w])\rangle, \tag{13.5}$$

where $g = \lim_{\alpha\to 0} g_\alpha$. The third step is concerned with the computation of g. Since Q_0 is a BGK operator, we can write

$$g_\alpha = -\frac{1}{\alpha}Q_0(w_\alpha) = -\alpha\partial_t w_\alpha - p\cdot\nabla_x w_\alpha - \theta[V]w_\alpha + \alpha Q_1(w_\alpha).$$

In the limit $\alpha \to 0$ we infer that

$$g = -p\cdot\nabla_x M[w] - \theta[V]M[w].$$

Inserting this expression into (13.5) gives the conclusion. □

The system of equations (13.4) is of diffusive type. Indeed, the second-order term can be written, since $M[w] = \operatorname{Exp}(\lambda\cdot\kappa)$, as

$$\operatorname{div}\left(\operatorname{div}\langle p\otimes p\kappa(p)M[w]\rangle\right) = \sum_{j,\ell=1}^{3}\sum_{m=0}^{N}\frac{\partial}{\partial x_j}\left(\langle p_j p_\ell\kappa_m\kappa\operatorname{Exp}(\lambda\cdot\kappa)\rangle\frac{\partial\lambda_m}{\partial x_\ell}\right)$$
$$= \operatorname{div}(D:\nabla\lambda),$$

where $D = \langle pp\kappa\kappa\mathrm{Exp}(\lambda \cdot \kappa)\rangle$ is a 4-tensor (the quantum "diffusion matrix") and the product ":" means summation over two indices. Setting $B = \langle \kappa \otimes \kappa M[w]\rangle$ and observing that

$$\partial_t m = \partial_t \langle \kappa \mathrm{Exp}(\lambda \cdot \kappa)\rangle = \langle \kappa \otimes \kappa \mathrm{Exp}(\lambda \cdot \kappa)\rangle \partial_t \lambda = B\partial_t \lambda,$$

we can formulate (13.4) in compact form as

$$B\partial_t \lambda - \mathrm{div}\,(D : \nabla\lambda) = f(\lambda),$$

where $f(\lambda)$ denotes the remaining terms. This formulation indicates (if D is positive definite) that the moment system (13.4) is of parabolic type. Unfortunately, it seems to be difficult to make the system more explicit. Moreover, due to their complexity, no mathematical results are currently available for such systems. In the case $N = 0$ and $\kappa_0(p) = 1$, we recover the quantum drift-diffusion equations studied in Chap. 12. In the following section, we discuss the case $N = 1$ and $\kappa = (1, \frac{1}{2}|p|^2)$.

13.2 The Quantum Energy-Transport Model

The quantum energy-transport equations are obtained from the general moment model (13.4) in the case $N = 1$ with the weight functions $\kappa(p) = (1, \frac{1}{2}|p|^2)$. The model was first derived by Degond, Méhats, and Ringhofer in [1]. We recall from Lemma 12.9 that

$$\langle \theta[V]w\rangle = 0, \quad \langle p\theta[V]w\rangle = -\langle w\rangle \nabla V \tag{13.6}$$

for all functions w. We also need to compute higher-order moments of the potential operator (12.3).

Lemma 13.2 (Moments of the potential operator). *It holds, for all functions $w = w(x,p,t)$,*

$$\left\langle \frac{1}{2}|p|^2\theta[V]w\right\rangle = -\langle pw\rangle \cdot \nabla V,$$

$$\left\langle \frac{1}{2}p|p|^2\theta[V]w\right\rangle = -\left(\langle p\otimes pw\rangle + \left\langle \frac{1}{2}|p|^2w\right\rangle \mathrm{Id}\right)\nabla V + \frac{\varepsilon^2}{8}\langle w\rangle \nabla\Delta V.$$

Proof. We recall the formula (12.11):

$$\phi(x,0,t) = \frac{1}{(2\pi)^3}\int_{\mathbb{R}^3\times\mathbb{R}^3} \phi(x,\eta,t)e^{ip\cdot\eta}\,d\eta\,dp$$

for all functions ϕ. Then, integrating by parts, we compute

$$\left\langle \frac{1}{2}|p|^2\theta[V]w \right\rangle = \int_{\mathbb{R}^9} (\delta V)(x,\eta,t)w(x,p',t)e^{-ip'\cdot\eta}(-\Delta_\eta e^{ip\cdot\eta})\frac{dp'\,d\eta\,dp}{(2\pi)^3(2\pi\varepsilon)^3}$$

$$= -\int_{\mathbb{R}^9} \Delta_\eta\left(\delta V e^{-ip'\cdot\eta}\right)w(x,p',t)e^{ip\cdot\eta}\frac{dp'\,d\eta\,dp}{(2\pi)^3(2\pi\varepsilon)^3}$$

$$= -\int_{\mathbb{R}^3} \Delta_\eta\left(\delta V e^{-ip'\cdot\eta}\right)\Big|_{\eta=0}w(x,p',t)\frac{dp'}{(2\pi\varepsilon)^3}.$$

Since $(\delta V)(x,0,t) = \Delta_\eta(\delta V)(x,0,t) = 0$ and $\nabla_\eta(\delta V)(x,0,t) = i\nabla V(x,t)$, the above expression becomes

$$\left\langle \frac{1}{2}|p|^2\theta[V]w \right\rangle = -2\int_{\mathbb{R}^3} p'\cdot\nabla V(x,t)w(x,p',t)\frac{dp'}{(2\pi\varepsilon)^3} = -\langle pw\rangle\cdot\nabla V.$$

In a similar way, we can prove the second identity. Employing $\nabla_\eta\Delta_\eta(\delta V)(x,0,t) = i(\varepsilon^2/4)\nabla\Delta V$, we obtain

$$\left\langle \frac{1}{2}p|p|^2\theta[V]w \right\rangle = -i\int_{\mathbb{R}^9} \nabla_\eta\Delta_\eta(\delta V e^{-ip'\cdot\eta})w(x,p',t)e^{ip\cdot\eta}\frac{dp'\,d\eta\,dp}{(2\pi)^3(2\pi\varepsilon)^3}$$

$$= \int_{\mathbb{R}^3} \left(\frac{\varepsilon^2}{4}\nabla\Delta V - 2(p'\otimes p')\nabla V - |p'|^2\nabla V\right)w(x,p',t)\frac{dp'}{(2\pi\varepsilon)^3}$$

$$= \frac{\varepsilon^2}{8}\nabla\Delta V\langle w\rangle - \langle p\otimes pw\rangle\nabla V - \langle\tfrac{1}{2}|p|^2w\rangle\nabla V.$$

This finishes the proof.   □

With the above result, we can simplify the moment system (13.4).

Theorem 13.3 (Nonlocal quantum energy-transport equations). *We assume that the operator Q_1 in (13.3) conserves mass, $\langle Q_1(w)\rangle = 0$ for all functions w. Let V be a given potential and w_α be a solution of the Wigner–Boltzmann equation (13.1) and (13.2). We choose the weight functions $\kappa(p) = (1, \frac{1}{2}|p|^2)$. Then, in the formal limit $\alpha \to 0$, w_α converges to $M[w]$ which solves the* nonlocal quantum energy-transport equations

$$\partial_t n - \mathrm{div}\,J_n = 0, \quad \partial_t(ne) - \mathrm{div}\,J_e + J_n\cdot\nabla V = W, \tag{13.7}$$

where $n = \langle M[w]\rangle$ is the particle density, $ne = \langle\frac{1}{2}|p|^2M[w]\rangle$ the energy density, $W = \langle\frac{1}{2}|p|^2Q_1(M[w])\rangle$ the averaged collision term, and the particle and energy current densities are given by, respectively,

$$J_n = \mathrm{div}\,P - n\nabla V, \quad J_e = \mathrm{div}\,U - (P + ne\,\mathrm{Id})\nabla V + \frac{\varepsilon^2}{8}n\nabla\Delta V, \tag{13.8}$$

where $P = \langle p\otimes pM[w]\rangle$ is the quantum stress tensor and $U = \langle\frac{1}{2}p\otimes p|p|^2M[w]\rangle$ is a fourth-order moment.

Proof. Using (13.6) and Lemma 13.2, the moment system (13.4) can be written as

$$\partial_t m_0 - \operatorname{div} J_n = 0, \quad \partial_t m_1 - \operatorname{div} J_e - \left\langle \frac{1}{2}|p|^2 \theta[V](p \cdot \nabla_x M[w] + \theta[V]M[w]) \right\rangle = W,$$
(13.9)

where $n = m_0 = \langle M[w] \rangle$, $ne = m_1 = \langle \frac{1}{2}|p|^2 M[w] \rangle$, and the current densities are defined by

$$J_n = \operatorname{div} \langle p \otimes p M[w] \rangle - \langle M[w] \rangle \nabla V = \operatorname{div} P - n \nabla V,$$

$$J_e = \operatorname{div} \left\langle \frac{1}{2} p \otimes p |p|^2 M[w] \right\rangle - \left(\langle p \otimes p M[w] \rangle + \left\langle \frac{1}{2}|p|^2 M[w] \right\rangle \operatorname{Id} \right) \nabla V$$

$$+ \frac{\varepsilon^2}{8} n \nabla \Delta V = \operatorname{div} U - (P + ne \operatorname{Id}) \nabla V + \frac{\varepsilon^2}{8} n \nabla \Delta V.$$

It remains to compute the integral in (13.9). With Lemma 13.2 and the second identity in (13.6), we obtain

$$\left\langle \frac{1}{2}|p|^2 \theta[V](p \cdot \nabla_x M[w] + \theta[V]M[w]) \right\rangle = -\langle p(p \cdot \nabla_x M[w] + \theta[V]M[w]) \rangle \cdot \nabla V$$

$$= -(\operatorname{div}_x \langle p \otimes p M[w] \rangle + \langle p \theta[V]M[w] \rangle) \cdot \nabla V$$

$$= -(\operatorname{div} P - n \nabla V) \cdot \nabla V = -J_n \cdot \nabla V,$$

which concludes the proof. □

The quantum Maxwellian is given by $M[w] = \operatorname{Exp}(A - |p|^2/2T)$, where A and T are the Lagrange multipliers related to n and ne by the nonlocal moment constraints

$$n = \frac{2}{(2\pi\varepsilon)^3} \int_{\mathbb{R}^3} \operatorname{Exp}\left(A - \frac{|p|^2}{2T}\right) dp, \quad ne = \frac{2}{(2\pi\varepsilon)^3} \int_{\mathbb{R}^3} \frac{|p|^2}{2} \operatorname{Exp}\left(A - \frac{|p|^2}{2T}\right) dp.$$

The variables A and T may be interpreted, in analogy to the classical case, as the "chemical potential" and "temperature", respectively. The quantum energy-transport model can be viewed equivalently as an evolution system for (n, ne) or for (A, T). The relations between (P, U) and (A, T) are also nonlocal in space.

The balance equations (13.7) for the particle and energy densities are the same as their semi-classical counterparts (6.10). However, the relations for the current densities (13.8) are significantly different from the semi-classical expressions. Indeed, the relation between (J_n, J_e) and (n, ne) is nonlocal in space, whereas in the semi-classical energy-transport equations, the fluxes are functions of the gradients of (n, ne). Furthermore, the quantum tensors P and U are generally not diagonal. In the semi-classical case, they are diagonal, since the classical Maxwellian is an even function with respect to each component p_j of the momentum vector p. In the quantum case, the Maxwellian $M[w]$ is even in p but not with respect to each component of p separately.

The quantum energy-transport model dissipates the quantum fluid entropy. To make this precise, we define the quantum fluid entropy similar as in Sect. 12.3 by

$$S = -\frac{2}{(2\pi\varepsilon)^3} \int_{\mathbb{R}^3 \times \mathbb{R}^3} w_0 (\text{Log}\, w_0 - 1)\, dp\, dx,$$

where $w_0 = \text{Exp}(A - |p|^2/2T)$. Inserting $\text{Log}\, w_0 = A - |p|^2/2T$, we infer that

$$S = -\frac{2}{(2\pi\varepsilon)^3} \int_{\mathbb{R}^3 \times \mathbb{R}^3} \left(A - \frac{|p|^2}{2T} - 1 \right) \text{Exp}\left(A - \frac{|p|^2}{2T} \right) dp\, dx$$

$$= -\int_{\mathbb{R}^3} \left(An + \frac{ne}{T} - n \right) dx. \tag{13.10}$$

The following result was shown in [1, Prop. 3.3].

Proposition 13.4. *Let (n, ne) be a solution of the quantum energy-transport model (13.7) and (13.8). Then the quantum fluid entropy (13.10) is nondecreasing in time:*

$$\frac{dS}{dt} \geq 0 \quad \text{for all } t \geq 0.$$

Proof. Let w_α be a solution of the Wigner–Boltzmann equation (13.1). Multiplying this equation by $\text{Log}\, w_\alpha$ and integrating over the phase space give

$$\int_{\mathbb{R}^3 \times \mathbb{R}^3} \partial_t w_\alpha \text{Log}\, w_\alpha\, dp\, dx + \alpha^{-1} \int_{\mathbb{R}^3 \times \mathbb{R}^3} (p \cdot \nabla_x w_\alpha + \theta[V]w_\alpha) \text{Log}\, w_\alpha\, dp\, dx$$

$$= \alpha^{-2} \int_{\mathbb{R}^3 \times \mathbb{R}^3} Q(w_\alpha) \text{Log}\, w_\alpha\, dp\, dx. \tag{13.11}$$

Since by Lemma 12.1, $\partial_t \text{Log}\, w_\alpha = (\partial_t w_\alpha)/w_\alpha$, the first integral of the above equation can be written as

$$\int_{\mathbb{R}^3 \times \mathbb{R}^3} \partial_t w_\alpha \text{Log}\, w_\alpha\, dp\, dx = \frac{d}{dt} \int_{\mathbb{R}^3 \times \mathbb{R}^3} w_\alpha (\text{Log}\, w_\alpha - 1)\, dp\, dx.$$

For the second integral, we observe that for $w_\alpha = W(\widehat{\rho}_\alpha)$,

$$p \cdot \nabla_x w_\alpha + \theta[V]w_\alpha = W([H, \widehat{\rho}_\alpha])$$

(this is a consequence of the results of Sect. 10.1). Thus, using the Parseval-type identity (12.22) and the cyclicity of the commutator,

$$\int_{\mathbb{R}^3 \times \mathbb{R}^3} (p \cdot \nabla_x w_\alpha + \theta[V]w_\alpha) \text{Log}\, w_\alpha\, dp\, dx = \frac{(2\pi\varepsilon)^3}{2} \int_{\mathbb{R}^3} \text{Tr}\,([H, \widehat{\rho}_\alpha]\log\widehat{\rho}_\alpha)\, dx$$

$$= \frac{(2\pi\varepsilon)^3}{2} \int_{\mathbb{R}^3} \text{Tr}\,(H[\widehat{\rho}_\alpha, \log\widehat{\rho}_\alpha])\, dx = 0,$$

since $\widehat{\rho}_\alpha$ commutes with any function of $\widehat{\rho}_\alpha$.

Finally, we claim that the right-hand side of (13.11) is nonpositive. Set $F(\widehat{\rho}) = \int \text{Tr}\,(\widehat{\rho}(\log\widehat{\rho} - 1))\, dx$. Its derivative is $DF(\widehat{\rho})(\widehat{\sigma}) = \int \text{Tr}\,(\log\widehat{\rho}\widehat{\sigma})\, dx$. We define for $\lambda \in [0, 1]$ the function $G(\lambda) = F(W^{-1}((1 - \lambda)M[w_\alpha] + \lambda w_\alpha))$. Employing again the identity (12.22), the derivative of G is given by

$$G'(\lambda) = DF\left(W^{-1}((1-\lambda)M[w_\alpha] + \lambda w_\alpha)\right)\left(W^{-1}(w_\alpha - M[w_\alpha])\right)$$

$$= \int_{\mathbb{R}^3} \text{Tr}\left(\log W^{-1}((1-\lambda)M[w_\alpha] + \lambda w_\alpha)W^{-1}(w_\alpha - M[w_\alpha])\right) dx$$

$$= \frac{2}{(2\pi\varepsilon)^3} \int_{\mathbb{R}^3 \times \mathbb{R}^3} \text{Log}\left((1-\lambda)M[w_\alpha] + \lambda w_\alpha\right)(w_\alpha - M[w_\alpha]) \, dp \, dx.$$

It can be seen that F is convex, and so is G. Therefore $G'(1) \geq G(1) - G(0)$, which is equivalent to

$$\frac{2}{(2\pi\varepsilon)^3} \int_{\mathbb{R}^3 \times \mathbb{R}^3} (\text{Log}\, w_\alpha)(w_\alpha - M[w_\alpha]) \, dp \, dx \geq F(W^{-1}(w_\alpha)) - F(W^{-1}(M[w_\alpha])).$$

The right-hand side is nonnegative since, by definition of the quantum Maxwellian, $W^{-1}(M[w_\alpha])$ is the maximizer of $-F$. Thus, (13.11) becomes

$$\frac{d}{dt} \int_{\mathbb{R}^3 \times \mathbb{R}^3} w_\alpha(\text{Log}\, w_\alpha - 1) \, dp \, dx \leq \alpha^{-2} \int_{\mathbb{R}^3 \times \mathbb{R}^3} (M[w_\alpha] - w_\alpha)\text{Log}\, w_\alpha \, dp \, dx \leq 0.$$

Passing to the limit $\alpha \to 0$, w_α converges to $w_0 = M[w] = \text{Exp}(A - |p|^2/2T)$, where (A, T) solves the quantum energy-transport equations, and thus,

$$\frac{d}{dt} \int_{\mathbb{R}^3 \times \mathbb{R}^3} w_0(\text{Log}\, w_0 - 1) \, dp \, dx \leq 0,$$

showing the assertion. □

It is possible to express $\text{div}\, P$ and $\text{div}\, U$ in terms of A and T and its derivatives [2]. If the "temperature" T is slowly varying, these expressions can be simplified and we obtain the following expansion. For a proof we refer to [2, Sect. 5.2].

Proposition 13.5. *For slowly varying variables T in the sense of $|\nabla T/T| = \delta \ll 1$, the following approximations hold:*

$$\text{div}\, P = n\nabla(A/T) + \mathcal{O}(\delta),$$

$$\text{div}\, U = (P + ne\,\text{Id}) \cdot \nabla(A/T) - \frac{\varepsilon^2}{8} n\nabla\Delta(A/T) + \mathcal{O}(\delta).$$

Moreover, the current densities can be approximated by

$$J_n = n\nabla(A/T - V) + \mathcal{O}(\delta),$$

$$J_e = (P + ne\,\text{Id}) \cdot \nabla(A/T - V) - \frac{\varepsilon^2}{8} n\nabla\Delta(A/T - V) + \mathcal{O}(\delta).$$

Unfortunately, no approximation of the quantum pressure P is known up to now. Furthermore, the mathematical structure of the above quantum energy-transport model is unclear. More explicit expressions are obtained in the $\mathcal{O}(\varepsilon^4)$ approxima-tion. For this, we need to expand P, U, and ne in terms of ε^2. If $\nabla \log T = \mathcal{O}(\varepsilon^2)$, some tedious computations lead to the following formulas:

$$P = nT \operatorname{Id} - \frac{\varepsilon^2}{12} n(\nabla \otimes \nabla) \log n + \mathcal{O}(\varepsilon^4), \quad ne = \frac{3}{2} nT - \frac{\varepsilon^2}{24} n \Delta \log n + \mathcal{O}(\varepsilon^4),$$

$$U = \frac{5}{2} nT^2 \operatorname{Id} - \frac{\varepsilon^2}{24} nT \left(\Delta \log n \operatorname{Id} + 7(\nabla \otimes \nabla) \log n \right) + \mathcal{O}(\varepsilon^4).$$

Equations (13.7) and (13.8) together with the above constitutive relations are called the *local quantum energy-transport equations*. Notice that the expressions for P and U differ from those presented in [1]. We expect that the local model possesses an entropic formulation similar to the classical energy-transport model (see Sect. 6.3) but currently, no entropic structure is known.

In Fig. 13.1 we illustrate the nonlocal and local quantum models derived in this and the previous chapter.

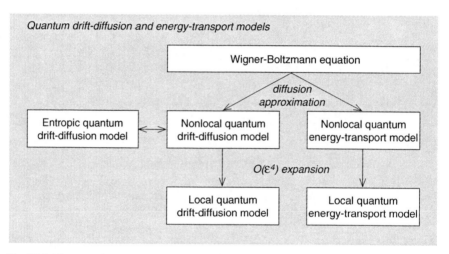

Fig. 13.1 Macroscopic quantum models and their relations

References

1. P. Degond, F. Méhats, and C. Ringhofer. Quantum energy-transport and drift-diffusion models. *J. Stat. Phys.* 118 (2005), 625–665.
2. P. Degond, S. Gallego, and F. Méhats. On quantum hydrodynamic and quantum energy trans-port models. *Commun. Math. Sci.* 5 (2007), 887–908.

Chapter 14
Quantum Hydrodynamic Equations

In the previous chapters, we have derived quantum macroscopic models from a Wigner–Boltzmann equation using a diffusion scaling. In this chapter, we show that, in analogy to the semi-classical situation, quantum hydrodynamic models can be derived by employing a hydrodynamic scaling. We present two derivations: one from the (mixed-state) Schrödinger equation and one from a Wigner–Boltzmann equation. This approach can be extended to general quantum moment hydrodynamics, presented in the final section.

14.1 Zero-Temperature Quantum Hydrodynamic Equations

It is well known since Madelung [1] that there exists a fluiddynamical description of the Schrödinger equation, also called the *Madelung hydrodynamic formulation* or *quantum fluid dynamics*. In this section we consider a single electron moving in a vacuum. Electron ensembles are studied in the following sections.

The quantum evolution of the particle with mass m is described by the single-state Schrödinger equation

$$i\hbar \partial_t \psi = -\frac{\hbar^2}{2m} \Delta \psi - qV(x,t)\psi, \quad t > 0, \quad \psi(\cdot,0) = \psi_I \quad \text{in } \mathbb{R}^3,$$

where ψ is the wave function and V the electric potential. The potential may be a given function or the solution of the Poisson equation. In the following, we assume that V is a given function. However, the following arguments do not change if V solves the Poisson equation. First we scale the equation by introducing reference values for the time τ, length λ, and potential U. We assume that the kinetic energy is of the same order as the electric energy,

$$m \left(\frac{\lambda}{\tau}\right)^2 = qU.$$

Jüngel, A.: *Quantum Hydrodynamic Equations*. Lect. Notes Phys. **773**, 283–308 (2009)
DOI 10.1007/978-3-540-89526-8_14

Then the scaled Schrödinger equation becomes

$$i\varepsilon\partial_t\psi = -\frac{\varepsilon^2}{2}\Delta\psi - V(x,t)\psi, \quad t>0, \quad \psi(\cdot,0)=\psi_I \quad \text{in } \mathbb{R}^3, \tag{14.1}$$

where the scaled Planck constant is the ratio between wave energy and kinetic energy:

$$\varepsilon = \frac{\hbar/\tau}{m(\lambda/\tau)^2} = \frac{\hbar\tau}{m\lambda^2}.$$

This is the same scaling as in Sect. 10.2 after choosing $U = k_B T_L/q$.

In order to derive a fluiddynamical formulation, we need to assume that the initial wave function is given in the WKB (Wentzel [2], Kramers [3], Brillouin [4]) state:

$$\psi_I = \sqrt{n_I}\exp(iS_I/\varepsilon), \tag{14.2}$$

where $n_I(x) \geq 0$ and $S_I(x) \in \mathbb{R}$ are some functions. Then, inserting the ansatz $\psi = \sqrt{n}\exp(iS/\varepsilon)$ in the Schrödinger equation leads to the following result.

Theorem 14.1 (Zero-temperature quantum hydrodynamic equations). *Let ψ be a solution of the initial-value problem* (14.1) *with initial datum* (14.2). *Then $n = |\psi|^2$, $J_n = -\varepsilon\,\text{Im}(\overline{\psi}\nabla\psi)$ are a solution of the zero-temperature quantum hydrodynamic or Madelung equations*

$$\partial_t n - \text{div}\,J_n = 0, \quad \partial_t J_n - \text{div}\left(\frac{J_n \otimes J_n}{n}\right) + n\nabla V + \frac{\varepsilon^2}{2}n\nabla\left(\frac{\Delta\sqrt{n}}{\sqrt{n}}\right) = 0, \tag{14.3}$$

$$n(\cdot,0)=n_I, \quad J_n(\cdot,0)=J_I \quad \text{in } \mathbb{R}^3,$$

where the initial data are given by $n_I = |\psi_I|^2$ and $J_I = -n_I\nabla S_I$, as long as $n > 0$ in $\mathbb{R}^3$. On the other hand, let (n,S) be a (smooth) solution of

$$\partial_t n + \text{div}(n\nabla S) = 0, \quad \partial_t S + \frac{1}{2}|\nabla S|^2 - V - \frac{\varepsilon^2}{2}\frac{\Delta\sqrt{n}}{\sqrt{n}} = 0, \quad t>0, \tag{14.4}$$

$$n(\cdot,0)=n_I, \quad S(\cdot,0)=S_I \quad \text{in } \mathbb{R}^3, \tag{14.5}$$

such that $n > 0$ in $\mathbb{R}^3$, $t > 0$. Then $\psi = \sqrt{n}\exp(iS/\varepsilon)$ solves the Schrödinger equation (14.1) *with initial datum* (14.2).

Proof. Let ψ be a solution of the initial-value problem (14.1) with initial datum (14.2). As long as $|\psi| > 0$, we can decompose $\psi = \sqrt{n}\exp(iS/\varepsilon)$, where $n = |\psi|^2$ and S is some phase function. Then

$$J_n = -\varepsilon\,\text{Im}(\overline{\psi}\nabla\psi) = -\varepsilon\,\text{Im}\left(\sqrt{n}\nabla\sqrt{n} + \frac{i}{\varepsilon}n\nabla S\right) = -n\nabla S.$$

Thus, n and J_n satisfy the initial conditions. Inserting the decomposition $\psi = \sqrt{n}\exp(iS/\varepsilon)$ into the Schrödinger equation (14.1) gives, after division by the factor $\exp(iS/\varepsilon)$,

$$\frac{i\varepsilon}{2}\frac{\partial_t n}{\sqrt{n}} - \sqrt{n}\partial_t S = -\frac{\varepsilon^2}{2}\left(\Delta\sqrt{n} + \frac{2i}{\varepsilon}\nabla\sqrt{n}\cdot\nabla S + \frac{i}{\varepsilon}\sqrt{n}\Delta S - \frac{\sqrt{n}}{\varepsilon^2}|\nabla S|^2\right) - \sqrt{n}V.$$

(14.6)

The imaginary part of this equation equals

$$\partial_t n = -2\sqrt{n}\nabla\sqrt{n}\cdot\nabla S - n\Delta S = -\text{div}(n\nabla S),$$

which is the first equation of (14.3). Dividing the real part of (14.6) by $\sqrt{n}$, then differentiating the resulting equation with respect to x and multiplying it by n, we infer, using the first equation in (14.3), that

$$n\nabla V + \frac{\varepsilon^2}{2}n\nabla\left(\frac{\Delta\sqrt{n}}{\sqrt{n}}\right) = n\partial_t(\nabla S) + \frac{1}{2}n\nabla|\nabla S|^2$$

$$= \partial_t(n\nabla S) - (\text{div}\,J_n)\nabla S + \frac{1}{2}n\nabla|\nabla S|^2$$

$$= -\partial_t J_n + \text{div}\left(\frac{J_n \otimes J_n}{n}\right).$$

(14.7)

For the last equality, we have employed the identity

$$\frac{1}{2}n\nabla|\nabla S|^2 = n((\nabla\otimes\nabla)S)\nabla S = \text{div}(n\nabla S\otimes\nabla S) - \text{div}(n\nabla S)\nabla S,$$

where $(\nabla\otimes\nabla)S$ denotes the Hessian of S. Equation (14.7) is the second equation in (14.3).

Let (n, S) be a solution of (14.4) and (14.5) with $n > 0$ and set $\psi = \sqrt{n}\exp(iS/\varepsilon)$. Then, differentiating ψ gives

$$i\varepsilon\partial_t\psi + \frac{\varepsilon^2}{2}\Delta\psi = e^{iS/\varepsilon}\left(i\varepsilon\frac{\partial_t n}{2\sqrt{n}} - \sqrt{n}\partial_t S + \frac{\varepsilon^2}{2}\Delta\sqrt{n} + i\varepsilon\nabla\sqrt{n}\cdot\nabla S\right.$$

$$\left. + \frac{i\varepsilon}{2}\sqrt{n}\Delta S - \frac{\sqrt{n}}{2}|\nabla S|^2\right)$$

$$= e^{iS/\varepsilon}\left(-\frac{i\varepsilon}{2}\frac{\text{div}(n\nabla S)}{\sqrt{n}} + i\varepsilon\nabla\sqrt{n}\cdot\nabla S + \frac{i\varepsilon}{2}\sqrt{n}\Delta S - \sqrt{n}V\right)$$

$$= -\sqrt{n}e^{iS/\varepsilon}V = -V\psi.$$

Thus, ψ solves the Schrödinger equation. □

The system (14.3) is the quantum analogue of the classical pressureless Euler equations of gas dynamics, which are obtained in the classical limit $\varepsilon \to 0$. This limit is made rigorous in some sense by Gasser and Markowich [5] (also see [6]). The second equation in (14.4) is also called a *quantum Hamilton–Jacobi equation* [7]. We notice that the above derivation requires an irrotational initial velocity J_n/n since $\text{curl}(J_n/n) = -\text{curl}(\nabla S) = 0$. The quantum term can be interpreted as a quantum self-potential with the so-called quantum or *Bohm potential* $\phi_B = \Delta\sqrt{n}/\sqrt{n}$ or as a quantum stress term:

$$\frac{\varepsilon^2}{2}n\nabla\left(\frac{\Delta\sqrt{n}}{\sqrt{n}}\right) = \frac{\varepsilon^2}{4}\mathrm{div}\left(n(\nabla\otimes\nabla)\log n\right),$$

where $P = (\varepsilon^2/4)n(\nabla\otimes\nabla)\log n$ is a nondiagonal stress tensor. The quantum hydrodynamic equations are employed in Bohmian mechanics and for the description of quantum trajectories [7]. They are also used, for instance, for simulations of photodissociation problems [8], superfluidity models [9], collinear chemical reactions [10], and for weakly interacting Bose gases [11].

There is a problem with the formulation (14.4) if vacuum occurs, i.e., if $n = 0$ locally. In this situation, the phase S is not defined which manifests in the fact that the Bohm potential may become singular at vacuum points. The appearance and properties of these vacuum points inside a nondissipative shock wave in the quantum hydrodynamic equations are studied by El et al. [12]. It is shown by Gasser and Markowich [5] that the formulation (14.3) has generally better mathematical properties than (14.4). In fact, it can be shown that $n\nabla\phi_B$ is an element of a Sobolev space with negative index and not just a distribution. A mathematical formulation of the quantum hydrodynamic equations based on polar factorization was given recently by Antonelli and Marcati [13].

Another problem is the reconstruction of the initial datum ψ_I in terms of the variables n_I and J_I. In the above theorem it is explicitly required that ψ_I is given in terms of n_I and S_I. This problem is connected to a more general important problem in physics, the so-called *Pauli problem* [14], regarding the possibility to reconstruct a pure quantum state by knowing a finite set of measurements of the state, in our case: the particle and current densities. Here, the possible existence of vacuum points generally makes this reconstruction impossible (see [14] and references therein).

14.2 Mixed-State Schrödinger Models and Quantum Hydrodynamics

The quantum hydrodynamic model of the previous section is derived for a single particle and therefore, it does not contain a temperature term. In order to include temperature, we consider now an electron ensemble represented by a mixed state (see Sect. 10.1). A mixed quantum state consists of a sequence of occupation probabilities $\lambda_j \geq 0$ ($j \in \mathbb{N}$) for the jth state ψ_j described by the scaled Schrödinger equation [15]

$$i\varepsilon\partial_t\psi_j = -\frac{\varepsilon^2}{2}\Delta\psi_j - V(x,t)\psi_j, \quad t > 0, \quad \psi_j(\cdot,0) = \psi_j^0 \quad \text{in } \mathbb{R}^3, \tag{14.8}$$

where the electric potential is assumed to be given. The occupation numbers satisfy $\sum_{j=1}^{\infty} \lambda_j = 1$, which means that the probability of finding the electron ensemble in any of the quantum states is one.

We define the single-state particle and current densities of the jth state as in the previous section as

$$n_j = |\psi_j|^2, \quad J_j = -\varepsilon \operatorname{Im}(\overline{\psi}_j \nabla \psi_j), \quad j \in \mathbb{N}.$$

We claim that, following [15], the total electron density n and current density J of the mixed state,

$$n = \sum_{j=1}^{\infty} \lambda_j |\psi_j|^2, \quad J = \sum_{j=1}^{\infty} \lambda_j J_j, \tag{14.9}$$

are a solution of the quantum hydrodynamic equations with a temperature tensor.

Theorem 14.2 (Quantum hydrodynamic equations). *Let ψ_j be single-state solutions of the Schrödinger equation (14.8) with occupation numbers λ_j of the jth quantum state. Then (n, J), defined in (14.9), is a solution of the quantum hydrodynamic equations*

$$\partial_t n - \operatorname{div} J = 0, \tag{14.10}$$

$$\partial_t J - \operatorname{div}\left(\frac{J \otimes J}{n} + n\theta\right) + n\nabla V + \frac{\varepsilon^2}{2} n \nabla\left(\frac{\Delta \sqrt{n}}{\sqrt{n}}\right) = 0, \tag{14.11}$$

where $x \in \mathbb{R}^3$ and $t > 0$, with initial conditions

$$n(\cdot, 0) = \sum_{j=1}^{\infty} \lambda_j |\psi_j^0|^2, \quad J(\cdot, 0) = -\varepsilon \sum_{j=1}^{\infty} \lambda_j \operatorname{Im}(\overline{\psi}_j^0 \nabla \psi_j^0) \quad \text{in } \mathbb{R}^3.$$

The temperature tensor θ is defined by $\theta = \theta_{\mathrm{cu}} + \theta_{\mathrm{os}}$, where the "current temperature" and "osmotic temperature" are given by, respectively,

$$\theta_{\mathrm{cu}} = \sum_{j=1}^{\infty} \lambda_j \frac{n_j}{n} (u_{\mathrm{cu},j} - u_{\mathrm{cu}}) \otimes (u_{\mathrm{cu},j} - u_{\mathrm{cu}}),$$

$$\theta_{\mathrm{os}} = \sum_{j=1}^{\infty} \lambda_j \frac{n_j}{n} (u_{\mathrm{os},j} - u_{\mathrm{os}}) \otimes (u_{\mathrm{os},j} - u_{\mathrm{os}}),$$

and the variables

$$u_{\mathrm{cu},j} = -\frac{J_j}{n_j}, \quad u_{\mathrm{cu}} = -\frac{J}{n}, \quad u_{\mathrm{os},j} = \frac{\varepsilon}{2} \nabla \log n_j, \quad u_{\mathrm{os}} = \frac{\varepsilon}{2} \nabla \log n$$

are called the "current velocities" and "osmotic velocities", respectively.

The notion "osmotic" comes from the fact that the quantum term can be written as the divergence of the quantum stress tensor $P = (\varepsilon^2/4) n (\nabla \otimes \nabla) \log n$ [15].

Proof. The pair (n_j, J_j) solves the single-state quantum hydrodynamic equations (14.3) with initial conditions

$$n_j(\cdot, 0) = |\psi_j^0|^2, \quad J_j(\cdot, 0) = -\varepsilon \operatorname{Im}(\overline{\psi}_j^0 \nabla \psi_j^0).$$

Multiplication of (14.3) by λ_j and summation over j yields

$$\partial_t n - \operatorname{div} J = 0,$$

$$\partial_t J - \sum_{j=1}^{\infty} \lambda_j \operatorname{div}\left(\frac{J_j \otimes J_j}{n_j}\right) + n\nabla V + \frac{\varepsilon^2}{2} \sum_{j=1}^{\infty} \lambda_j n_j \nabla\left(\frac{\Delta\sqrt{n_j}}{\sqrt{n_j}}\right) = 0. \qquad (14.12)$$

We rewrite the second and fourth term of the second equation. With the definitions of the "current temperature" and "current velocity", we obtain

$$\sum_{j=1}^{\infty} \lambda_j \operatorname{div}\left(\frac{J_j \otimes J_j}{n_j}\right) = \sum_{j=1}^{\infty} \lambda_j \operatorname{div}\left(n_j u_{\mathrm{cu},j} \otimes u_{\mathrm{cu},j}\right)$$

$$= \sum_{j=1}^{\infty} \lambda_j \operatorname{div}\left(n_j(u_{\mathrm{cu},j} - u_{\mathrm{cu}}) \otimes (u_{\mathrm{cu},j} - u_{\mathrm{cu}}) + 2n_j u_{\mathrm{cu},j} \otimes u_{\mathrm{cu}}\right) - \operatorname{div}\left(n u_{\mathrm{cu}} \otimes u_{\mathrm{cu}}\right)$$

$$= \operatorname{div}\left(n\theta_{\mathrm{cu}}\right) + 2 \sum_{j=1}^{\infty} \operatorname{div}\left(\lambda_j J_j \otimes \frac{J}{n}\right) - \operatorname{div}\left(\frac{J \otimes J}{n}\right)$$

$$= \operatorname{div}\left(n\theta_{\mathrm{cu}}\right) + \operatorname{div}\left(\frac{J \otimes J}{n}\right).$$

Furthermore, employing the definitions of the "osmotic temperature" and "osmotic velocity", we compute

$$\frac{\varepsilon^2}{2} \sum_{j=1}^{\infty} \lambda_j n_j \nabla\left(\frac{\Delta\sqrt{n_j}}{\sqrt{n_j}}\right) = \frac{\varepsilon^2}{4} \sum_{j=1}^{\infty} \lambda_j \operatorname{div}\left((\nabla \otimes \nabla)n_j - \frac{\nabla n_j \otimes \nabla n_j}{n_j}\right)$$

$$= \frac{\varepsilon^2}{4} \sum_{j=1}^{\infty} \lambda_j \operatorname{div}\left((\nabla \otimes \nabla)n_j + \frac{n_j}{n}\frac{\nabla n \otimes \nabla n}{n} - 2\frac{\nabla n \otimes \nabla n_j}{n}\right.$$

$$\left. - n_j\left(\frac{\nabla n_j}{n_j} - \frac{\nabla n}{n}\right) \otimes \left(\frac{\nabla n_j}{n_j} - \frac{\nabla n}{n}\right)\right)$$

$$= \frac{\varepsilon^2}{4} \operatorname{div}\left((\nabla \otimes \nabla)n - \frac{\nabla n \otimes \nabla n}{n}\right) - \operatorname{div}\left(n\theta_{\mathrm{os}}\right)$$

$$= \frac{\varepsilon^2}{2} n\nabla\left(\frac{\Delta\sqrt{n}}{\sqrt{n}}\right) - \operatorname{div}\left(n\theta_{\mathrm{os}}\right).$$

Inserting these expressions into (14.12) gives (14.11). □

The temperature tensor cannot be expressed in terms of the total particle and current densities without further assumptions, and as in the derivation of the semi-classical hydrodynamic equations, we need a closure condition to obtain a closed set of equations. In the literature, the following closures were employed.

Assume that the temperature tensor is diagonal with equal entries on the diagonal, $\theta = T\,\mathrm{Id}$, where T is a scalar temperature and Id the identity matrix. Then we can close Eqs. (14.10) and (14.11) by taking T to be constant and refer to this case as the *isothermal quantum hydrodynamic model*. The isothermal model was first proposed by Grubin and Kreskovsky in the context of semiconductor modeling [16]. If T is given by $T(n) = T_0 n^\alpha$ for some $\alpha > 0$, we refer to (14.10) and (14.11) as the *isentropic quantum hydrodynamic models*.

Another closure was proposed by Gasser et al. [17] using small temperature and small (scaled) Planck constant asymptotics. In this work, the initial wave function is written as

$$\psi_j^0 = \sqrt{n_j^0}\exp\left(iS_j^0/\varepsilon\right)$$

and the initial phase functions are assumed to satisfy

$$S_j^0 = S_I + \mathscr{O}\left(\sqrt{\delta}\right) + \mathscr{O}\left(\varepsilon^2\right), \quad j \in \mathbb{N},$$

where $\delta > 0$ is a small parameter. An initial state satisfying this condition is called "almost coherent". Initially, the current velocity $u_{\mathrm{cu},j} = -J_j/n_j = \nabla S_j$ satisfies

$$u_j^0 = \nabla S_j^0 = \nabla S_I + \mathscr{O}\left(\sqrt{\delta}\right) + \mathscr{O}\left(\varepsilon^2\right).$$

It can be shown that then for all times,

$$u_{\mathrm{cu},j} = -\frac{J_j}{n_j} = \nabla S + \mathscr{O}\left(\sqrt{\delta}\right) + \mathscr{O}\left(\varepsilon^2\right).$$

This implies that the difference $J_j/n_j - J/n$ is of the order $\mathscr{O}(\sqrt{\delta}) + \mathscr{O}\left(\varepsilon^2\right)$ and hence, the current and osmotic temperatures, defined in Theorem 14.2, can be written as

$$\theta_{\mathrm{cu}} = \mathscr{O}(\delta) + \mathscr{O}\left(\sqrt{\delta}\varepsilon^2\right) + \mathscr{O}\left(\varepsilon^4\right), \quad \theta_{\mathrm{os}} = \mathscr{O}\left(\varepsilon^2\right).$$

Then the temperature $\theta = \theta_{\mathrm{cu}} + \theta_{\mathrm{os}}$ satisfies

$$\theta = \theta_{\mathrm{os}} + \mathscr{O}(\delta) + \mathscr{O}\left(\sqrt{\delta}\varepsilon^2\right) + \mathscr{O}\left(\varepsilon^4\right).$$

Now, we multiply as in the proof of Theorem 14.2 the single-state Schrödinger equations (14.8) by λ_j and sum over j. After similar calculations as in the proof of the previous theorem, we arrive at Eqs. (14.10) and (14.11) for the particle density n and the current density J, where now the temperature tensor θ solves the energy equation

$$\partial_t E_{j\ell} + \sum_{m=1}^{3} \frac{\partial}{\partial x_m}\left(\frac{J_m}{n}E_{j\ell} + \frac{J_j}{2n}P_{\ell m} + \frac{J_\ell}{2n}P_{jm}\right) + \frac{1}{2}\left(J_j\frac{\partial V}{\partial x_\ell} + J_\ell\frac{\partial V}{\partial x_j}\right)$$
$$- \frac{\varepsilon^2}{8}\sum_{m=1}^{3}\frac{\partial}{\partial x_m}\left(n\frac{\partial^2}{\partial x_j \partial x_\ell}\left(\frac{J_m}{n}\right)\right) = \mathscr{O}(\delta) + \mathscr{O}(\sqrt{\delta}\varepsilon^2) + \mathscr{O}\left(\varepsilon^4\right), \quad (14.13)$$

and the energy tensor $E = (E_{j\ell})$ and stress tensor $P = (P_{j\ell})$ are given by

$$E = \frac{1}{2}\left(\frac{J \otimes J}{n} + n\theta - \frac{\varepsilon^2}{4}n(\nabla \otimes \nabla)\log n\right), \quad P = n\theta - \frac{\varepsilon^2}{4}n(\nabla \otimes \nabla)\log n.$$

Thus, for "almost coherent" initial states with $\sqrt{\delta} \ll \varepsilon$, the continuity equation (14.10), the momentum equation (14.11), and the energy equation (14.13) are a closed set of equations which are correct up to order $\mathcal{O}(\delta) + \mathcal{O}(\sqrt{\delta}\varepsilon^2) + \mathcal{O}(\varepsilon^4)$. In the case of exactly coherent initial states, the parameter δ vanishes, and the quantum hydrodynamic equations are valid up to order $\mathcal{O}(\varepsilon^4)$. It is mentioned in [17] that the standard single-state closure is not valid for *fixed* (scaled) Planck constant. The above derivation, however, is based on small temperature *and* small Planck constant asymptotics.

When the quantum hydrodynamic equations are considered in a bounded domain, some boundary conditions are needed. In the literature, the following boundary conditions for irrotational flows were suggested [18]. We consider the steady-state isothermal equations

$$\text{div}\, J = 0, \quad -\text{div}\left(\frac{J \otimes J}{n}\right) - \nabla n + n\nabla V + \frac{\varepsilon^2}{6}n\nabla\left(\frac{\Delta\sqrt{n}}{\sqrt{n}}\right) = 0 \quad \text{in } \Omega,$$

and we assume that the velocity J/n is irrotational and that there exists a velocity potential S such that $J = -n\nabla S$. Since $\text{div}\,(J \otimes J/n) = \frac{1}{2}n\nabla|\nabla S|^2$, we can write the momentum equation as

$$n\nabla\left(\frac{1}{2}|\nabla S|^2 + \log n - V - \frac{\varepsilon^2}{6}\frac{\Delta\sqrt{n}}{\sqrt{n}}\right) = 0.$$

Supposing that $n > 0$, we infer that

$$\frac{1}{2}|\nabla S|^2 + \log n - V - \frac{\varepsilon^2}{6}\frac{\Delta\sqrt{n}}{\sqrt{n}} = 0. \tag{14.14}$$

The integration constant can be assumed to vanish by choosing a reference point for the electric potential. This formulation allows us to write the stationary quantum hydrodynamic equations, together with a self-consistent electric potential, as a system of second-order differential equations:

$$\text{div}\,(n\nabla S) = 0, \quad \lambda_D^2\Delta V = n - C(x),$$
$$\frac{\varepsilon^2}{6}\Delta\sqrt{n} = \frac{1}{2}\sqrt{n}|\nabla S|^2 + \sqrt{n}\log n - \sqrt{n}V.$$

Thus, we need boundary conditions for the functions S, n, and V.

The boundary $\partial\Omega$ is assumed to consist of two parts: the Dirichlet part Γ_D and the insulating part Γ_N, where $\Gamma_D \cup \Gamma_N = \partial\Omega$ and $\Gamma_D \cap \Gamma_N = \emptyset$. We suppose that the normal derivatives of the variables vanish on the insulating boundary,

$$\nabla n \cdot \eta = \nabla S \cdot \eta = \nabla V \cdot \eta = 0 \quad \text{on } \Gamma_N,$$

where η denotes the exterior unit normal vector on $\partial\Omega$. On the Dirichlet part, the boundary data are assumed to be the superposition of the thermal equilibrium functions (n_{eq}, S_{eq}, V_{eq}) and the applied potential U:

$$n = n_{eq}, \quad S = S_{eq} + U, \quad V = V_{eq} + U \quad \text{on } \Gamma_D.$$

The thermal equilibrium is defined by $J = 0$ or (as n is positive) $S = \text{const}$. By fixing the reference point for S, we can suppose that $S_{eq} = 0$. We assume further that

- the total space charge $C - n_{eq}$ vanishes on the boundary and
- no quantum effects occur on the boundary, i.e., $\Delta\sqrt{n_{eq}}/\sqrt{n_{eq}} = 0$.

Then we obtain from (14.14)

$$\frac{1}{2}|\nabla S_{eq}|^2 + \log n_{eq} - V_{eq} = 0$$

and, since $S_{eq} = 0$, $V_{eq} = \log n_{eq}$ on Γ_D. Therefore, the Dirichlet boundary conditions are given by

$$n = C, \quad S = U, \quad V = \log C + U \quad \text{on } \Gamma_D.$$

These conditions were employed to prove the existence and uniqueness of solutions for subsonic flow in [18].

14.3 Wigner–Boltzmann Equations and Quantum Hydrodynamics

The quantum hydrodynamic model of the previous section does not include collisional effects since the Schrödinger equation only models ballistic transport. In order to allow for collisional phenomena, we employ a (scaled) Wigner–Boltzmann equation

$$\partial_t w + p \cdot \nabla_x w + \theta[V]w = Q(w), \quad t > 0, \quad w(x,p,0) = w_I(x,p), \quad (x,p) \in \mathbb{R}^6$$

(see Sect. 11.3 for a discussion of Wigner–Boltzmann models). The electric potential V is assumed to be a solution of the Poisson equation

$$\lambda_D^2 \Delta V = \frac{2}{(2\pi\varepsilon)^3}\int_{\mathbb{R}^3} w\,dp - C(x), \quad x \in \mathbb{R}^3. \tag{14.15}$$

The following presentation is based on [19] and [20]. A derivation of macroscopic quantum models from a collisional Liouville–von Neumann equation, leading to similar results as those presented below, can be found in [21]. We assume that the collision operator $Q(w)$ is the sum of two operators Q_0 and Q_1. We employ as in Sect. 9.1 a hydrodynamic scaling and replace x by x/α and t by t/α, where $\alpha > 0$

is the ratio of the mean free paths corresponding to Q_0 and Q_1, respectively. We assume that Q_0 models collisions which occur more frequently than those described by Q_1, thus implying that $\alpha \ll 1$. Then we can write the Wigner–Boltzmann equation as

$$\alpha \partial_t w + \alpha (p \cdot \nabla_x w + \theta[V]w) = Q_0(w) + \alpha Q_1(w), \quad w(\cdot,\cdot,0) = w_I. \qquad (14.16)$$

The collisions modeled by Q_0 are supposed to conserve mass, momentum, and energy,

$$\langle Q_0(w)\kappa_j(p)\rangle = 0 \quad \text{for all } w, \ j = 0,1,2, \qquad (14.17)$$

where the brackets are defined by $\langle g(p)\rangle = 2(2\pi\varepsilon)^{-3}\int_{\mathbb{R}^3} g(p)\,dp$ and

$$\kappa_0(p) = 1, \quad \kappa_1(p) = p, \quad \kappa_2(p) = \frac{1}{2}|p|^2$$

are the weight functions. The collision operator Q_1 remains unspecified, but we will consider some examples below (see (14.43) and (14.47)). For the moment, we only suppose that Q_1 conserves mass:

$$\langle Q_1(w)\rangle = 0 \quad \text{for all } w. \qquad (14.18)$$

We also assume that the kernel of Q_0 consists exactly of the quantum Maxwellians. We need to specify this notion.

Let w be given. By Lemma 12.2, the maximizer of the quantum entropy

$$S(w) = -\frac{2}{(2\pi\varepsilon)^3} \int_{\mathbb{R}^3 \times \mathbb{R}^3} w\left(\mathrm{Log}\,w - 1 + \frac{|p|^2}{2} - V\right) dx\,dp$$

under the constraints

$$\frac{2}{(2\pi\varepsilon)^3} \int_{\mathbb{R}^3} w(x,p,t) \begin{pmatrix} 1 \\ p \\ \frac{1}{2}|p|^2 \end{pmatrix} dp = \begin{pmatrix} n \\ nu \\ ne \end{pmatrix} (x,t), \quad x \in \mathbb{R}^3, \ t > 0, \qquad (14.19)$$

where $n = \langle w\rangle$, $nu = \langle pw\rangle$, and $ne = \langle \frac{1}{2}|p|^2 w\rangle$, is (if it exists) given by

$$M[w](x,t) = \mathrm{Exp}\left(A(x,t) - \frac{|p - v(x,t)|^2}{2T(x,t)}\right), \qquad (14.20)$$

where Log and Exp are the quantum logarithm and the quantum exponential, respectively, defined in Sect. 12.1. The Lagrange multipliers A, v, and T are uniquely determined by the moments of w. In the classical setting, they correspond to the logarithm of the particle density, the velocity, and the temperature, respectively. Thus, the assumption on the kernel of Q_0 can be formulated as

$$Q_0(w) = 0 \quad \text{if and only if} \quad w = M[w]. \qquad (14.21)$$

Derivation of the quantum hydrodynamic equations. The quantum hydrodynamic equations are now derived in the formal limit $\alpha \to 0$. Let w_α be a solution of the Wigner–Boltzmann equation (14.16) and V_α be a solution of the Poisson equation (14.15). The limit $\alpha \to 0$ in (14.16) gives $Q_0(w) = 0$, where $w = \lim_{\alpha \to 0} w_\alpha$, and hence, by assumption, $w = M[w]$. Multiplying (14.16) by the weight functions $\kappa_j(p)$ and integrating over $p \in \mathbb{R}^3$ yield the moment equations

$$\partial_t \langle \kappa_j w_\alpha \rangle + \mathrm{div}_x \langle p \kappa_j w_\alpha \rangle + \langle \kappa_j \theta[V] w_\alpha \rangle = \langle \kappa_j Q_1(w_\alpha) \rangle,$$

since $\langle \kappa_j Q_0(w_\alpha) \rangle = 0$ by assumption (14.17). The limit $\alpha \to 0$ in the moment equations then gives, taking into account the constraints (14.19), the property (14.18) on Q_1, and the moments of the potential operator, see (13.6) and Lemma 13.2,

$$\partial_t n - \mathrm{div}\, J_n = 0, \tag{14.22}$$

$$\partial_t J_n - \mathrm{div} \langle p \otimes p M[w] \rangle + n \nabla V = -\langle p Q_1(M[w]) \rangle, \tag{14.23}$$

$$\partial_t (ne) + \mathrm{div} \left\langle \frac{1}{2} |p|^2 p M[w] \right\rangle + J_n \cdot \nabla V = \left\langle \frac{1}{2} |p|^2 Q_1(M[w]) \right\rangle, \tag{14.24}$$

where $J_n = -nu$ is the current density. Defining the quantum stress tensor P and the quantum heat flux q by

$$P = \langle (p - u) \otimes (p - u) M[w] \rangle, \quad q = \left\langle \frac{1}{2} (p - u) |p - u|^2 M[w] \right\rangle, \tag{14.25}$$

we can simplify the integrals $\langle p \otimes p M[w] \rangle$ and $\langle \frac{1}{2} p |p|^2 M[w] \rangle$ slightly:

$$\langle p \otimes p M[w] \rangle = P + \frac{J_n \otimes J_n}{n}, \quad \left\langle \frac{1}{2} p |p|^2 M[w] \right\rangle = -(P + ne\,\mathrm{Id}) \frac{J_n}{n} + q.$$

The result is summarized in the following theorem.

Theorem 14.3 (Nonlocal quantum hydrodynamic equations). *Let the collision operator satisfy assumptions (14.17), (14.18), and (14.21). Let (w_α, V_α) be a solution of the Wigner–Boltzmann–Poisson system (14.15) and (14.16). Then, formally, as $\alpha \to 0$, $w_\alpha \to w$ and $V_\alpha \to V$, where $w = \mathrm{Exp}(A - |p - v|^2 / 2T)$, and (A, v, T, V) is a solution of the* quantum hydrodynamic equations

$$\partial_t n - \mathrm{div}\, J_n = 0, \tag{14.26}$$

$$\partial_t J_n - \mathrm{div} \left(\frac{J_n \otimes J_n}{n} + P \right) + n \nabla V = -\langle p Q_1(w) \rangle, \tag{14.27}$$

$$\partial_t (ne) - \mathrm{div} \left((P + ne\,\mathrm{Id}) J_n - q \right) + J_n \cdot \nabla V = \left\langle \frac{1}{2} |p|^2 Q_1(w) \right\rangle, \tag{14.28}$$

$$\lambda_D^2 \Delta V = n - C(x),$$

where the quantum stress tensor P and quantum heat flux q are introduced in (14.25). The initial data are given by

$$n(\cdot,0) = \langle w_I \rangle, \quad J_n(\cdot,0) = -\langle pw_I \rangle, \quad (ne)(\cdot,0) = \left\langle \frac{1}{2}|p|^2 w_I \right\rangle,$$

and the Lagrange multipliers (A,v,T) are determined by

$$\begin{pmatrix} n \\ nu \\ ne \end{pmatrix} = \frac{2}{(2\pi\varepsilon)^3} \int_{\mathbb{R}^3} \mathrm{Exp}\left(A - \frac{|p-v|^2}{2T}\right) \begin{pmatrix} 1 \\ p \\ \frac{1}{2}|p|^2 \end{pmatrix} dp,$$

where $J_n = -nu$ is the current density.

The quantum hydrodynamic model of the above theorem is rather involved and not easy to handle numerically. It can be made more explicit if we consider the isothermal situation. Below, we will expand the integrals in powers of ε^2 which yields another simplification.

Isothermal quantum hydrodynamic equations. An isothermal quantum hydro-dynamic model can be derived from the Wigner–Boltzmann equation (14.16) by employing the quantum Maxwellian

$$M[w] = \mathrm{Exp}\left(A - \frac{1}{2}|p-v|^2\right),$$

which follows from (14.20) after setting $T = 1$. For simplicity, we consider here only the case $Q_1 = 0$, as some choices of Q_1 are discussed below. Similar to Sect. 12.4, we introduce the modified Hamiltonian $H(A,v) = W^{-1}(A - |p-v|^2/2)$, i.e., the quantum Maxwellian reads as $M[w] = \exp(-H(A,v))$. More explicitly, we find that

$$H(A,v) = \frac{1}{2}(i\varepsilon\nabla + v)^2 + A.$$

If $v = 0$, $H(A,0) = -(\varepsilon^2/2)\Delta + A$, which is the Hamiltonian considered in Sect. 12.4. We assume that $H(A,v)$ has a complete set of orthonormal eigenfunction–eigenvalue pairs (ψ_j, λ_j). Then the particle and current densities can be computed from the formulas (also see (12.32))

$$n = \sum_{j=1}^{\infty} e^{-\lambda_j} |\psi_j|^2, \quad J_n = -\varepsilon \sum_{j=1}^{\infty} e^{-\lambda_j} \mathrm{Im}(\overline{\psi}_j \nabla \psi_j). \qquad (14.29)$$

By exploiting Gauge invariance properties and arguing similarly as in Sect. 12.4, Degond et al. [22] have formulated the *isothermal quantum hydrodynamic equations*

$$\partial_t n - \mathrm{div}\, J_n = 0,$$
$$\partial_t J_n + \mathrm{div}\,(J_n \otimes v) + (\nabla v)(J_n + nv) + n\nabla(V - A) = 0,$$

where the densities n and J_n are related to the Lagrange multipliers A and v by (14.29) through the modified Hamiltonian $H(A,v)$. The velocity $u = -J_n/n$ and the Lagrange multiplier v are linked by the relation $nv = nu + \mathcal{O}(\varepsilon^2)$. Thus, the convective term becomes $\operatorname{div}(J_n \otimes v) = -\operatorname{div}(J_n \otimes J_n/n) + \mathcal{O}(\varepsilon^2)$. It is shown in [22] that in the irrotational case, $u = v$ holds and hence, the convective term is the same as in the classical hydrodynamic equations (see Sect. 9.1).

An approximate isothermal model can be derived by expanding the quantum Maxwellian in powers of ε^2. After some computations, which are detailed in [22, 23], the following quantum hydrodynamic model up to terms of order $\mathcal{O}(\varepsilon^4)$ is obtained:

$$\partial_t n - \operatorname{div} J_n = 0, \tag{14.30}$$

$$\partial_t J_n - \operatorname{div}\left(\frac{J_n \otimes J_n}{n}\right) - \nabla n + n\nabla V + \frac{\varepsilon^2}{6}n\nabla\left(\frac{\Delta\sqrt{n}}{\sqrt{n}}\right) = \frac{\varepsilon^2}{12}\operatorname{div}(nU), \tag{14.31}$$

where U is a tensor with the components

$$U_{j\ell} = \sum_{m=1}^{3}\left(\frac{\partial u_m}{\partial x_j} - \frac{\partial u_j}{\partial x_m}\right)\left(\frac{\partial u_m}{\partial x_\ell} - \frac{\partial u_\ell}{\partial x_m}\right). \tag{14.32}$$

The right-hand side of the momentum equation can be written in terms of the vorticity $\omega = \operatorname{curl} u$ [22, Sect. 3.1]:

$$\operatorname{div}(nU) = \omega \times (\operatorname{curl}(n\omega)) + \frac{1}{2}n\nabla(|\omega|^2).$$

The vorticity satisfies the equation

$$\partial_t \omega + \operatorname{curl}(\omega \times v) = 0,$$

which shows that the flow is irrotational for all time if it does so initially.

Equations (14.30) and (14.31), without the vorticity term, correspond to the quantum hydrodynamic equations (14.10) and (14.11) with a diagonal temperature tensor except for the factor of the quantum potential term which is $\frac{1}{3}$ of the factor in (14.11). We remark that this factor is not related to the space dimension since it appears also when working in any dimension. In [24] it is argued that the factor $\frac{1}{3}$ is a statistical factor coming from the expansion of the quantum potential to leading order in ε^2.

Local quantum hydrodynamic equations. The quantum hydrodynamic model (14.26), (14.27), and (14.28) is nonlocal since the stress tensor and heat flux depend implicitly on the moments through the Lagrange multipliers. In the following, we will expand these expressions in powers of ε^2 in order to derive a local version of the model.

An expansion of the quantum Maxwellian was given in Sect. 12.1, see Proposition 12.5. Inserting the expansion into the definition of the moments, we obtain after some computations (see [20, Lemma 3.4] for details):

$$n = 2(2\pi\varepsilon^2)^{-3/2}e^A - \frac{\varepsilon^2}{12T}(2\pi\varepsilon^2)^{-3/2}e^A \left(-2\Delta A - |\nabla A|^2 \right. \tag{14.33}$$

$$\left. + \nabla \log T \cdot \nabla A - 2\Delta \log T + \frac{1}{4}|\nabla \log T|^2 + \frac{1}{2T}\mathrm{Tr}(R^\top R) \right) + \mathcal{O}\left(\varepsilon^4\right),$$

$$ne = \frac{3}{2}nT + \frac{1}{2}n|u|^2 - \frac{\varepsilon^2}{24}n\left(\Delta \log n - \frac{1}{T}\mathrm{Tr}(R^\top R) - \frac{3}{2}|\nabla \log T|^2 \right.$$

$$\left. - \Delta \log T - \nabla \log T \cdot \nabla \log n \right) + \mathcal{O}\left(\varepsilon^4\right), \tag{14.34}$$

where the vorticity matrix $R = (R_{j\ell})$ is the anti-symmetric part of the velocity derivatives, $R_{j\ell} = \partial u_j/\partial x_\ell - \partial u_\ell/\partial x_j$ and it relates to (14.32) by $U = R^\top R$. Furthermore, the quantum stress tensor and quantum heat flux can be expanded as follows (see [20, Lemma 3.5]):

$$P = nT\,\mathrm{Id} + \frac{\varepsilon^2}{12}n\left(\frac{5}{2}\nabla \log T \otimes \nabla \log T - \nabla \log T \otimes \nabla \log n - \nabla \log n \otimes \nabla \log T \right.$$

$$\left. - (\nabla \otimes \nabla)\log(nT^2) + \frac{1}{T}R^\top R \right) + \frac{\varepsilon^2}{12}T\mathrm{div}\left(\frac{n}{T}\nabla \log T \right) + \mathcal{O}\left(\varepsilon^4\right), \tag{14.35}$$

$$q = -\frac{\varepsilon^2}{24}n\left(5R\nabla \log T + 2\mathrm{div}R + 3\Delta u\right) + \mathcal{O}\left(\varepsilon^4\right). \tag{14.36}$$

The expansions simplify if we assume that the temperature varies slowly in the sense of $\nabla \log T = \mathcal{O}\left(\varepsilon^2\right)$. Then the expressions $\varepsilon^2 \nabla \log T$ are of order $\mathcal{O}\left(\varepsilon^4\right)$ and can be neglected in our approximation. We obtain, up to order $\mathcal{O}\left(\varepsilon^4\right)$:

$$P = nT\,\mathrm{Id} - \frac{\varepsilon^2}{12}n\left((\nabla \otimes \nabla)\log n - \frac{1}{T}R^\top R \right),$$

$$q = -\frac{\varepsilon^2}{24}n(2\mathrm{div}R + 3\Delta u),$$

$$ne = \frac{3}{2}nT + \frac{1}{2}n|u|^2 - \frac{\varepsilon^2}{24}n\left(\Delta \log n - \frac{1}{T}\mathrm{Tr}(R^\top R) \right).$$

Equations (14.26), (14.27), and (14.28), together with the above expansions, form a closed set of equations. The stress tensor consists of the classical pressure nT on the diagonal, the quantum pressure $(\varepsilon^2/12)n(\nabla \otimes \nabla)\log n$, and the vorticity correction $(\varepsilon^2/12)nR^\top R/T$. In the isothermal case, this correction coincides with the expression $(\varepsilon^2/12)nU$ in (14.31). The quantum heat flux depends on the second derivatives of the velocity. Finally, the energy density consists of the thermal energy, kinetic energy, and quantum energy including a vorticity correction.

Further simplifications are obtained by assuming that the vorticity tensor R is small in the sense of $R = \mathcal{O}\left(\varepsilon^2\right)$. Notice that in one space dimension, this term vanishes.

Theorem 14.4 (Local quantum hydrodynamic equations). *Let the assumptions of Theorem 14.3 hold. We assume further that the temperature variations and the vorticity tensor are small in the sense of $\nabla \log T = \mathcal{O}\left(\varepsilon^2\right)$ and $R = \mathcal{O}\left(\varepsilon^2\right)$. Then the moments (n, J_n, ne) of the limit quantum Maxwellian solve the* quantum hydro-dynamic equations

$$\partial_t n - \operatorname{div} J_n = 0, \tag{14.37}$$

$$\partial_t J_n - \operatorname{div}\left(\frac{J_n \otimes J_n}{n}\right) - \nabla(nT) + n\nabla V + \frac{\varepsilon^2}{6} n\nabla\left(\frac{\Delta\sqrt{n}}{\sqrt{n}}\right) = -\langle pQ_1(w)\rangle, \tag{14.38}$$

$$\partial_t(ne) - \operatorname{div}\left((P + ne\,\mathrm{Id})u\right) - \frac{\varepsilon^2}{8}\operatorname{div}(n\Delta u) + J_n \cdot \nabla V = \left\langle \frac{1}{2}|p|^2 Q_1(w)\right\rangle, \tag{14.39}$$

where the energy density ne and the quantum stress tensor P are given by

$$P = nT\,\mathrm{Id} - \frac{\varepsilon^2}{12} n(\nabla \otimes \nabla)\log n, \quad ne = \frac{3}{2}nT + \frac{1}{2}n|u|^2 - \frac{\varepsilon^2}{24}n\Delta\log n. \tag{14.40}$$

The initial conditions for n, J_n, and ne are as in Theorem 14.3.

Notice that the above energy equation is scalar in contrast to the energy tensor equation (14.13) mentioned in the previous section. The quantum heat flux q (but not the vorticity tensor R) also appears in other quantum hydrodynamic derivations. Gardner has derived it from a mixed-state Wigner model and interpreted it as a dispersive heat flux (see formula (36) in [25]). Moreover, it shows up in the quantum hydrodynamic equations of Gardner and Ringhofer [26] involving a "smoothed" potential, derived from a Wigner–Boltzmann equation by a Chapman–Enskog expansion. Numerical results in [20] indicate that the dispersive term stabilizes the quantum hydrodynamic system numerically.

The above equations can be written in unscaled form as follows:

$$\partial_t n - \frac{1}{q}\operatorname{div} J_n = 0, \quad x \in \mathbb{R}^3,\ t > 0,$$

$$\partial_t J_n - \operatorname{div}\left(\frac{J_n \otimes J_n}{qn}\right) - \frac{qk_B}{m^*}\nabla(nT) + \frac{q^2}{m^*}n\nabla V + \frac{q\hbar^2}{6(m^*)^2}n\nabla\left(\frac{\Delta\sqrt{n}}{\sqrt{n}}\right)$$
$$= -\frac{q}{\tau}\langle pQ_1(w)\rangle,$$

$$\partial_t(ne) - \operatorname{div}\left((P + ne\,\mathrm{Id})u\right) - \frac{\hbar^2}{8m^*}\operatorname{div}(n\Delta u) + J_n \cdot \nabla V$$
$$= \frac{1}{\tau}\left\langle \frac{1}{2}|p|^2 Q_1(w)\right\rangle,$$

where the unscaled pressure tensor and energy density read as

$$P = nk_B T \operatorname{Id} - \frac{\hbar^2}{12m^*} n(\nabla \otimes \nabla) \log n, \quad ne = \frac{3}{2} nk_B T + \frac{m^*}{2} n|u|^2 - \frac{\hbar^2}{24m^*} n\Delta \log n,$$

and the current density is given by $J_n = -qnu$.

Comparison with other quantum hydrodynamic equations. A quantum hydrodynamic model including an energy equation was derived first by Ferry and Zhou from the Bloch equation for the density matrix [27]. A derivation from the Wigner equation was proposed by Gardner [28]. He obtained the equations (14.37) and (14.38) except for the dispersive velocity term $(\varepsilon^2/8)\operatorname{div}(n\Delta u)$. The origin of this difference lies in the different choice of the quantum Maxwellian. In both approaches, closure of the moment equations is obtained by assuming that the Wigner function is in equilibrium. However, the notion of equilibrium is different. In the following, we explain this difference.

A quantum system, which is characterized by its energy operator $H = W^{-1}(h)$, where W^{-1} is the inverse of the Wigner transform and $h(p) = |p|^2/2 + V$ the Hamiltonian, attains its maximum of the relative von Neumann entropy in the mixed state with Wigner function $w_Q = \operatorname{Exp}(-h)$, as shown in Sect. 10.1. This state represents the *unconstrained* quantum equilibrium. The expansion of w_Q in terms of ε^2 reads, according to Wigner [29], as

$$w_Q = e^{-h(x,p)} \left(1 + \varepsilon^2 g(x, p)\right) + \mathcal{O}\left(\varepsilon^4\right)$$

with an appropriate function g. As a definition of the quantum equilibrium *with* moment constraints, Gardner employed this expansion of w_Q and modified it, mimicking the momentum shift of the equilibrium distribution in the classical situation (see Example 2.1):

$$\widetilde{w}_Q = n(x)e^{-h(x,p)/T(x)} \left(1 + \varepsilon g(x, p - v(x))\right) + \mathcal{O}\left(\varepsilon^4\right).$$

In contrast to the classical case, $\widetilde{w}_Q$ is *not* the constrained maximizer for the quantum entropy. The quantum Maxwellian $M[w] = \operatorname{Exp}(A - |p - v|^2/2T)$ is a genuine maximizer of the quantum entropy with respect to the given moments.

Both approaches coincide if the temperature is constant and if only the particle density is prescribed as a constraint. In order to see this, we write Gardner's momentum-shifted quantum Maxwellian more explicitly as (see Remark 12.7)

$$\widetilde{w}_Q = e^{V/T - |p|^2/2T} \left[1 + \frac{\varepsilon^2}{8T} \left(\Delta V + \frac{1}{3T}|\nabla V|^2 - \frac{1}{3T} p^\top (\nabla \otimes \nabla) V p\right)\right] + \mathcal{O}\left(\varepsilon^4\right). \tag{14.41}$$

The quantum Maxwellian obtained from entropy maximization with given particle density equals (see Corollary 12.6)

$$M[w] = e^{A - |p|^2/2} \left[1 + \frac{\varepsilon^2}{8} \left(\Delta A + \frac{1}{3}|\nabla A|^2 - \frac{1}{3} p^\top (\nabla \otimes \nabla) A p\right)\right] + \mathcal{O}\left(\varepsilon^4\right).$$

Using the expansion

$$n = 2(2\pi\varepsilon)^{-3} \int_{\mathbb{R}^3} \widetilde{w}_Q \, dp = (2\pi T)^{3/2} e^{V/T} + \mathcal{O}\left(\varepsilon^2\right) \tag{14.42}$$

and assuming that the temperature varies slowly, Gardner has substituted $\nabla V = T \nabla \log n + \mathcal{O}\left(\varepsilon^2\right)$ into the formula for $\widetilde{w}_Q$ in order to avoid second-order derivatives of the electric potential. This substitution yields $M[w]$ since, by (12.17), $\nabla A = \nabla \log n + \mathcal{O}\left(\varepsilon^2\right)$ and thus, both expansions of $\widetilde{w}_Q$ and $M[w]$ coincide if $T = 1$.

The quantum hydrodynamic model is used for the simulation of quantum devices, like the resonant tunneling diode [16, 28, 30] which consists of different materials. At the interface of the materials (heterojunctions), the (mean-field) potential is calibrated by a barrier potential which models the gap between the conduction bands of each material. The barrier potential is modeled by a given function which is constant inside each material. Thus, the sum of the (mean-field) potential and the barrier potential is discontinuous. However, in the derivation of Gardner's model, the approximation $\partial^2 \log n / \partial x_j \partial x_\ell = T^{-1} \partial^2 V / \partial x_j \partial x_\ell + \mathcal{O}\left(\varepsilon^2\right)$, which follows from (14.42) for slowly varying temperature, was employed. This approximation is not valid for discontinuous potentials. Gardner and Ringhofer [24] have overcome this problem by deriving so-called *"smooth" quantum hydrodynamic equations*. More precisely, they obtain in the Born approximation to the Bloch equation the model (14.37), (14.38), (14.39) and (14.40) (without the dispersive velocity term) in which the terms

$$\frac{\varepsilon^2}{6} n \nabla \left(\frac{\Delta \sqrt{n}}{\sqrt{n}} \right) \quad \text{and} \quad \frac{\varepsilon^2}{12} (\nabla \otimes \nabla) \ln n$$

are replaced by

$$\frac{\varepsilon^2}{4} \operatorname{div} (n(\nabla \otimes \nabla) \overline{V}) \quad \text{and} \quad \frac{\varepsilon^2}{4} (\nabla \otimes \nabla) \overline{V},$$

and $\overline{V} = \overline{V}(x, T)$ depends nonlocally on x and T (see [24] for details). The quantum hydrodynamic equations (14.37), (14.38), (14.39), and (14.40) are recovered in the $\mathcal{O}\left(\varepsilon^2\right)$ approximation

$$\overline{V} = \frac{1}{3} V + \mathcal{O}\left(\varepsilon^2\right), \quad \nabla \log n = \frac{\nabla V}{T} + \mathcal{O}\left(\varepsilon^2\right),$$

if n, J, and T are varying very slowly.

Finally, we mention the approach of Frosali et al. [31] to derive *high-field quantum hydrodynamic equations*. Starting from the rescaled Wigner–Boltzmann equation

$$\partial_t w_\alpha + p \cdot \nabla_x w_\alpha + \frac{1}{\alpha} \theta[V] w_\alpha = \frac{1}{\alpha} Q(w_\alpha),$$

where the collision operator is of relaxation-time type, in the formal limit $\alpha \to 0$ and employing Gardner's quantum Maxwellian in the $\mathcal{O}\left(\varepsilon^2\right)$ approximation (14.41), the following quantum hydrodynamic model is obtained:

$$\partial_t n - \operatorname{div} J_n = 0, \quad \partial_t J_n - \operatorname{div}\left(\frac{J_n \otimes J_n}{n}\right) - \operatorname{div} P = 0,$$

$$\partial_t (ne) - \operatorname{div}\left((P + ne\,\mathrm{Id})u\right) + \operatorname{div} q = 0,$$

where the high-field stress tensor P and quantum heat flux q are given by

$$P = nT\,\mathrm{Id} - \frac{\varepsilon^2}{12} n (\nabla \otimes \nabla) V + n u_w \otimes u_w, \quad q = -\frac{\varepsilon^2}{8} n \Delta u_w + n(u_w \otimes u_w) u_w$$

and $u_w = \nabla V$ is a kind of velocity coming from the formula

$$n u_w = \frac{2}{(2\pi\varepsilon)^3} \int_{\mathbb{R}^3} p w\, dp,$$

where $w = \lim_{\alpha \to 0} w_\alpha$ is the solution of the Wigner–Boltzmann equation in the $\alpha \to 0$ limit, i.e., w solves $\theta[V]w = Q(w)$. Interestingly, the first expression in the heat flux resembles the quantum heat flux in (14.39).

Dissipative quantum hydrodynamic equations. By specifying the collision operator Q_1, we can make explicit the right-hand sides of (14.38) and (14.39). In Sect. 11.3 we have discussed various quantum collision models. For instance, with a variant of the Caldeira–Leggett operator (11.21) from Sect. 11.3,

$$Q_1(w) = \frac{1}{\tau}(\Delta_p w + \operatorname{div}_p(pw)), \tag{14.43}$$

we obtain $\langle Q_1(w) \rangle = 0$ and, by integration by parts,

$$-\langle p Q_1(M[w]) \rangle = \frac{1}{\tau}\frac{2}{(2\pi\varepsilon)^3} \int_{\mathbb{R}^3} (\nabla_p M[w] + p M[w])\, dp = -\frac{J_n}{\tau}, \tag{14.44}$$

$$\left\langle \frac{1}{2}|p|^2 Q_1(M[w]) \right\rangle = -\frac{1}{\tau}\frac{2}{(2\pi\varepsilon)^3} \int_{\mathbb{R}^3} p \cdot (\nabla_p M[w] + p M[w])\, dp \tag{14.45}$$

$$= \frac{1}{\tau}\frac{2}{(2\pi\varepsilon)^3} \int_{\mathbb{R}^3} (3M[w] - |p|^2 M[w])\, dp = -\frac{2}{\tau}\left(ne - \frac{3}{2}n\right).$$

Thus, in the space homogeneous case and in the absence of external forces, equations (14.38) and (14.39) become

$$\partial_t J_n = -\frac{J_n}{\tau}, \quad \partial_t (ne) = -\frac{2}{\tau}\left(ne - \frac{3}{2}n\right),$$

and for $t \to \infty$, the current and energy densities converge to their equilibrium values $J_n = 0$ and $ne = \frac{3}{2}n$. Therefore, the expressions (14.44) and (14.45) are referred to as *relaxation-time terms* and the parameter τ as the *relaxation time*. In the literature, the quantum hydrodynamic model is usually considered with these terms. The existence of solutions of the stationary isentropic quantum hydrodynamic equations in bounded domains with physically motivated boundary conditions was shown in [18] under a smallness condition on the applied voltage, which corresponds to a subsonic condition on the velocity for the underlying Euler system. In [32], the non-existence of weak solutions was proved for particular boundary conditions. The semi-classical and inviscid limits were analyzed in [33]. Furthermore, in [34], current–voltage characteristics for simplified quantum hydrodynamic equations were derived. The existence of local-in-time solutions was proved in [35]. For the time-dependent model in the whole space, a global-in-time existence result for arbitrary velocities was shown in [13].

Interestingly, the quantum hydrodynamic equations with relaxation-time terms are formally equivalent to a nonlinear Schrödinger equation. More precisely, let us consider the isothermal quantum hydrodynamic model

$$\partial_t n - \text{div} J_n = 0, \quad \partial_t J_n - \text{div}\left(\frac{J_n \otimes J_n}{n}\right) - \nabla n + n\nabla V + \frac{\varepsilon^2}{2}n\nabla\left(\frac{\Delta\sqrt{n}}{\sqrt{n}}\right) = -\frac{J_n}{\tau}.$$

(14.46)

Then, if the quantum flow is irrotational and we can write $J_n = -n\nabla S$ for some phase function S, and setting $\psi = \sqrt{n}\exp(iS/\varepsilon)$, the above system of equations is formally equivalent to the *Schrödinger–Langevin equation*

$$i\varepsilon\partial_t\psi = -\frac{\varepsilon^2}{2}\Delta\psi - V\psi + \log(|\psi|^2)\psi - \frac{i\varepsilon}{\tau}\log\frac{\psi}{\overline{\psi}}.$$

This equation was derived by Kostin [36] from a Heisenberg–Langevin equation modeling linearly coupled harmonic oscillators. The fluid formulation of this equation in terms of the velocity J_n/n instead of the current density J_n was investigated by Nassar [37].

In the so-called relaxation-time limit, the solutions of the quantum hydrodynamic model converge to a solution of the quantum drift-diffusion model. More precisely, we consider the isothermal model (14.46) and rescale the time by $t \mapsto t/\tau$ and the current density by $J_n \mapsto \tau J_n$. Then the rescaled equations become

$$\tau\partial_t n - \tau\text{div} J_n = 0,$$

$$\tau^2\partial_t J_n - \tau^2\text{div}\left(\frac{J_n \otimes J_n}{n}\right) - \nabla n + n\nabla V + \frac{\varepsilon^2}{2}n\nabla\left(\frac{\Delta\sqrt{n}}{\sqrt{n}}\right) = -J_n,$$

and the formal limit $\tau \to 0$ gives the quantum drift-diffusion model

$$\partial_t n - \text{div} J_n = 0, \quad J_n = \nabla n - n\nabla V - \frac{\varepsilon^2}{2}n\nabla\left(\frac{\Delta\sqrt{n}}{\sqrt{n}}\right),$$

studied in Chap. 11. This limit was made rigorous in the whole space for smooth solutions in [38].

When the collision operator Q_1 is given by the Fokker–Planck operator (11.20),

$$Q_1(w) = D_{pp}\Delta_p w + 2\gamma\text{div}_p(pw) + D_{qq}\Delta_x w + 2D_{pq}\text{div}_x(\nabla_p w),$$

(14.47)

where D_{pp}, D_{pq}, D_{qq}, and γ are nonnegative parameters, we obtain additional viscous terms. For this, we compute for $w = M[w]$

$$\langle Q_1(w)\rangle = D_{qq}\Delta_x n, \tag{14.48}$$

$$-\langle pQ_1(w)\rangle = 4(2\pi\varepsilon)^{-3}\int_{\mathbb{R}^3}(\gamma pw + D_{pq}\nabla_x w)\,dp - 2D_{qq}(2\pi\varepsilon)^{-3}\Delta_x\int_{\mathbb{R}^3}pw\,dp$$

$$= -2\gamma J_n + 2D_{pq}\nabla_x n + D_{qq}\Delta_x J_n, \tag{14.49}$$

$$\left\langle \frac{1}{2}|p|^2 Q_1(w)\right\rangle = -2(2\pi\varepsilon)^{-3}\int_{\mathbb{R}^3}\left(D_{pp}p\cdot\nabla_p w + 2\gamma|p|^2 w + 2D_{pq}p\cdot\nabla_x w\right)dp$$

$$+ 2D_{qq}(2\pi\varepsilon)^{-3}\Delta_x\int_{\mathbb{R}^3}\frac{1}{2}|p|^2 w\,dp$$

$$= -2\left(2\gamma ne - \frac{3}{2}D_{pp}n\right) + 2D_{pq}\mathrm{div}_x J_n + D_{qq}\Delta_x(ne). \tag{14.50}$$

The spatial second-order expressions involving the parameter D_{qq} can be interpreted as viscous terms. The corresponding model is called the *viscous quantum hydrodynamic equations*. They were first suggested in [39]. Existence results for the stationary isothermal model in one space dimension were shown in [40], and the transient equations were analyzed in [41–43]. The long-time behavior of the solutions is studied in [44, 45]; numerical results are presented in [40, 46]. Notice that the mass conservation equation now reads as

$$\partial_t n - \mathrm{div}\,J_n = D_{qq}\Delta n,$$

which can be written in conservative form as

$$\partial_t n - \mathrm{div}\,(J_n + D_{qq}\nabla n) = 0.$$

Then the quantity $J_n + D_{qq}\nabla n$ can be interpreted as an effective current density.

Conserved quantities. We consider the local quantum hydrodynamic equations (14.26), (14.27), and (14.28) without scattering terms, coupled to the Poisson equation (14.15) and together with the relations (14.35) and (14.36) for the quantum stress tensor and the heat flux, neglecting the $\mathscr{O}(\varepsilon^4)$ terms. Clearly, the mass is conserved but not the momentum due to the electric force given by $\int n\nabla V\,dx$. We show that also the energy is conserved [20].

Proposition 14.5. *Let $Q_1 = 0$ and let (n, J_n, ne, V) be a solution of* (14.26), (14.27), *and* (14.28), (14.35), *and* (14.36) *(without the $\mathscr{O}(\varepsilon^4)$ terms), and the Poisson equation* (14.15). *Then the energy*

$$E(t) = \int_{\mathbb{R}^3}\left(ne + \frac{\lambda_D^2}{2}|\nabla V|^2\right)dx,$$

where ne is defined in (14.34) *(without the $\mathscr{O}(\varepsilon^4)$ terms), is conserved, i.e., $dE(t)/dt = 0$ for all $t > 0$. Furthermore, the energy can be written as*

$$E(t) = \int_{\mathbb{R}^3} \left(\frac{3}{2}nT + \frac{1}{2}n|u|^2 + \frac{\lambda_D^2}{2}|\nabla V|^2 + \frac{\varepsilon^2}{6}|\nabla\sqrt{n}|^2 \right.$$
$$\left. + \frac{\varepsilon^2}{16}n|\nabla\log T|^2 + \frac{\varepsilon^2}{24T}n\,\mathrm{Tr}(R^\top R) \right) dx \geq 0. \tag{14.51}$$

Proof. We differentiate the energy with respect to time and employ (14.15) and (14.28),

$$\frac{dE}{dt} = \int_{\mathbb{R}^3} \left(\partial_t(ne) + \lambda_D^2 \nabla V \cdot \nabla \partial_t V \right) dx = \int_{\mathbb{R}^3} \left(-J_n \cdot \nabla V - \lambda_D^2 V \partial_t \Delta V \right) dx$$
$$= \int_{\mathbb{R}^3} \left((\mathrm{div}\, J_n)V - V\partial_t n \right) dx = 0.$$

It remains to prove formula (14.51). By (14.34), the integral of the energy density ne can be written as

$$\int_{\mathbb{R}^3} ne\,dx = \int_{\mathbb{R}^3} \left(\frac{3}{2}nT + \frac{1}{2}n|u|^2 + \frac{\varepsilon^2}{16}n|\nabla\log T|^2 + \frac{\varepsilon^2}{24T}n\,\mathrm{Tr}(R^\top R) \right) dx$$
$$+ \frac{\varepsilon^2}{24} \int_{\mathbb{R}^3} \left(-n\Delta\log n + n\Delta\log T + n\nabla\log T \cdot \nabla\log n \right) dx.$$

The last integral equals, after an integration by parts,

$$\frac{\varepsilon^2}{24} \int_{\mathbb{R}^3} \left(4|\nabla\sqrt{n}|^2 - \nabla n \cdot \nabla\log T + n\nabla\log T \cdot \nabla\log n \right) dx = \frac{\varepsilon^2}{6} \int_{\mathbb{R}^3} |\nabla\sqrt{n}|^2\,dx,$$

which shows the claim. □

The energy (14.51) consists of, in this order, the thermal energy, the kinetic energy, the electric energy, and the energy of the Bohm potential. The remaining two terms represent additional field quantum energies associated with spatial variations of the temperature and the vorticity. In the case of the local quantum hydrodynamic equations of Theorem 14.4, the energy is given by (14.51) except the last two terms.

In the presence of the relaxation-time terms (14.44) and (14.45) or the viscous terms (14.48), (14.49), and (14.50), the energy is no longer conserved due to the dissipative effects but it is bounded; see for instance, the estimates in [44].

The quantum hydrodynamic models of this and the previous sections are summarized in Fig. 14.1.

14.4 Extended Quantum Hydrodynamic Equations

The strategy of the previous section can be used to derive general quantum moment equations. Let w_α be a solution of the Wigner–Boltzmann equation in the hydrodynamic scaling (14.16),

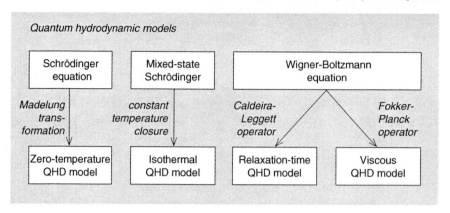

Fig. 14.1 Overview of quantum hydrodynamic (QHD) equations

$$\alpha \partial_t w_\alpha + \alpha \left(p \cdot \nabla_x w_\alpha + \theta[V] w_\alpha \right) = Q(w_\alpha), \quad w_\alpha(x,p,0) = w_I, \qquad (14.52)$$

where $x, p \in \mathbb{R}^3$, $t > 0$, and the potential V_α solves the Poisson equation

$$\lambda_D^2 \Delta V_\alpha = \langle w_\alpha \rangle - C(x), \quad x \in \mathbb{R}^3. \qquad (14.53)$$

Furthermore, let $\kappa(p) = (\kappa_0(p), \ldots, \kappa_N(p))$ be weight functions with $\kappa_0(p) = 1$. We assume as in Sect. 14.3 that the collision operator is the sum of two operators Q_0 and Q_1,

$$Q(w) = Q_0(w) + \alpha Q_1(w),$$

where Q_0 satisfies the properties

$$\langle \kappa(p) Q_0(w) \rangle = 0 \text{ for all } w, \quad Q_0(w) = 0 \text{ if and only if } w = M[w], \qquad (14.54)$$

and Q_1 conserves mass,

$$\langle Q_1(w) \rangle = 0 \quad \text{for all } w. \qquad (14.55)$$

A simple example of a collision operator satisfying (14.54) is the BGK operator $Q_0(w) = M[w] - w$. If $\kappa(p) = (1, p, \frac{1}{2}|p|^2)$, the first condition in (14.54) expresses the conservation of mass, momentum, and energy.

For given w, the quantum Maxwellian $M[w]$ is the maximizer of the quantum entropy

$$S(\widetilde{w}) = -\frac{2}{(2\pi\varepsilon)^3} \int_{\mathbb{R}^3} \widetilde{w} \left(\text{Log}\,\widetilde{w} - 1 + \frac{|p|^2}{2} - V \right) dx\,dp$$

under the constraints $\langle \kappa_j(p) w \rangle = \langle \kappa_j(p) \widetilde{w} \rangle$, $j = 0, \ldots, N$. We call $m_j = \langle \kappa_j(p) w \rangle$ the *jth moment* of w. By Lemma 12.2, the quantum Maxwellian equals

$$M[w] = \text{Exp}(\lambda \cdot \kappa),$$

where $\lambda(x,t) = (\lambda_0,\ldots,\lambda_N)$ are the Lagrange multipliers of the constrained extremal problem.

The *moment equations* are obtained by multiplying (14.52) by $\kappa(p)/\alpha$, integrating over the momentum space, and employing (14.54):

$$\partial_t\langle\kappa_j(p)w_\alpha\rangle + \mathrm{div}_x\langle p\kappa_j(p)w_\alpha\rangle + \langle\kappa_j(p)\theta[V_\alpha]w_\alpha\rangle = \langle\kappa_j(p)Q_1(w_\alpha)\rangle.$$

The formal limit $\alpha \to 0$ in (14.52) gives $Q_0(w) = 0$, where $w = \lim_{\alpha\to 0} w_\alpha$, and hence, $w = M[w]$. The same limit in the above moment equations leads to the following result which is due to [19].

Theorem 14.6 (Quantum moment hydrodynamic equations). *Let the collision operator $Q(w) = Q_0(w) + \alpha Q_1(w)$ satisfy (14.54) and (14.55) and let (w_α, V_α) be a solution of the Wigner–Poisson system (14.52) and (14.53). Then, in the formal limit $\alpha \to 0$, (w_α, V_α) converges to $(M[w], V)$, solving the quantum moment hydrodynamic model*

$$\partial_t m_j + \mathrm{div}_x\langle p\kappa_j(p)M[w]\rangle + \langle\kappa_j(p)\theta[V]M[w]\rangle = \langle\kappa_j(p)Q_1(M[w])\rangle, \tag{14.56}$$

$$\lambda_D^2\Delta V = m_0 - C(x), \quad x \in \mathbb{R}^3,\ t > 0, \tag{14.57}$$

where the quantum Maxwellian is given by $M[w] = \mathrm{Exp}(\lambda \cdot \kappa)$ and $m_j = \langle\kappa_j M[w]\rangle$, $j = 0,\ldots,N$. The initial conditions are $m_j(\cdot,0) = \langle\kappa_j w_I\rangle$.

In the classical gas-dynamics case, Levermore [47] has shown that the moment equations are symmetrizable and hyperbolic. In the present situation, this concept of hyperbolicity cannot be used since (14.56) is not a partial differential equation but a differential equation with nonlocal operators of the type $\lambda \mapsto \langle\mathrm{Exp}(\lambda \cdot \kappa(p))\rangle$. Nevertheless, it is possible to prove that there exists a bound on the energy

$$E(t) = \frac{1}{(2\pi\varepsilon)^3}\int_{\mathbb{R}^3\times\mathbb{R}^3}\left(\left\langle\frac{1}{2}|p|^2 M[w]\right\rangle + \frac{\lambda_D^2}{2}|\nabla V|^2\right)\mathrm{d}x\,\mathrm{d}p.$$

Proposition 14.7. *Let (w,V) be a solution of (14.56) and (14.57). Furthermore, let 1 and $\frac{1}{2}|p|^2$ be included in the set of weight functions. We assume that Q_1 dissipates energy, i.e., $\langle\frac{1}{2}|p|^2 Q_1(w)\rangle \le 0$ for all functions w. Then*

$$\frac{\mathrm{d}E}{\mathrm{d}t}(t) \le 0.$$

Proof. We recall that for all w, by Lemmas 12.9 and 13.2,

$$\langle\theta[V]w\rangle = 0, \quad \langle p\theta[V]w\rangle = -\langle w\rangle\nabla_x V, \quad \left\langle\frac{1}{2}|p|^2\theta[V]w\right\rangle = -\langle pw\rangle\cdot\nabla_x V.$$

Thus, from the moment equations

$$\partial_t\langle M[w]\rangle + \mathrm{div}_x\langle pM[w]\rangle = 0,$$

$$\partial_t \left\langle \frac{1}{2}|p|^2 M[w] \right\rangle + \mathrm{div}_x \left\langle \frac{1}{2}p|p|^2 M[w] \right\rangle - \langle pM[w] \rangle \cdot \nabla_x V$$

$$= \left\langle \frac{1}{2}|p|^2 Q_1(M[w]) \right\rangle \le 0$$

and the Poisson equation (14.53), we obtain

$$\frac{\mathrm{d}E}{\mathrm{d}t} \le \int_{\mathbb{R}^3} \left(\langle pM[w] \rangle \cdot \nabla_x V + \lambda_D^2 \nabla_x V \cdot \partial_t \nabla_x V \right) \mathrm{d}x$$

$$= \int_{\mathbb{R}^3} \left(\langle pM[w] \rangle \cdot \nabla_x V - V \partial_t \langle M[w] \rangle \right) \mathrm{d}x$$

$$= \int_{\mathbb{R}^3} \left(\langle pM[w] \rangle \cdot \nabla_x V + V \mathrm{div}_x \langle pM[w] \rangle \right) \mathrm{d}x = 0,$$

which shows the monotonicity of the energy. □

It is shown in [19] that the quantum fluid entropy,

$$S = -\frac{2}{(2\pi\varepsilon)^3} \int_{\mathbb{R}^3 \times \mathbb{R}^3} \mathrm{Exp}(\lambda \cdot \kappa)(\lambda \cdot \kappa - 1) \mathrm{d}x \mathrm{d}p,$$

is nondecreasing along the solutions of (14.56). These dissipation properties may indicate local well-posedness of the quantum moment hydrodynamic equations. However, due to the complicated mathematical structure, it seems to be difficult to prove the existence of solutions, even locally in time, and to solve the model numerically.

On the other hand, the system of Eq. (14.56) possesses some interesting properties also present in classical thermodynamic systems. It is shown in [19] that the entropy $S = S(m)$ is concave. Denoting by $\Sigma(\lambda)$ the Legendre dual of α, the mapping $\lambda \mapsto m$, which relates the extensive variables m_j to the intensive variables λ_j, can be inverted by means of the functionals S and Σ according to

$$\frac{\mathrm{d}S}{\mathrm{d}m} = \lambda, \quad \frac{\mathrm{d}\Sigma}{\mathrm{d}\lambda} = m.$$

This relation also holds in classical nonequilibrium thermodynamics.

Finally, we notice that recently, an extended quantum hydrodynamic model based on the entropy maximization closure and a variant of Wigner's quantum equilibrium function was proposed by Romano [48]. This model consists of the balance equations for the particle, current, energy density, and, additionally, the energy flux.

References

1. E. Madelung. Quantentheorie in hydrodynamischer Form. *Z. Physik* 40 (1927), 322–326.
2. G. Wentzel. Eine Verallgemeinerung der Quantenbedingungen für die Zwecke der Wellenmechanik. *Z. Physik* 38 (1926), 518–529.
3. H. Kramers. Wellenmechanik und halbzählige Quantisierung. *Z. Physik* 39 (1926), 828–840.

4. L. Brillouin. La méchanique ondulatoire de Schrödinger: une méthode générale de resolution par approximations successives. *C. R. Acad. Sci.* 183 (1926), 24–26.
5. I. Gasser and P. Markowich. Quantum hydrodynamics, Wigner transforms and the classical limit. *Asympt. Anal.* 14 (1997), 97–116.
6. P. Gérard, P. Markowich, N. Mauser, and F. Poupaud. Homogenization limits and Wigner transforms. *Commun. Pure Appl. Math.* 50 (1997), 323–379.
7. R. Wyatt. *Quantum Dynamics with Trajectories.* Springer, New York, 2005.
8. F. Sales Mayor, A. Askar, and H. Rabitz. Quantum fluid dynamics in the Lagrangian representation and applications to photodissociation problems. *J. Chem. Phys.* 111 (1999), 2423–2435.
9. M. Loffredo and L. Morato. On the creation of quantum vortex lines in rotating He II. *Il nouvo cimento* 108B (1993), 205–215.
10. R. Wyatt. Quantum wave packet dynamics with trajectories: application to reactive scattering. *J. Chem. Phys.* 111 (1999), 4406–4413.
11. J. Grant. Pressure and stress tensor expressions in the fluid mechanical formulation of the Bose condensate equations. *J. Phys. A: Math., Nucl. Gen.* 6 (1973), L151–L153.
12. G. El, V. Geogjaev, A. Gurevich, and A. Krylov. Decay of an initial discontinuity in the defocusing NLS hydrodynamics. *Physica D* 87 (1995), 185–192.
13. P. Antonelli and P. Marcati. On the finite energy weak solutions to a system in quantum fluid dynamics. To appear in *Commun. Math. Phys.* 2009.
14. S. Weigert. How to determine a quantum state by measurements: the Pauli problem for a particle with arbitrary potential. *Phys. Rev. A* 53 (1996), 2078–2083.
15. I. Gasser, P. Markowich, D. Schmidt, and A. Unterreiter. Macroscopic theory of charged quantum fluids. In: *Mathematical Problems in Semiconductor Physics*, Pitman Res. Notes Math. Ser. 340, 42–75. Longman, Harlow, 1995.
16. H. Grubin and J. Kreskovsky. Quantum moment balance equations and resonant tunneling structures. *Solid State Electr.* 32 (1989), 1071–1075.
17. I. Gasser, P. Markowich, and C. Ringhofer. Closure conditions for classical and quantum moment hierarchies in the small-temperature limit. *Transp. Theory Stat. Phys.* 25 (1996), 409–423.
18. A. Jüngel. A steady-state quantum Euler-Poisson system for semiconductors. *Commun. Math. Phys.* 194 (1998), 463–479.
19. P. Degond and C. Ringhofer. Quantum moment hydrodynamics and the entropy principle. *J. Stat. Phys.* 112 (2003), 587–628.
20. A. Jüngel, D. Matthes, and J.-P. Milišić. Derivation of new quantum hydrodynamic equations using entropy minimization. *SIAM J. Appl. Math.* 67 (2006), 46–68.
21. I. Burghardt and L. Cederbaum. Hydrodynamic equations for mixed quantum states. I. General formulation. *J. Chem. Phys.* 115 (2001), 10303–10311.
22. P. Degond, S. Gallego, and F. Méhats. Isothermal quantum hydrodynamics: derivation, asymptotic analysis, and simulation. *SIAM Multiscale Model. Simul.* 6 (2007), 246–272.
23. A. Jüngel and D. Matthes. A derivation of the isothermal quantum hydrodynamic equations using entropy minimization. *Z. Angew. Math. Mech.* 85 (2005), 806–814.
24. C. Gardner and C. Ringhofer. Smooth quantum potential for the hydrodynamic model. *Phys. Rev. E* 53 (1996), 157–168.
25. C. Gardner. Resonant tunneling in the quantum hydrodynamic model. *VLSI Design* 3 (1995), 201–210.
26. C. Gardner and C. Ringhofer. The Chapman–Enskog expansion and the quantum hydrodynamic model for semiconductor devices. *VLSI Design* 10 (2000), 415–435.
27. D. Ferry and J.-R. Zhou. Form of the quantum potential for use in hydrodynamic equations for semiconductor device modeling. *Phys. Rev. B* 48 (1993), 7944–7950.
28. C. Gardner. The quantum hydrodynamic model for semiconductor devices. *SIAM J. Appl. Math.* 54 (1994), 409–427.
29. E. Wigner. On the quantum correction for the thermodynamic equilibrium. *Phys. Rev.* 40 (1932), 749–759.

30. Z. Chen, B. Cockburn, C. Gardner, and J. Jerome. Quantum hydrodynamic simulation of hysteresis in the resonant tunneling diode. *J. Comput. Phys.* 117 (1995), 274–280.
31. G. Borgioli, G. Frosali, and C. Manzini. Derivation of a quantum hydrodynamic model in the high-field case. In: N. Manganaro et al. (eds.), *Proceedings WASCOMP 2007 – 14th Conference on Waves and Stability in Continuous Media*, 60–65. World Scientific, Singapore, 2008.
32. I. Gamba and A. Jüngel. Positive solutions to singular and third order differential equations for quantum fluids. *Arch. Rat. Mech. Anal.* 156 (2001), 183–203.
33. I. Gamba and A. Jüngel. Asymptotic limits in quantum trajectory models. *Commun. Part. Diff. Eqs.* 27 (2002), 669–691.
34. A. Jüngel. A note on current-voltage characteristics from the quantum hydrodynamic equations for semiconductors. *Appl. Math. Letters* 10 (1997), 29–34.
35. H.-L. Li and P. Marcati. Existence and asymptotic behavior of multi-dimensional quantum hydrodynamic model for semiconductors. *Commun. Math. Phys.* 245 (2004), 215–247.
36. M. Kostin. On the Schrödinger-Langevin equation. *J. Chem. Phys.* 57 (1972), 3589–3591.
37. A. Nassar. Fluid formulation of a generalised Schrödinger-Langevin equation. *J. Phys. A: Math. Gen.* 18 (1985), L509–L511.
38. A. Jüngel, H.-L. Li, and A. Matsumura. The relaxation-time limit in the quantum hydrodynamic equations for semiconductors. *J. Diff. Eqs.* 225 (2006), 440–464.
39. M. Gualdani and A. Jüngel. Analysis of the viscous quantum hydrodynamic equations for semiconductors. *Europ. J. Appl. Math.* 15 (2004), 577–595.
40. A. Jüngel and J.-P. Milišić. Physical and numerical viscosity for quantum hydrodynamics. *Commun. Math. Sci.* 5 (2007), 447–471.
41. L. Chen and M. Dreher. The viscous model of quantum hydrodynamics in several dimensions. *Math. Models Meth. Appl. Sci.* 17 (2007), 1065–1093.
42. M. Dreher. The transient equations of viscous quantum hydrodynamics. *Math. Meth. Appl. Sci.* 31 (2008), 391–414.
43. I. Gamba, A. Jüngel, and A. Vasseur. Global existence of solutions to one-dimensional viscous quantum hydrodynamic equations. Preprint, Vienna University of Technology, Austria, 2009.
44. M. Gualdani, A. Jüngel, and G. Toscani. Exponential decay in time of solutions of the viscous quantum hydrodynamic equations. *Appl. Math. Letters* 16 (2003), 1273–1278.
45. B. Liang and S. Zheng. Exponential decay to a quantum hydrodynamic model for semiconductors. *Nonlin. Anal.: Real World Appl.* 9 (2008), 326–337.
46. A. Jüngel and S. Tang. Numerical approximation of the viscous quantum hydrodynamic model for semiconductors. *Appl. Numer. Math.* 56 (2006), 899–915.
47. C. Levermore. Moment closure hierarchies for kinetic theories. *J. Stat. Phys.* 83 (1996), 1021–1065.
48. V. Romano. Quantum corrections to the semiclassical hydrodynamical model of semiconductors based on the maximum entropy principle. *J. Math. Phys.* 48 (2007), 123504.

Index